D0930238

HANDBOOK OF THE AGING BRAIN

HANDBOOK OF THE AGING BRAIN

EUGENIA WANG

The Bloomfield Centre for Research in Aging
Lady Davis Institute for Medical Research
Jewish General Hospital, and
Department of Medicine, McGill University
Montréal, Québec, Canada

D. STEPHEN SNYDER

National Institute on Aging
National Institutes of Health
Bethesda, Maryland

ACADEMIC PRESS

San Diego   London   Boston   New York   Sydney   Tokyo   Toronto

This book is printed on acid-free paper.

CONTENTS

3

___

SPATIAL COGNITION AND FUNCTIONAL ALTERATIONS
OF AGED RAT HIPPOCAMPUS

C. A. BARNES

4

___

IDENTIFICATION OF MOLECULAR AND
CELLULAR MECHANISMS OF LEARNING AND MEMORY:
THE IMPACT OF GENE TARGETING

ALCINO J. SILVA, KARL PETER GIESE, AND PAUL W. FRANKLAND

5

NORMAL AGING AND ALZHEIMER'S DISEASE

BRADLEY T. HYMAN AND TERESA GÓMEZ-ISLA

6

TRANSGENIC MOUSE MODELS WITH NEUROFILAMENT-INDUCED PATHOLOGIES

JEAN-PIERRE JULIEN

7

TRANSGENIC MODELS OF AMYOTROPHIC LATERAL SCLEROSIS AND ALZHEIMER'S DISEASE

PHILIP C. WONG, DAVID R. BORCHELT, MICHAEL K. LEE, GOPAL THINAKARAN, SANGRAM S. SISODIA, AND DONALD L. PRICE

8

TOWARD A GENETIC ANALYSIS OF UNUSUALLY SUCCESSFUL NEURAL AGING

GEORGE M. MARTIN

9

THE ROLE OF THE PRESENILINS IN ALZHEIMER'S DISEASE

RUDOLPH E. TANZI

10

MECHANISMS OF NEURON DEATH IN NEURODEGENERATIVE DISEASES OF THE ELDERLY: ROLE OF THE LEWY BODY

JOHN Q. TROJANOWSKI, JAMES E. GALVIN, M. LUISE SCHMIDT,
PANG-HSIEN TU, TAKESHI IWATSUBO, AND VIRGINIA M.-Y. LEE

11

MICROTUBULE-ASSOCIATED PROTEIN TAU: BIOCHEMICAL MODIFICATIONS, DEGRADATION, AND ALZHEIMER'S DISEASE

SHU-HUI YEN, PARIMALA NACHARAJU, LI-WEN KO,
AGNES KENESSEY, AND WAN-KYNG LIU

12

A NOVEL GENE IN THE ARMADILLO FAMILY INTERACTS WITH PRESENILIN 1

KENNETH S. KOSIK, CAROLE HO, UDAYA LIYANGE, CYNTHIA LEMERE,
MIGUEL MEDINA, AND JIANHUA ZHOU

13

PUTATIVE LINKS BETWEEN SOME OF THE KEY PATHOLOGICAL FEATURES OF THE ALZHEIMER'S BRAIN

RÉMI QUIRION, DANIEL AULD, UWE BEFFERT, JUDES POIRIER,
AND SATYABRATA KAR

14

UNRAVELING THE CONTROVERSY OF HUMAN PRION PROTEIN DISEASES

ANDRÉA LEBLANC

15

TRANSLATIONAL CONTROL, APOPTOSIS, AND THE AGING BRAIN

EUGENIA WANG

16

ASTROCYTE SENESCENCE AND THE PATHOGENESIS
OF PARKINSON'S DISEASE

HYMAN M. SCHIPPER

CONTRIBUTORS

Numbers in parentheses indicate the pages on which the authors' contributions begin.

Marilyn S. Albert (1) Massachusetts General Hospital, Charlestown, Massachusetts 02129

Daniel Auld (181) Douglas Hospital Research Centre, McGill Center for Studies on Aging and Department of Psychiatry, McGill University, Montréal, Québec, Canada H4H 1R3

C. A. Barnes (51) Departments of Psychology and Neurology and Arizona Research Laboratories Division of Neural Systems, Memory and Aging, University of Arizona, Tucson, Arizona 85724

Uwe Beffert (181) Douglas Hospital Research Centre, McGill Center for Studies on Aging and Department of Psychiatry, McGill University, Montréal, Québec, Canada H4H 1R3

David R. Borchelt (107) Department of Pathology and the Division of Neuropathology, The Johns Hopkins University School of Medicine, Baltimore, Maryland 21205

Paul W. Frankland (67) Cold Spring Harbor Laboratory, Cold Spring Harbor, New York 11724

James E. Galvin (143) Center for Neurodegenerative Disease Research, Department of Pathology and Laboratory Medicine, University of Pennsylvania School of Medicine, and Department of Neurology, Allegheny University of the Health Sciences, Philadelphia, Pennsylvania 19104

Karl Peter Giese (67) Cold Spring Harbor Laboratory, Cold Spring Harbor, New York 11724

Teresa Gómez-Isla (83) Neurology Department, Massachusetts General Hospital, Boston, Massachusetts 02114

Carole Ho (171) Center for Neurologic Diseases, Brigham and Women's Hospital and Harvard Medical School, Boston, Massachusetts 02115

Bradley T. Hyman (83) Neurology Department, Massachusetts General Hospital, Boston, Massachusetts 02114

Takeshi Iwatsubo (143) Department of Neuropathology and Neuroscience, University of Tokyo, 113-0052 Tokyo, Japan

Jean-Pierre Julien (93) Centre for Research in Neuroscience, McGill University, The Montréal General Hospital Research Institute, Montréal, Québec, Canada H3G 1A4

Satyabrata Kar (181) Douglas Hospital Research Centre, McGill Center for Studies on Aging and Department of Psychiatry, McGill University, Montréal, Québec, Canada H4H 1R3

Agnes Kenessey (153) Department of Pathology, Albert Einstein College of Medicine, Bronx, New York 10461

Li-wen Ko (153) Department of Pathology, Albert Einstein College of Medicine, Bronx, New York 10461

Kenneth S. Kosik (171) Center for Neurologic Diseases, Brigham and Women's Hospital and Harvard Medical School, Boston, Massachusetts 02115

Andréa LeBlanc (201) The Bloomfield Center for Research in Aging, Lady Davis Institute for Medical Research, Sir Mortimer B. Davis Jewish General Hospital, and Department of Neurology and Neurosurgery, McGill University, Montréal, Québec, Canada H3T 1E2

Michael K. Lee (107) Department of Pathology and the Division of Neuropathology, The Johns Hopkins University School of Medicine, Baltimore, Maryland 21205

Virginia M.-Y. Lee (143) Center for Neurodegenerative Disease Research, Department of Pathology and Laboratory Medicine, University of Pennsylvania School of Medicine, Philadelphia, Pennsylvania 19104

Cynthia Lemere (171) Center for Neurologic Diseases, Brigham and Women's Hospital and Harvard Medical School, Boston, Massachusetts 02115

Wan-Kyng Liu (153) Department of Pathology, Albert Einstein College of Medicine, Bronx, New York 10461

Udaya Liyange (171) Center for Neurologic Diseases, Brigham and Women's Hospital and Harvard Medical School, Boston, Massachusetts 02115

Sonia J. Lupien (19) Aging Research Center, Douglas Hospital Research Center, McGill University, Montréal, Québec, Canada H4H 1R3

George M. Martin (125) Departments of Pathology and Genetics and the Alzheimer's Disease Research Center, University of Washington, Seattle, Washington 98195

Michael J. Meaney (19) Aging Research Center, Douglas Hospital Research Center, McGill University, Montréal, Québec, Canada H3W 1W5

Miguel Medina (171) Center for Neurologic Diseases, Brigham and Women's Hospital and Harvard Medical School, Boston, Massachusetts 02115

Parimala Nacharaju (153) Department of Pathology, Albert Einstein College of Medicine, Bronx, New York 10461

Judes Poirier (181) Douglas Hospital Research Centre, McGill Center for Studies on Aging and Department of Psychiatry, McGill University, Montréal, Québec, Canada H4H 1R3

Donald L. Price (107) Departments of Pathology, Neurology, and Neuroscience and Division of Neuropathology, The Johns Hopkins University School of Medicine, Baltimore, Maryland 21205

Rémi Quirion (181) Douglas Hospital Research Centre, McGill Center for Studies on Aging and Department of Psychiatry, McGill University, Montréal, Québec, Canada H4H 1R3

Hyman M. Schipper (243) The Bloomfield Centre for Research in Aging, Lady Davis Institute for Medical Research, Sir Mortimer B. Davis Jewish General Hospital, and Departments of Neurology and Neurosurgery and Medicine (Division of Geriatrics), McGill University, Montréal, Québec, Canada H3T 1E2

M. Luise Schmidt (143) Center for Neurodegenerative Disease Research, Department of Pathology and Laboratory Medicine, University of Pennsylvania School of Medicine, Philadelphia, Pennsylvania 19104

Alcino J. Silva (67) Cold Spring Harbor Laboratory, Cold Spring Harbor, New York 11724

Sangram S. Sisodia (107) Departments of Pathology, Neuroscience, and the Division of Neuropathology, The Johns Hopkins University School of Medicine, Baltimore, Maryland 21205

Rudolph E. Tanzi (135) Genetics and Aging Unit, Massachusetts General Hospital and Harvard Medical School, Charlestown, Massachusetts 02129

Gopal Thinakaran (107) Department of Pathology and the Division of Neuropathology, The Johns Hopkins University School of Medicine, Baltimore, Maryland 21205

John Q. Trojanowski (143) Center for Neurodegenerative Disease Research, Department of Pathology and Laboratory Medicine, University of Pennsylvania School of Medicine, Philadelphia, Pennsylvania 19104

Pang-Hsien Tu (143) Center for Neurodegenerative Disease Research, Department of Pathology and Laboratory Medicine, University of Pennsylvania School of Medicine, Philadelphia, Pennsylvania 19104

Eugenia Wang (223) The Bloomfield Center for Research in Aging, Lady Davis Institute for Medical Research, Sir Mortimer B. Davis Jewish General Hospital, and Department of Medicine, McGill University, Montréal, Québec, Canada H3T 1E2

Philip C. Wong (107) Department of Pathology and the Division of Neuropathology, The Johns Hopkins University School of Medicine, Baltimore, Maryland 21205

Shu-Hui Yen (153) Department of Research, Mayo Clinic Jacksonville, Jacksonville, Florida 32224

Jianhua Zhou (171) Center for Neurologic Diseases, Brigham and Women's Hospital and Harvard Medical School, Boston, Massachusetts 02115

PREFACE

Some age-dependent diseases may derive from a lifetime accumulation of either genetic or epigenetic events, and often both. With no small amount of effort, considerable progress has been made in uncovering the genetic underpinnings of a few neurodegenerative diseases. However, the progress has usually been restricted to "familial" disorders that constitute the smaller fraction of a given disease entity. Disorders that arise "sporadically" present an even greater challenge in attempts to determine cause. Advances in genetic epidemiology will likely have a role to play here and will aid in the discovery of heretofore unrecognized epigenetic risk factors and polymorphisms for age-dependent neurodegenerative diseases.

To further the exploration and understanding of these events and processes, this book brings together preeminent scientists whose research focus is on how the brain changes with age and how these molecular and biochemical changes affect behavior and cognition in the elderly. It is assumed that such changes contribute to the neurodegenerative processes not infrequently observed with increasing age and also to processes occurring in normal aging. The goal of this book is to exchange findings, discuss ideas, and assess what is currently known of the aging brain. Contributors represent disciplines across the spectrum of neurobiology from the morphological to the molecular and cellular as well as the cognitive. The chapters attempt to integrate the growing knowledge of molecular changes in the aging brain to cognitive patterns in later life. The participation by investigators working in overlapping fields, in areas requiring markedly different backgrounds and skills, made possible discussion, interpretation, and understanding of the significance of discrete findings on issues ranging from molecules to systems to the whole organism.

The contents of this book should provide unique opportunities for cross-fertilization and initiate a promising beginning for future advances in our understanding of how the brain functions and, possibly more importantly, how it degenerates in diseases associated with aging. This *Handbook of the Aging Brain* should provide an overview and a starting point for the reader interested in acquiring an integrated perspective on CNS aging and late-life changes in both the presence and the absence of disease.

D. Stephen Snyder
Eugenia Wang

1

NORMAL AND ABNORMAL MEMORY

AGING AND ALZHEIMER'S DISEASE

MARILYN S. ALBERT

Psychiatry/Gerontology
Massachusetts General Hospital
Charlestown, Massachusetts 02129

Changes in memory occur with age, but memory changes are also the earliest cognitive change seen in Alzheimer's disease (AD). Recent studies indicate that there is a considerable difference between the nature of the memory changes in aging and in AD and the underlying neurobiology responsible for those changes. Neuropathological and neuroimaging studies comparing persons with no evidence of AD during life with those in the very early stages of AD indicate which neurobiological changes are likely to be responsible for the memory changes seen in healthy aging and in the early stages of AD. The present chapter will review the changes in memory seen in healthy individuals across the age range and those that appear to characterize the early stages of AD and suggest their neurobiological correlates.

AGE-RELATED CHANGES IN MEMORY

There are substantial changes with age in explicit secondary memory, in contrast to the minimal age changes in sensory and primary memory (see Craik[1] and Poon[2] for a review). The age at which changes in secondary memory occur depends upon the methods that are used to test the memory store. Difficult

1

explicit memory tasks (e.g., delayed recall) demonstrate statistically significant differences by subjects in their 50s, in comparison to younger individuals.[3] Age decrements are greater on recall than recognition tasks. This is true whether words or pictures are used. Cueing during encoding or retrieval also alters the appearance of an age decline. Cueing at both encoding and retrieval produces the smallest age differences, whereas no cueing at either stage of the task maximizes age differences.[4] However, even with cued recall and recognition, there are often declines. Rabinowitz[5] reported a 33% age-related decrement in cued recall and an 11% age-related decrement in recognition when comparing young and old subjects (mean age 19 versus 68).

Figure 1.1 shows the performance of subjects across the age range on delayed recall of two lengthy paragraphs. That is, each subject is read two lengthy paragraphs; immediately after hearing each one and then again after 20 min, the subject is asked to state what he or she can recall of the paragraphs.

A close examination of these data indicates that the older individuals are not more rapidly forgetting what they learned but rather they are taking longer to learn the new information. For example, if one compares the difference between immediate and delayed recall over the life span, there are no statistically significant age differences.[6] Thus, if one allows older subjects to learn material well (i.e., to the point where few errors are made), they do not forget what they have learned more rapidly than the young. However, if older subjects are not given the ability to learn material to the same level of proficiency as younger individuals, after a delay, less information will be retained by the average older person.

However, there is considerable variability among older subjects on tasks of this sort. There are many healthy older subjects who have test scores that overlap those of subjects many years younger than themselves [e.g., about one-third

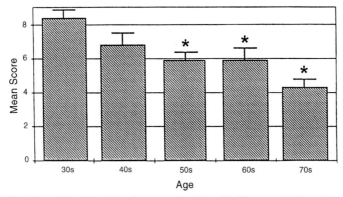

FIGURE 1.1   Delayed recall performance of subjects 30–80 years old. The subjects are asked to report what they remember of two lengthy paragraphs after a 15-min delay.

of healthy 70-year-old humans have delayed recall scores that overlap those of 30-year-olds (equated for education)].

Similar findings have emerged from studies in monkeys. Since free recall cannot be easily tested in monkeys, considerable effort has gone into developing a memory task that uses recognition but determines the quantity of information aging monkeys can retain across varying delay intervals. The delayed nonmatching to sample task (DNMS) is the most widely used method for assessing recognition memory in the nonhuman primate.[7–12] This task relies upon a two-alternative forced choice paradigm in which the monkey is required to discriminate which of two objects was recently presented. By using the delayed nonmatching to sample task, many studies have shown that aged monkeys are impaired at learning the nonmatching principle but are, at best, only mildly impaired across the delay conditions of the test.[13–18] Thus, like humans, the old monkeys take longer to learn something new but do not appear to forget this information more rapidly over lengthening delays than younger monkeys.

Another task, the Delayed Recognition Span Test (DRST),[19, 20] has demonstrated similar findings. This test requires a monkey to identify a novel stimulus from an increasing array of previously presented stimuli. The task has been administered to monkeys with two types of conditions: a spatial version and an object version. In both conditions the monkey first sees a board with 18 positions (food wells) and a stimulus object on one position. A second stimulus is added, while the board is obscured from view, and the animal is required to point to the new stimulus. On the first trial of the spatial version, the animal's task is to indicate which of two disks is occupying the new position on the board (i.e., a nonmatching to position task). On the first trial of the object version, the animal's task is to indicate which of two objects on the board is the new one (i.e., in essence, a nonmatching to sample task). However, since new disks or objects are then added in series to the previous ones on the board, the task becomes increasingly difficult. For example, young adult monkeys often achieve a span of five or greater before committing an error.

Middle-aged monkeys (16–23 years) are impaired on the spatial version of the DRST but not on the color version[21] (Fig. 1.2) This finding is similar to the declines in performance on delayed recall seen in middle-aged humans. Aged monkeys (25–27 years) are impaired relative to young adults (5–7 years) under both conditions of the task. This suggests that the performance of monkeys on the spatial version of the DRST may be functionally equivalent to the performance of humans on difficult delayed recall tasks. In addition, the difference between performance on the DNMS and the spatial condition of the DRST appears to be similar to what one sees between recall and recognition paradigms in aging humans.

Like aging humans, there is also considerable variability in the performance of aging monkeys. Among the oldest animals there are subjects who perform within the range of younger animals, even though the mean performance of the

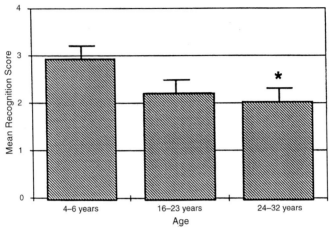

FIGURE 1.2   Spatial delayed recognition span scores by young adult monkeys (aged 5–7 years), middle-aged monkeys (aged 25–27 years), and elderly monkeys (aged 25–27 years).

group declines significantly. For example, on DNMS, the number of trials to criterion ranges from 50 to 220 and the number of errors range from 29 to 60. Among the older monkeys the trials to criterion range from 200 to 516 and the number of errors range from 50 to 115.

AGE-RELATED CHANGES IN BRAIN STRUCTURE AND FUNCTION

There is substantial evidence that as people age they show significant increases in cerebrospinal fluid (CSF) and decreases in brain tissue.[22, 23] That is, there is increasing atrophy with age. This alteration becomes statistically significant when subjects are in their 70s.[22]

However, recent studies in humans suggest that the decrease in brain tissue observed is primarily the result of decreases in white matter with age. For example, a magnetic resonance imaging (MRI) study measuring volumes of white matter, gray matter and CSF[24] found that decreases in gray matter are only marginally statistically significant, whereas the age-related decrease in the volume of white matter was substantial ($p \leq .001$) (Fig. 1.3). These findings are consistent with recent data in humans,[25–27] indicating that with advancing age neuronal loss in the cortex is either not significant or not as extensive as earlier reports (i.e., reports prior to 1984) had suggested.[28–33]

There are comparable data in monkeys. A recent study of MRIs in young and old monkeys demonstrates significant decreases in white matter volume with age

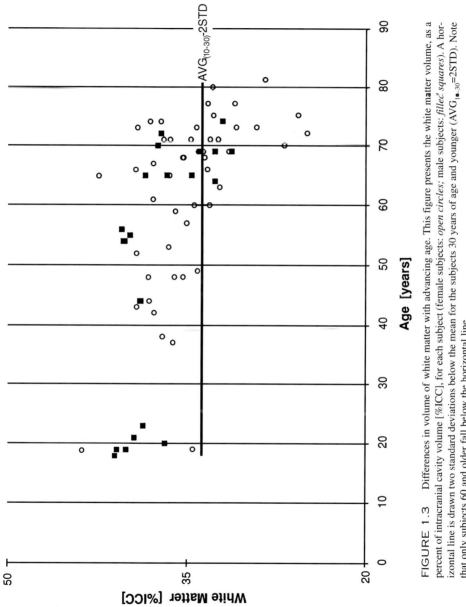

FIGURE 1.3   Differences in volume of white matter with advancing age. This figure presents the white matter volume, as a percent of intracranial cavity volume [%ICC], for each subject (female subjects: *open circles*; male subjects: *filled squares*). A horizontal line is drawn two standard deviations below the mean for the subjects 30 years of age and younger (AVG$_{(0-30)}$=2STD). Note that only subjects 60 and older fall below the horizontal line.

but minimal decreases in gray matter volume.[34] Neuronal counts of numerous brain regions in young and old monkeys also demonstrate minimal neuronal cortical loss with age, including the striate cortex,[35] motor cortex,[36] frontal cortex,[37] and the entorhinal cortex.[38] (It should be noted that these findings do not exclude focal changes with age in small areas of gray matter, such as selected areas of cortex or small subcortical nuclei, as discussed further below.)

Data in monkeys and humans also indicate that neuronal loss is highly selective within the hippocampal formation. For example, the subiculum shows a significant age-related loss in humans and a similar trend in monkeys; however, the CA1, CA2, and CA3 fields show no evidence of an age-related neuronal loss.[38–40]

A recent positron emission tomography (PET) study also suggests that the hippocampal formation is more highly functional with age than previously thought.[41] In this investigation, young and elderly subjects attempted to learn two types of word lists. The testing conditions were such that high levels of recall were achieved with one set of word lists, whereas low levels of recall were achieved with the other. Analyses of the PET data from both the young and older adults revealed blood flow increases in the hippocampal formation in comparisons in which the high recall condition was emphasized; bilateral blood flow increases were seen in the High Recall Minus Baseline comparison, and right unilateral increases were seen in the High Recall Minus Low Recall comparison. The pattern of blood flow increases were similar in both the young and older subjects, in both z-score units and location of change. This suggests that the features of recollection indexed by hippocampal activity operate in a similar manner in old and young.

These PET studies, however, also demonstrate differences in brain activation between young and older subjects. These were evident in comparisons emphasizing the low recall condition. Most striking was the difference between the groups in the Low Recall Minus Baseline comparison, which produced blood flow increases bilaterally in the anterior frontal lobe (centering on Area 10) for the young subjects but not the older ones. By contrast, the older subjects demonstrated unilateral blood flow increases in the right posterior frontal lobe (centering on Area 45) and the right motor area (centering on Area 4/6). These differences in frontal activation between the young and older adults may reflect differences in retrieval strategies when subjects are attempting to recall information that is not well learned. Based on the reports of preserved neuronal populations in the cortex cited above, it seems unlikely that these frontal activation differences between young and old are the result of substantial cortical neuronal loss. Age-related alterations in subcortical nuclei that project to the cortex seem to be the more likely explanation for these findings, since there is substantial neuronal loss in selected subcortical regions involved in neurotransmitter systems important for memory function, such as the basal forebrain and the locus coeruleus.[39, 42] For example, in humans and monkeys there is approximately a 50% neuronal loss with age in the basal forebrain and a 35–40% loss in the locus coeruleus and dorsal raphe.[43] This compares with an approximate loss of 5% in CA1 of the hippocampus. Although neuronal loss appears to be minimal in the hippocampus with age, recent reports suggest alterations in specific receptor types (e.g., N-methyl-D-aspartate [NMDA] receptors) that may play a role in memory function.[44]

In addition, there is an age-related decrease in dopaminergic binding sites in the caudate nucleus[45] and the substantia nigra and a loss of neurons in the substantia nigra of about 6% per year.[46] This loss of dopamine is thought to be responsible for many neurological symptoms that increase in frequency with age, such as decreased armswing and increased rigidity.[47] Changes in dopamine levels may also cause age-related changes in cognitive flexibility. This is suggested by the fact that patients with Parkinson's disease (a disorder associated with a loss of cells in the substantia nigra and a severe decline in dopamine levels) have cognitive deficits that have variably been described as "mental inflexibility,"[48] a disorder of the "shifting attitude,"[49] an "instability of cognitive set,"[50] and difficulty with "set formation, maintenance and shifting."[51] An age-related functional loss of dopamine has been demonstrated in monkeys,[52] making it possible to study this hypothesis in nonhuman primates.

MEMORY CHANGES IN EARLY ALZHEIMER'S DISEASE

The alterations in memory associated with early AD differ in important ways from those associated with age-related changes in memory. Difficulty with the acquisition of new information is generally the first and most salient symptom to emerge in patients with AD. When clinical neuropsychological tests are used to evaluate memory in AD patients, it is clear that recall and recognition performance are impaired in both the verbal and nonverbal domain.[53, 54]

Experimental studies have examined AD patients to determine whether the manner in which information is lost over brief delays is unique in any way to this patient group. The results of these studies suggest that a comparison of immediate and delayed recall performance may be a useful diagnostic measure for identifying patients with AD.

The first such study was conducted by Moss et al.[55] They compared patients with AD to a group of amnesic patients who had alcoholic Korsakoff's syndrome (KS), a group of dementing patients with Huntington's disease (HD), and a group of normal control subjects (NC). All of the subjects were administered the DRST, the task mentioned above with respect to studies of memory in monkeys. The recognition portion of the task is entirely comparable to that used in monkeys. Disks are employed on which are placed a variety of stimuli (words, colors, faces, patterns, etc.). During the recognition portion of the task, the disks are placed on a board one at a time (there are 16 disks in all). As each disk is added, the board is hidden from view. The subject is then asked to point to the disk that was added during the delay interval. In order to do this the subject must keep track of an increasingly long series of disks. The disks are then added one at a time, until the subject makes an error. This yields a delayed recognition span for each of the stimuli sets. All of the patient groups were impaired in their recognition performance with respect to the control subjects, but there is overlap among the patient groups. There was no significant difference among the three patient groups in their ability to recognize new spatial, color, pattern, or facial stimuli; patients with HD performed significantly better than the other two groups when verbal stimuli were used.

However, when the DRST was given to humans, a unique recall portion was added to the verbal recognition span paradigm. Fifteen seconds and two minutes after completion of the last verbal recognition trial, the subject was asked to recall the words that had been on the disks. In this condition the AD patients differ considerably from the other patients. They recall significantly fewer words over this brief delay interval (2 min) than either HD or KS patients. Although all three patient groups were equally impaired relative to normal control subjects at the 15-sec interval, patients with AD recalled significantly fewer words than either the HD or KS groups at the 2-min interval; in fact, only the AD group performed significantly worse at the longer interval as compared with the shorter interval (Fig. 1.4). It is notable that by the end of the 2-min interval, 11 of the 12 patients in the AD group could recall fewer than 3 of the 16 words presented repeatedly during recognition testing. Of these 11 patients, 7 were unable to recall any of the 16 words at the longer interval. Whereas the KS, HD, and normal control subjects lost an average of 10–15% of the verbal information between the 15-sec and 2-min delay intervals, patients with AD lost an average of 75% of the material. This pattern of recall performance demonstrated for the first time that patients with AD lose more information over a brief delay than other patients with amnesic or dementing disorders.

A similar pattern of results has been reported by numerous other investigators.[56–59] The findings of Hart et al.[57] are particularly notable. They administered a continuous recognition task to AD patients and control subjects and equated both groups of subjects for retention 90 sec after the task was completed. They then retested the subjects at 10 min, 2 hr and 48 hr after completion of the task. The AD patients showed a greater loss of information than the control subjects between the 90-sec interval and the 10-min interval, but not between the 10-min and 2-hr or 48-hr intervals, suggesting that intervals of 10 min or less may be optimal for differentiating AD patients from other patient groups and from control subjects.

Since these findings were first reported, additional patient groups have been compared to AD patients on tasks of this nature. They, likewise, appear to recall more information after a delay than patients with AD. Milberg and Albert[60] compared the performance of AD patients with that of progressive supranuclear palsy (PSP) patients. The two groups were equated for overall level of impairment on

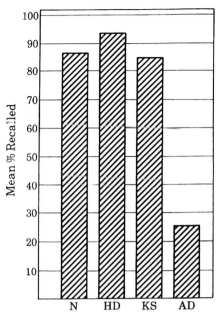

FIGURE 1.4   The difference between immediate and delayed recall on the verbal recall portion of the Delayed Recognition Span Test (DRST). The groups compared are normal control subjects *(N)*, patients with Alzheimer's disease *(AD)*, patients with Huntington's disease *(HD)*, and patients with alcoholic Korsakoff's syndrome *(KS)*.

the basis of the Mattis Dementia Rating Scale[61] and were equivalent in years of education. There was no difference between the patient groups on most of the tasks administered (e.g., vocabulary, digit span forward, similarities, block design). There was, however, a striking difference between the groups on both of the memory tasks. The AD patients were significantly impaired in comparison to the PSP patients on tests of both verbal and nonverbal memory.

A comparison of patients with AD and patients with frontotemporal dementia (FTD)[13] also demonstrates the severe recall deficits of the AD patients. Here again, patients with AD and patients with FTD, equated for overall level of cognitive impairment, were administered the DRST described earlier. As in the earlier study, the difference in total recall between the 15-sec and the 2-min delay interval (i.e., the savings score) differentiated the groups. The retention of the FTD patients over this delay interval approaches normality, whereas the AD patients lose a substantial amount of information.

In general, these findings suggest that the nature and severity of the AD patients' memory disturbance in relation to delays spanning the first 10 min after encoding is likely to be the result of a unique pattern of neuropathological and/or neurochemical dysfunction.

CHANGES IN BRAIN
STRUCTURE IN
EARLY ALZHEIMER'S DISEASE

The most likely explanation for the abnormalities in memory that character-ize the early stage of AD pertains to the damage to the hippocampal formation seen in AD patients.[62, 63] In the hippocampal formation, neuronal loss and abnor-mal formations with the cells (e.g., neurofibrillary tangles and neuritic plaques) are seen primarily in the entorhinal cortex and subiculum, the primary pathways that convey information into and out of the hippocampus. It has been suggested that abnormalities in these regions produce a functional isolation of the hip-pocampus.[63, 64] These findings suggest that neuropathological damage to medial temporal lobe structures may be responsible for the marked short-term memory impairment evident in the early stages of AD.

These results were first observed in patients with end-stage disease; however, they have recently been extended to patients with very mild disease.[65] Most strik-ing is the fact that the entorhinal cortex has neuronal loss of approximately 60% and 40% in layers 2 and 4, respectively. The normal control subjects examined in this investigation spanned the ages of 60 to 90. They were carefully screened to exclude individuals with AD and demonstrated no significant neuronal loss in layers 2 and 4 of the entorhinal cortex. Since this region is known to be critical-ly important for the acquisition and retention of new information,[66] abnormalities in these regions are likely to be responsible for the severe anterograde memory loss evident early in the course of AD.

MRI studies focusing on mildly impaired AD patients are entirely consistent with the neuropathological data cited above. They have uniformly reported sig-nificant and striking differences between MRI measures of the medial temporal lobe in AD patients and control subjects. Measures of the hippocampal forma-tion, the parahippocampal gyrus, the amygdalohippocampal complex, and the temporal horn of the lateral ventricles have demonstrated significant differences between mildly impaired AD patients and control subjects across a wide range of studies, using a variety of techniques.[67–74] These studies are consistent with computed tomography (CT) studies showing suprasellar cistern–temporal horn abnormalities in AD.[75–77]

A recent study[78] indicates that in order to identify AD in the prodromal phase (i.e., when subjects had evidence of recent declines in memory but were still "questionable"), very specific medial temporal lobe measures are needed. In this study, MRI measures were obtained when the subjects were "questionable" and the measures that predicted subsequent "conversion" to AD were examined. Two sets of MRI measures were employed. The first set pertained to regions of interest (ROI) used in previous studies with mild AD patients and control sub-jects. These ROIs were as follows and represented the total volume of each region: the hippocampal formation, the amygdala, the basal forebrain, the ante-rior cingulate gyrus, the head of the caudate nucleus, the temporal horn, the lateral ventricles, and the third ventricle. In the previous study a discriminant function analysis, in which measurements of the hippocampal formation and the temporal horn were combined, differentiated 100% of the AD patients from control subjects.[79]

For the second set of MRI measures, two ROIs were examined: (1) the entorhinal cortex, representative of a limbic structure known to have pathological changes early in the disease process (as described above), and (2) the superior temporal sulcus region,[65] representative of high-order association cortex, which is believed to be affected later in the disease process.[80]

As in previous studies, the first set of ROIs (i.e., the hippocampal formation, the amygdala, the basal forebrain, the anterior cingulate gyrus, the head of the caudate nucleus, the temporal horn, the lateral ventricles, and the third ventricle) revealed significant differences between the mildly impaired AD patients and normal control subjects. However, somewhat surprisingly, these measures were only minimally able to differentiate individuals who were control subjects from individuals who went on to develop AD during the longitudinal follow-up period ("converters").

The second group of ROIs, measures of the entorhinal cortex and the banks of the superior temporal sulcus region, readily distinguished control subjects from patients with mild AD. Importantly, however, these measures were also able to demonstrate substantial differences between control subjects and individuals who at the time of the MRI did not have sufficient clinical difficulties to warrant a diagnosis of dementia but subsequently converted over the follow-up period to a diagnosis of AD. The converters had a 30% smaller volume of the entorhinal cortex in comparison to the control subjects, which was highly statistically significant. The AD patients demonstrated a 50% loss in comparison to the control subjects. The measurement of the banks of the superior temporal gyrus in the converters showed a less substantial decrease in size (about 20%) in comparison to control subjects, but this was also highly significant. The AD patients demonstrated a 36% loss in comparison to the control subjects. Using these two MRI measures, a discriminant function analysis (adjusted for age and gender) demonstrated an overall accuracy of 93%. These findings suggest that measures of the entorhinal cortex and the banks of the superior temporal sulcus are likely to be highly useful in differentiating subjects with AD in prodromal phase from control subjects who are not likely to develop dementia.

It is particularly interesting to note that the degree of loss in the measure of the entorhinal cortex between the control subjects and the "questionable" subjects was extremely comparable to that seen in neuropathological studies of this region in similar subjects. In a study reporting the neuropathological examination of 20 cases with varying degrees of dementia at the time of death, stereological cell counts in "questionable" cases in the entorhinal cortex demonstrated profound loss of neurons, with a 60% and 40% loss from layers 2 and 4, respectively, and average loss in the entire region of 32%.[65] By contrast, in severe AD patients, layer 2 lost over 90% of its neurons, and overall loss was over 60%.

These figures compare closely with overall volume loss of 30% in "questionable" cases and 50% in individuals with severe AD.

In the same cases the banks of superior temporal gyrus showed a 10% loss, which was not significant in the "questionable" cases, but about a 50% loss in established moderate and severe AD.[65] The comparable figures from the current structural MRI study are 20% and 36% for volume loss based on MRI scans in "questionable" and established AD. These data support the hypothesis that the present MRI findings are likely to be a reflection of underlying AD neuropathology and that especially the measure of the entorhinal cortex will be highly useful in preclinical prediction of AD.

Recent functional neuroimaging data support the conclusions based on neuropathological and structural neuroimaging data. In addition, the functional data suggests that a distributed brain network may be involved in early AD.

Functional imaging techniques, such as single photon emission computed tomography (SPECT) and PET, have demonstrated regional abnormalities in brain perfusion in patients with AD (for a review see Jagust[81]). Temporoparietal abnormalities have emerged as the most consistent functional alteration seen in mild-to-moderately impaired AD patients. Among studies that have compared AD patients at different stages of disease, the temporal region appears to be the most sensitive and specific in distinguishing mildly impaired AD patients from control subjects, consistent with the structural findings described above.

Like the studies of prodromal AD patients, functional neuroimaging techniques also identify selected brain regions that are affected in the earliest stages of AD. A recently completed study using SPECT successfully identified over 80% of the "questionable" subjects who would later progress to the point where they met criteria for AD.[82]

Interestingly, the SPECT findings suggested that four brain regions were selectively involved in these early cases: the hippocampal formation, the posterior cingulate gyrus, the anterior thalamus, and the anterior cingulate gyrus. As described above, the hippocampal formation has been implicated for some time in the early stages of AD. However, the other brain regions that were significantly associated with the development of AD in this study have received relatively less attention. Nevertheless, a review of the literature suggests that they may also be involved in the very early stages of disease. For example, decreased perfusion in the posterior cingulate gyrus has been reported in two recent PET studies related to the development of AD,[83,84] and cingulate activation is frequently found in "activation" studies of memory.[85]

In addition, three of the four brain regions that were important in the present study for discriminating converters from normal control subjects are thought to be important in memory function in general. Recent memory studies in rodents indicate that the hippocampal formation, the posterior cingulate gyrus, and the anterior thalamus comprise a memory system critical for learning the relationships among cues (e.g., spatial, temporal, etc.).[86, 87] Neuroanatomical studies in monkeys also demonstrate strong interconnections between these three regions.[88–90] There is also increasing neuropathological evidence that all three sites are affected early in the course of AD.[79] The SPECT findings thus suggest that there are alterations in a distributed brain network with multiple nodes that may underly the earliest stages of AD.

CONCLUSION

These findings indicate that the memory changes that occur with age differ in important and significant ways from those seen in the early stages of AD and that selected, and differing, alterations in the brain are responsible for them. Understanding the nature of these cognitive changes and the brain alterations associated with them is the first step in developing methods of changing them.

ACKNOWLEDGMENTS

The preparation of this chapter was supported in part by funds from grant P01-AG04953 from the National Institute on Aging and by a grant from the Dana Foundation.

REFERENCES

1. Craik, F. (1977). Age differences in human memory. In J. E. Birren & K. W. Schaie (Eds.), *Handbook of the psychology of aging* (pp. 384–420). New York: Van Nostrand Reinhold.
2. Poon, L. (1985). Differences in human memory with aging. In J. E. Birren & K. W. Schaie (Eds.), *Handbook of the psychology of aging* (pp. 427–462). New York: Van Nostrand Reinhold.
3. Albert, M., Duffy, F., & Naeser, M. (1987). Non-linear changes in cognition with age and neurophysiological correlates. *Canad. J. Psychol., 41,* 141–157.
4. Craik, E., Byrd, M., & Swanson, J. (1987). Patterns of memory loss in three elderly samples. *Psychol. Aging, 2,* 79–86.
5. Rabinowitz, J. (1986). Priming in episodic memory. *J. Gerontol., 41,* 204–213.
6. Petersen, R., Smith, G., Kokmen, E., Ivnik, R., & Tangalos, E. (1991). Memory function in normal aging. *Neurol., 42,* 396–401.
7. Bachevalier, J., & Mishkin, M. (1989). Mnemonic and neuropathological effects of occluding the posterior cerebral artery in *Macaca mulatta. Neuropsyhologia, 27,* 83–105.
8. Murray, E., & Mishkin, M. (1984). Severe tactual as well as visual memory deficits follow combined removal of the amygdala and hippocampus in monkeys. *J. Neurosci., 4,* 2565–2580.
9. Murray, E., & Mishkin, M. (1986). Visual recognition in monkeys following rhinal cortical ablations combined with either amygdalectomy or hippocampectomy. *J. Neurosci., 6,* 1991–2003.
10. Mishkin, M. (1978). Memory in monkeys severely impaired by combined but not separate removal of amygdala and hippocampus. *Nature, 273,* 297–298.
11. Mahut, H., Zola-Morgan, S., & Moss, M. (1982). Hippocampal resections impair associative learning and recognition memory in the monkey. *J. Neurosci., 2,* 1214–1229.
12. Zola-Morgan, S., & Squire, L. (1984). Preserved learning in monkeys with medial temporal lesions: Sparing of motor and cognitive skills. *J. Neurosci., 4,* 1072–1085.
13. Moss, M., & Albert, M. (1988). Alzheimer's disease and other dementing disorders. In M. Albert & M. Moss (Eds.), *Geriatric neuropsychology* (pp. 145–178). New York: Guilford Press.

14. Moss, M., Rosene, D., & Peters, A. (1988). Effects of aging on visual recognition memory in the rhesus monkey. *Neurobiol. Aging, 9,* 495–502.

15. Arnsten, A., & Goldman-Rakic, P. (1990). Analysis of alpha-2 adrenergic agonist effects on the delayed non-matching-to-sample performance of aged rhesus monkeys. *Neurobiol. Aging, 11,* 583–590.

16. Bachevalier, J., Landis, L., Walker, M., Brickson, M., Mishkin, M., Price, D., & Cork, L. (1991). Aged monkeys exhibit behavioral deficits indicative of widespread cerebral dysfunction. *Neurobiol. Aging, 12,* 99–111.

17. Presty, S., Bachevalier, J., Walker, J., Struble, R., Price, D., Mishkin, M., & Cork, L. (1987). Age differences in recognition memory of the rhesus monkey *(Macaca mulatta). Neurobiol. Aging, 8,* 435–440.

18. Rapp, P., & Amaral, D. (1989). Evidence for task-dependent memory dysfunction in the aged monkey. *J. Neurosci., 9,* 3568–3576.

19. Rehbein, L. (1983). *Long-term effects of early hippocampectomy in the monkey.* Unpublished doctoral dissertation. Boston: Northeastern University.

20. Moss, M. (1983). Assessment of memory in amnesic and dementia patients: Adaptation of behavioral tests used with non-human primates. *INS Bull. 5,* 15.

21. Moss, M., Killiany, R., Herndon, J., Lai, A., & Herndon, J. (submitted). Recognition span in aged monkeys.

22. Stafford, J., Albert, M., Naeser, M., Sandor, T., & Garvey, A. (1988). Age-related differences in computed tomographic scan measurements. *Arch. Neurol., 45,* 405–419.

23. Zatz, L., Jernigan, T., & Ahumada, A. (1982). Changes in computed cranial tomography with aging: Intracranial fluid volume. *Am. J. Neurorad., 3,* 1–11.

24. Guttmann, C., Jolesz, F., Kikinis, R., Killiany, R., Moss, M., Sandor, T., & Albert, M. (in press). White matter and gray matter differences with age. *Neurology.*

25. Haug, H. (1984). Macroscopic and microscopic morphometry of the human brain and cortex. A survey in the light of new results. *Brain Pathol., 1,* 123–149.

26. Terry, R., Deteresa, R., & Hansen, L. (1987). Neocortical cell counts in normal human adult aging. *Ann. Neurol., 21,* 530–539.

27. Leuba, G., & Garey, L. (1989). Comparison of neuronal and glial numerical density in primary and secondary visual cortex. *Exp. Brain. Res., 77,* 31–38.

28. Brody, H. (1955). Organization of cerebral cortex. III. A study of aging in the human cerebral cortex. *J. Comp. Neurol., 102,* 511–556.

29. Brody, H. (1970). Structural changes in the aging nervous system. *Interdiscipl. Top. Gerontol., 7,* 9–21.

30. Colon, E. (1972). The elderly brain. A quantitative analysis of the cerebral cortex in two cases. *Psychiatr. Neurol. Neurochir., 75,* 261–270.

31. Shefer, V. (1973). Absolute number of neurons and thickness of the cerebral cortex during aging, senile and vascular dementia, and Pick's and Alzheimer's diseases. *Neurosci. Behav. Physiol., 6,* 319–324.

32. Henderson, G., Tomlinson, B., & Gibson, P. (1988). Cell counts in human cerebral cortex in normal adults throughout life, using an image analysing computer. *J. Neurol. Sci., 46,* 113–136.

33. Anderson, J., Hubbard, B., Coghill, G., & Slidders, W. (1983). The effect of advanced old age on the neurone content of the cerebral cortex. Observations with an automatic image analyser point counting method. *J. Neurol. Sci., 58,* 233–244.

34. Rosene, D., Lai, Z., Killiany, R., Moss, M., Jolesz, D., Sandor, T., & Albert, M. (in press). Age-related loss of white matter with preservation of gray matter in the forebrain of the rhesus monkey: An MRI study. *Neurobiol. Aging.*

35. Vincent, S., Peters, A., & Tigges, J. (1989). Effects of aging on neurons within area 17 of rhesus monkey cerebral cortex. *Anat. Rec., 223,* 329–341.

36. Tigges, J., Herndon, J., & Peters, A. (1992). Neuronal population of area 4 during life span of rhesus monkeys. *Neurobiol. Aging, 11,* 201–208.

37. Peters, A., Leahu, D., Moss, M., & McNally, K. (1994). The effects of aging on area 46 of the frontal cortex of the rhesus monkey. *Cerebr. Cortex, 6,* 621–635.
38. Amaral, D. (1993). Morphological analyses of the brains of behaviorally characterized aged non-human primates. *Neurobiol. Aging, 14,* 671–672.
39. Rosene, D. (1993). Comparing age-related changes in the basal forebrain and hippocampus of the rhesus monkey. *Neurobiol. Aging, 14,* 669–670.
40. West, M., Coleman, P., Flood, D., & Troncoso, J. (1994). Differences in the pattern of hippocampal neuronal loss in normal ageing and Alzheimer's disease. *Lancet, 344,* 769–772.
41. Schacter, D., Savage, C., Alpert, N., Rauch, S., & Albert, M. (1996). The role of hippocampus and frontal cortex in age-related memory changes: A PET study. *NeuroReport, 7,* 1165–1169.
42. Chan-Palay, V., & Asan, E. (1989). Quantitation of catecholamine neurons in the locus ceruleus in human brains of normal young and older adults in depression. *J. Comp. Neurol., 287,* 357–372.
43. Kemper, T. (1993). The relationship of cerebral cortical changes to nuclei in the brainstem. *Neurobiol. Aging, 14,* 659–660.
44. Gazzaley, A., Siegel, S., Kordower, J., Mufson, E., & Morrison, J. (1996). Circuit-specific alterations of N-methyl-D-aspartate receptor subunit 1 in the dentate gyrus of aged monkeys. *Proc. Natl. Acad. Sci., 93,* 3121–3125.
45. Severson., J., Marcusson, J., Winblad., B., & Finch, C. (1982). Age-correlated loss of dopaminergic binding sites in human basal ganglia. *J. Neurochem., 39,* 1623–1631.
46. McGeer, P., McGeer, E., & Suzuki, J. (1977). Aging and extrapyramidal function. *Arch. Neurol., 34,* 33–35.
47. Odenheimer, G., Funkenstein, H., Beckett, L., Chown, M., Pilgrim, D., Evans, D., & Albert, M. (1994). Comparison of neurologic changes in successfully aging persons vs the total aging population. *Arch. Neurol., 51,* 573–580.
48. Lees, A., & Smith, E. (1983). Cognitive deficits in the early stages of Parkinson's disease. *Brain, 106,* 257–270.
49. Cools, A., van den Bercken, J., Horstink, M., van Spaendonck, K., & Berger, H. (1984). Cognitive and motor shifting aptitude disorder in Parkinson's disease. *J. Neurol. Neurosurg. Psychiat., 47,* 443–453.
50. Flowers, K., & Robertson, C. (1985). The effect of Parkinson's disease on the ability to maintain mental set. *J. Neurol. Neurosurg. Psychiat., 48,* 517–529.
51. Taylor, A., Saint-Cyr, J., & Lang, A. (1987). Parkinson's disease: Cognitive changes in relation to treatment response. *Brain, 110,* 35–51.
52. Arnsten, A., Cai, J., Steere, J., & Goldman-Rakic, P. (1994). Dopamine D2 receptor mechanisms in the cognitive performance of young adult and aged monkeys. *Psychopharm., 116,* 143–151.
53. Wilson, R., Bacon, L., Fox, J., & Kaszniak, A. (1983). Primary and secondary memory in dementia of the Alzheimer type. *J. Clin. Neuropsychol., 5,* 337–344.
54. Storandt, M., & Hill, R. (1989). Very mild senile dementia of the Alzheimer type II. Psychometric test performance. *Arch. Neurol., 46,* 383–386.
55. Moss, M., Albert, M., Butters, N., & Payne, M. (1986). Differential patterns of memory loss among patients with Alzheimer's disease, Huntington's disease, and alcoholic Korsakoff's syndrome. *Arch. Neurol., 43,* 239–246.
56. Butters, N., Salmon, D., Heindel, W., & Granholm, E. (1988). Episodic, semantic and procedural memory: Some comparisons of Alzheimer's and Huntington's disease patients. In R. D. Terry (Ed.), *Aging and the brain* (pp. 63–87). New York: Raven Press.
57. Hart, R., Kwentus, J., Harkins, S., & Taylor, J. (1988). Rate of forgetting in mild Alzheimer's-type dementia. *Brain Cognition, 7,* 31–38.
58. Welsh, K., Butters, N., Hughes, J., Mohs, R., & Heyman, A. (1991). Detection of abnormal memory decline in mild cases of Alzheimer's disease using CERAD neuropsychological measures. *Arch. Neurol., 48,* 278–281.
59. Tierney, M., Nores, A., Snow, W., Reid, D., & Zorzitto, M. (1994). Utility of the Rey Auditory

Verbal Learning Test in the differentiation of normal aging from Alzheimer's and Parkinson's disease. *Psychol. Assess., 6,* 129–134.

60. Milberg, W., & Albert, M. (1989). Cognitive differences between patients with PSP and Alzheimer's disease. *J. Clin. Exper. Neuropsychol., 11,* 605–614.

61. Mattis, S. (1976). Mental status examination for organic mental syndrome in the elderly patient. In L. Bellack & T. Karasu (Eds.), *Geriatric psychiatry* (pp. 71–121). New York: Grune & Stratton.

62. Ball, M. (1977). Neuronal loss, neurofibrillary tangles and granulovacuolar degeneration in the hippocampus with aging and dementia. *Acta Neuropathol., 37,* 111–118.

63. Hyman, G., Van Hoesen, G., Kromer, C., & Damasio, A. (1985). Alzheimer's disease: Cell specific pathology isolates the hippocampal formation. *Science, 225,* 1168–1170.

64. Hyman, B., Van Hoesen, G., Kromer, L., & Damasio, A. (1986). Perforant pathway changes and the memory impairment of Alzheimer's disease. *Ann. Neurol., 20,* 472–481.

65. Gómez-Isla, T., Price, J., McKeel, D., Morris, J., Growdon, J., & Hyman, B. (1996). Profound loss of layer II entorhinal cortex occurs in very mild Alzheimer's disease. *J. Neurosci., 16,* 4491–4500.

66. Zola-Morgan, S., Squire, L., & Ramos, S. (1994). Severity of memory impairment in monkeys as a function of locus and extent of damage within the medial temporal lobe memory system. *Hippocampus, 4,* 483–495.

67. Seab, J. P., Jagust, W. J., Wong, S. T., Roos, M. S., Reed, B. R., & Budinger, T. F. (1988). Quantitative NMR measurements of hippocampal atrophy in Alzheimer's disease. *Mag. Res. Med., 8,* 200–208.

68. Kesslak, J., Nalcioglu, O., & Cotman, C. (1991). Quantification of magnetic resonance scans for hippocampal and parahippocampal atrophy in Alzheimer's disease. *Neurology, 41,* 51–54.

69. Jack, C., Petersen, R., O'Brien, P., & Tangalos, E. (1992). MR-based hippocampal volumetry in the diagnosis of Alzheimer's disease. *Neurology, 42,* 183–188.

70. Killiany, R. J., Moss, M. B., Albert, M. S., Sandor, T., Tieman, J., & Jolesz, F. (1993). Temporal lobe regions on magnetic resonance imaging identify patients with early Alzheimer's disease. *Arch. Neurol., 50,* 949–954.

71. Convit, A., DeLeon, M., Golomb, J., George, A., Tarshish, C., Bobinski, M., Tsui, W., DeSanti, S., Weigel, J., & Wisniewski, H. (1993). Hippocampal atrophy in early Alzheimer's disease: Anatomic specificity and validation. *Psychiat. Qtr., 64.,* 371–387.

72. Lehericy, S. Baulac, M., Chiras, J., Pierot, L., Martin, N., Pillon, B., Deweer, B., Dubois, B., & Marsault, C. (1994). Amygdalohippocampal MR volume measurements in the early stages of Alzheimer's disease. *Am. J. Neuroradiol., 15,* 929–937.

73. Ikeda M., Tanabe, H., Nakagawa, Y., et al. (1994). MRI-based quantitative assessment of the hippocampal region in very mild to moderate Alzheimer's disease. *Neuroradiology, 36,* 7–10.

74. Laasko, M., Soininen, H., Partanen, K., Hallikainen, Lehtovirta, M., Hanninen, T., Vainio, P., & Reikkinen, P. (1995). The interuncal distance in Alzheimer's disease and age-associated memory impairment. *Am. J. Neuroradiol., 16,* 727–734.

75. Sandor, T., Albert, M., Stafford, J., & Harpley, S. (1988). Use of computerized CT analysis to discriminate between Alzheimer patients and normal control subjects. *Am. J. Neuroradiol., 9,* 1181–1187.

76. George, A., de Leon, M., Stylopoulos, L., Miller, J., Kluger, A., Smith, G., & Miller, D. (1991). CT diagnostic features of Alzheimer disease: Importance of the choroidal/hippocampal fissure complex. *Am. J. Neuroradiol., 12,* 583–584.

77. Sandor, T., Jolesz, F., Tieman, J., Kikinis, R., Jones, K., & Albert M. (1992). Comparative analysis of CT and MRI scans in Alzheimer patients and controls. *Arch. Neurol., 49,* 381–384.

78. Killiany, R., Gómez-Isla, T., Moss, M., Kikinis, R., Jolesz, F., Sandor, T., Hyman, B., & Albert, M. (submitted). Preclinical prediction of Alzheimer's disease using structural MRI measures of the entorhinal cortex.

79. Killiany, R. J., Moss, M. B., Albert, M. S., Sandor, T., Tieman, J., & Jolesz, F. (1993). Temporal

lobe regions on magnetic resonance imaging identify patients with early Alzheimer's disease. *Arch. Neurol., 50,* 949–954.

80. Braak, H., & Braak, E. (1993). Alzheimer neuropathology and limbic circuits. In B. Vogt & M. Gabriel (Eds.), *Neurobiology of cingulate cortex and limbic thalamus: A comprehensive handbook* (pp. 606–626). Boston: Birkhauser.

81. Jagust, W. (1996). Functional imaging patterns in Alzheimer's disease. *Annal NY Acad. Sci. 777,* 30–36.

82. Johnson, K., Jones, K., Holman, B. L., Becker, A., Spiers, P., Satlin, A., & Albert, M. (in press). Preclinical prediction of Alzheimer's disease using SPECT. *Neurology.*

83. Reiman, E., Caselli, R., Yun, L., Chen, K., Bandy, D., Minoshima, S., Thibideau, S., & Osbourne, D. (1996). Preclinical evidence of a genetic risk factor Alzheimer's disease in apolipoprotein E type homozygotes using positron emission tomography. *New Engl. J. Med., 334,* 752–758.

84. Minoshima, S., Frey, K., Koeppe, R., Foster, N., & Kuhl, D. (1995). A diagnostic approach to Alzheimer's disease using three-dimensional stereotactic surface projections of Fluorine-18-FDG PET. *J. Nucl. Med., 36,* 1238–1248.

85. Ungerleider, L. (1995). Functional brain imaging studies of cortical mechanisms for memory. *Science, 270,* 769–775.

86. Sutherland, R., & Rudy, J. (1989). Configurational association theory: The role of the hippocampal formation in learning, memory and amnesia. *Psychobiology, 17,* 129–144.

87. Sutherland, R., & Rudy, J. (1991). Exceptions to the rule of space. *Hippocampus, 1,* 250–252.

88. Vogt, B., Pandya, D., & Rosene, D. (1987). Cingulate cortex of the rhesus monkey: I. Cytoarchitecture and thalamic afferents. *J. Comp. Neurol., 262,* 256–270.

89. Van Hoesen, G. (1991). Hippocampal cortical and subcortical neural systems: Their origins and targets in the monkey. In L. Squire, M. Mishkin, & A. Shimamura (Eds.), *Learning and memory, discussions in neuroscience:* Vol. 6 (pp. 20–28). Amsterdam: Elsevier.

90. Pandya, D., Van Hoesen, G., & Mesulam, M.-M. (1981). Efferent connections of the cingulate gyrus in the rhesus monkey. *Exp. Brain. Res. 42,* 319–330.

2

STRESS, GLUCOCORTICOIDS, AND HIPPOCAMPAL AGING IN RAT AND HUMAN

SONIA J. LUPIEN*† AND
MICHAEL J. MEANEY*

*Aging Research Center
Douglas Hospital Research Center
McGill University, Montréal, Québec, Canada H4H 1R3
†Research Center, Montreal Geriatric Institute
Montréal, Québec, Canada H3W 1W5

Perhaps the most prominent feature of human aging is the variability to which there occurs a decline in intellectual process. The objective of our research program is to understand the basis of these individual differences in neurological function amongst the elderly. We are examining whether individual differences in hypothalamic-pituitary-adrenal (HPA) activity might contribute to the variability in neuropsychological function in elderly humans. We propose that prolonged exposure to elevated levels of cortisol, the principle glucocorticoid in humans, compromises hippocampal integrity leading to impairments in cognitive functions that depend upon hippocampal function.

This hypothesis first emerged from studies on brain aging in rodents. In the rat, as in humans,[1] increased HPA activity and the resulting increase in circulating glucocorticoid levels are not inevitable consequences of aging. Rather, increased glucocorticoid levels are selectively associated with neuropathology and cognitive impairments in aged rats.[2] About 25–35% of Long-Evans rats show pronounced memory impairments, and these animals also show increased HPA activity. In contrast, about 40% of the animals show no evidence of either memory impairments or HPA dysfunction. Subsequent studies have shown that basal plasma corticosterone levels among aged rats are significantly correlated with hippocampal dysfunction and degeneration and spatial learning deficits.[2–5] Moreover, adrenalectomy at midlife, with low-level corticosterone replacement, attenuates hippocampal degeneration and cognitive decline in the rat,[6] suggesting that the elevated glucocorticoid levels directly contribute to the hippocampal dysfunction and the development of cognitive impairments. Elevated levels of glucocorticoids compromise hippocampal function through effects on Ca^{2+} and glucose metabolism, morphology, and, in the extreme, survival. Hippocampal long-term potentiation describes a long-lasting enhancement in synaptic efficacy that occurs in response to high-frequency stimulation (an electrophysiological model of memory). Thus, high basal levels of glucocorticoids have been shown to impair long-term potentiation, which suggests that they may have detrimental effects on memory. In recent longitudinal studies with elderly humans we have found evidence for a similar relationship between individual differences in HPA function on the one hand and hippocampal morphology and cognitive function on the other.

HPA AXIS

The HPA axis, along with the sympathoadrenal system, governs metabolic responses to the slings and arrows of everyday life, as well as to the beleaguering demands that prevail under conditions of chronic, severe stress. These hormones, along with the adrenal catecholamines, support a range of metabolic responses[7, 8] that serve to ensure (1) the availability of sufficient energy substrates in circulation during periods when there is inevitably increased cellular activity in vital organs and (2) an accompanying increase in the requirement for fuel. This increase in energy substrates occurs in response to the stress-induced release of catecholamines and glucocorticoids from the adrenal gland, as well as a variety of other glucoregulators, which produce lipolysis, glycogenolysis, and protein catabolism, producing increased levels of fatty acids and triglycerides, and also resulting directly or indirectly (through gluconeogenesis) in the ability of the body to defend blood glucose levels. These actions assist the organism during stress, in part, by increasing the availability of energy substrates at the expense of existing energy stores. In addition, there is a transient inhibition of metabolically costly and nonessential anabolic processes such as growth and reproduction. These responses are absolutely essential for survival. In even a briefly fasted state, adrenalectomized animals seem unable to survive serious stress.

Under most conditions, the HPA axis lies under the dominion of specific releasing factors secreted by neurons located in the paraventricular nucleus (PVNh) of the hypothalamus. Most notable among these releasing factors is corticotropin-releasing hormone (CRH). CRH neurons from the parvocellular cells of the PVNh project extensively to the portal capillary zone of the median eminence, furnishing a potent excitatory signal for the synthesis and release of adrenocorticotropic hormone (ACTH) from the anterior pituitary. CRH, along with co-secretagogues such as arginine vasopressin (AVP) and oxytocin provide the signals by which neural inputs are coded into endocrine signals through their actions on pituitary corticotrophes.[9–11] Dynamic changes in ACTH levels occurring during the circadian peak or in response to stress are associated with changes in the secretion of one or more of these secretagogues from hypothalamic neurons into the portal system of the anterior pituitary, enhancing the synthesis and release of ACTH. Elevated ACTH levels, in turn, increase the synthesis and release of glucocorticoids from the adrenal gland. Although glucocorticoid responses to stress are essential for survival, prolonged exposure to elevated "stress hormone" levels can present a serious health risk, leading to a suppression of anabolism, decreased sensitivity to insulin and a risk of steroid-induced diabetes, hypertension, hyperlipidemia, hypercholesterolemia, arterial disease, amenorrhea, and the impairment of growth and tissue repair, as well as immunosuppression.[8, 12] Herein lies the dilemma: The same "stress hormones" that are the basis for survival during periods of stress can, in excess, produce illness. Indeed, the glucocorticoids and catecholamines, along with CRH are primary mediators of stress-induced illness. It is clearly in the organism's best interest to limit such metabolically costly responses.

The responsivity of the HPA axis to stress is, in part, determined by the ability of the glucocorticoids to regulate ACTH release (i.e., glucocorticoid negative feedback). Circulating glucocorticoids feedback onto the pituitary gland and specific brain regions to inhibit the secretion of releasing factors from hypothalamic neurons and pituitary ACTH.[13–19] In addition to pituitary and hypothalamic sites there is now considerable evidence for the importance of the limbic system, the hippocampus, and the frontal cortex in the regulation of HPA activity.[15–17, 20, 21] Both the hippocampus[19] and the frontal cortex[20, 22] contain high levels of corticosteroid receptors.

GLUCOCORTICOIDS AND AGING

There is now considerable evidence for the idea that glucocorticoids are involved in the occurrence of hippocampal pathology and/or the exacerbation of existing hippocampal pathology in later life in rodents. This work with nonhuman species has shown that increased exposure to glucocorticoids in later life is associated with the disruption of electrophysiological function, atrophy, and ultimately the death of hippocampal neurons, all of which can lead to the emergence of severe cognitive deficits. Together, these features are common to "unsuccessful" central nervous system (CNS) aging.

Plasma glucocorticoid and ACTH levels have generally been found to increase with age in the rat under both basal and poststress conditions.[3, 23–33] There has been some controversy here, with certain groups having reported no differences in basal corticosterone levels between young and aged rats. Sapolsky[4] noted that those studies in which young and aged rats did not differ in plasma corticosterone levels, reported AM basal corticosterone levels of 10 μg/dl or higher. This value is at least 3–4 times higher than one would expect. This suggests an inappropriate method of basal sampling and is likely due to the fact that many of these studies were not expressly designed to examine basal hormone levels. Each of those studies with more realistic AM basal corticosterone levels (i.e., <10 µg/dl) found that corticosterone levels were generally higher in the aged rats. As described below, much of the conflicting data may also refer to naturally occurring individual differences in HPA function in aged rats.

Negative feedback appears to be impaired in the aged rat.[34, 35] In later life, certain extrahypothalamic regions, which are critically involved in the HPA negative-feedback process, become less sensitive to circulating glucocorticoids (the signal that initiates negative feedback[36]). This is due, in part at least, to an apparent loss of intracellular corticosteroid receptors,[27, 31, 32, 37–39] decreasing the sensitivity of these structures to circulating glucocorticoids and dampening the inhibitory control over ACTH release.

Circulating glucocorticoids bind with high affinity to two corticosteroid receptor subtypes; the mineralocorticoid (or type I) and glucocorticoid (or type II) receptors. Both receptor subtypes have been implicated in mediating glucocorticoid feedback effects.[13, 15] These receptors exhibit a subtle, but important, difference in their affinity for corticosterone (or cortisol in humans). Thus, mineralocorticoid receptors bind corticosterone with an affinity that is about 2–5 times higher than that of the glucocorticoid receptor (i.e., 0.5–1.0 nM versus 2.0–5.0 nM). The significance of this difference in affinity is apparent under basal corticosterone levels. Low basal corticosterone levels characteristic of the inactive phase of the cycle (daytime in rodents, nighttime in humans) serve to activate largely mineralocorticoid receptors, whereas the elevated glucocorticoid levels characteristic of waking activate both mineralocorticoid and glucocorticoid receptors. Thus, the evidence suggests a stronger role for the mineralocorticoid receptor in the regulation of ACTH release during periods when basal glucocorticoid levels are low and both receptors during periods when basal HPA activity is highest.

The age-related decrease in corticosteroid receptors involves both the mineralocorticoid and glucocorticoid receptor subtypes (see Meaney et al.[40] for a review), and this decrease in corticosteroid receptors is largely specific to the hippocampus; there is no age-related loss of corticosteroid receptors in such obvious target sites as the hypothalamus and the pituitary gland.[27, 31, 32, 37–39, 41] This finding has lead to the idea that the loss of corticosteroid receptors leads to a decreased sensitivity of relevant, extrahypothalamic sites to circulating glucocorticoids and thus reduced feedback sensitivity, which results in increased HPA activity with age.[34] There is now abundant evidence for the importance of the hippocampus in feedback regulation of HPA function (see Sapolsky et al.,[34] McEwen et al.,[19] and Jacobson and Sapolsky[17] for reviews). Lesions to the hippocampus result in elevated corticosterone levels under basal and poststress conditions,[42–44] whereas glucocorticoid implants into the hippocampus normalize ACTH levels in adrenalectomized rats.[45] Hippocampectomized animals show reduced suppression of HPA activity following exogenous glucocorticoid administration.[46] Finally, hippocampectomy results in increased CRH mRNA levels in the parvocellular region of the PVN (a region which contains neurons whose axons terminate in the median eminence[47]) and increased portal concentrations of ACTH secretagogues.[4] The negative-feedback effects of glucocorticoids at the level of the hippocampus appear to involve both corticosteroid receptor subtypes. Thus, implants of either both RU 28318, an antagonist selective for the mineralocorticoid sites, or RU 38486, an antagonist selective for the glucocorticoid site, increase basal and poststress HPA activity (i.e., antagonize endogenous glucocorticoid negative feedback.[45] Sapolsky et al.[48] have shown that levels of glucocorticoid receptor occupancy in the hippocampus and hypothalamus are negatively correlated with portal concentrations of CRH and AVP. Finally, both mineralocorticoid and glucocorticoid receptor agonists suppress ACTH release when implanted directly into the hippocampus.[45]

There is a decrease in both mineralocorticoid and glucocorticoid receptors in the hippocampus of the aging rat. We have also shown that by 24 months of age the loss of hippocampal glucocorticoid receptors is paralleled by a decrease in glucocorticoid receptor mRNA expression.[39] The finding of reduced mineralocorticoid receptors in aged rats has proven very robust, less so the situation with glucocorticoid receptors. Some groups have reported finding no differences in hippocampal glucocorticoid receptor density between young and old rats. We feel that the answer to this problem lies in the naturally occurring individual variation in HPA dysfunction in aged rats (see the following).

CONSEQUENCES OF ELEVATED GLUCOCORTICOID LEVELS

This increase in circulating corticosterone has important consequences for neuronal function. Elevated levels of glucocorticoids have been found to compromise hippocampal function in a number of ways, and in each case there is reason to assume that these effects are associated with a loss of function. Thus, exposure to elevated glucocorticoid levels has been found to (1) produce atrophy of dendritic processes, (2) dampen electrophysiological signal conduction within hippocampal neurons in response to excitatory signals, (3) compromise the survival of hippocampal pyramidal cells, and (4) dampen compensatory responses (e.g., reactive synaptogenesis) of surviving hippocampal neurons to the loss of cortical inputs (i.e., decreased synaptic plasticity).

GLUCOCORTICOIDS AND HIPPOCAMPAL ELECTROPHYSIOLOGY

It has been known for some time that glucocorticoids can decrease unit activity in the dorsal hippocampus.[49] Joels and DeKloet[50] and Kerr et al.[51] have demonstrated that in young adult rats glucocorticoids increase both the amplitude and the duration of Ca^{2+} dependent slow after hyperpolarization (AHP) in dorsal hippocampus slice preparations; this is due to a direct, glucocorticoid effect on Ca^{2+} influx.[52] The effect is mediated by glucocorticoid receptors, and the time course is consistent with that of a genomic effect. Such AHPs dampen the neuron's response to subsequent excitatory inputs. Interestingly, reduced AHPs are normally associated with the acquisition of new information. Glucocorticoids also decrease the norephinephrine-evoked augmentation of depolarization-induced neuronal activity in CA1 neurons. Together, these effects decrease overall excitability in hippocampal neurons. Kerr et al.[51] demonstrated that the same pattern of effects is apparent in aged versus young rats, and the difference is reversed when the aged animals are adrenalectomized.

Kerr et al.[51] also provided evidence that the increased slow AHP was mediated by an increase in Ca^{2+} conductance (slow AHPs are related to slow Ca^{2+}-dependent potassium conductance). The effect on Ca^{2+} homeostasis is of considerable interest considering the wealth of evidence demonstrating the cytotoxic effects of elevated intracellular Ca^{2+}. Indeed, there may occur a continuum of glucocorticoid effects: In the early stages, the elevated glucocorticoid signal might be associated with elevated Ca^{2+}-dependent AHPs, compromising hippocampal function, and continued exposure to elevated glucocorticoid levels might then result in neuronal atrophy and potential death.[53] The studies from Landfield's group have shown that increased Ca^{2+} influx through voltage-sensitive channels at the soma (producing increased AHPs) is a major feature of hippocampal aging in the rat.[52]

These studies provide clear evidence of short-term effects of glucocorticoids on hippocampal function. These findings also suggest that cognitive processes dependent on hippocampal function would be compromised in the presence of elevated glucocorticoid levels (again, independent of neuron loss). Evidence for such an effect has emerged from both in vivo and in vitro experiments using the long-term potentiation model of synaptic plasticity (an electrophysiological model for learning and memory). Elevated levels of glucocorticoids dampen long-term potentiation (LTP) formation in the hippocampus in a concentration-dependent manner.[54, 55] Interestingly, in a lower range of concentrations corticosterone actually enhances LTP formation, resulting in an inverted U-shaped function between steroid levels and potentiation.[56] Overall, low-moderate basal levels of corticosterone, which activate principally mineralocorticoid receptors,[30, 57] facilitate LTP formation, whereas high basal to stress levels, which activate glucocorticoid receptors, impair LTP formation. Indeed, acute stress has been shown to disrupt hippocampal LTP formation,[58, 59] and this effect appears to be mediated by both increased adrenal release of corticosterone and enkephalins.[56, 60] The studies of Joels and DeKloet[50, 61] have shown that glucocorticoid receptor activation increases the magnitude of AHP; mineralocorticoid receptor activation produces the opposite effect. Since an increase in hippocampal AHP would be expected to decrease potentiation, it is likely that this finding provides a mechanism for the reduced LTP associated with elevated glucocorticoid levels. Conversely, low corticosterone levels might facilitate LTP formation by decreasing hippocampal AHP. These findings underscore the importance of an efficient modulation of corticosterone levels for the organism.

These data demonstrate that glucocorticoids can impair electrophysiological signal processing in the hippocampus and provide a mechanism for the impaired cognitive function associated with exposure to high corticoid levels. A cautionary note should be added here. Each of these studies has focused on the acute effects of elevated glucocorticoids. In the context of the aged rat, glucocorticoid levels are chronically elevated, and variations in the duration of exposure to elevated steroid levels can result in varying effects on neuronal excitability.[62] However, the clinical literature concerning the effects of long-term exposure to elevated glucocorticoid levels on cognition is certainly consistent with the idea that these steroids can disrupt information processing within certain neural systems (see the following). Reports on the effects of chronic exposure to stress in the rat suggest that longer exposure to stress can result in electrophysiological[53] and cognitive[63] impairments that are not related to neuron loss, but do persist even after the cessation of stress. Likewise, mid-aged rats administered high levels of corticosterone for a period of 3 months show greatly reduced hippocampal LTP and impaired spatial memory even 1–2 months following the end of the treatment.[63] The effects of stress become more pronounced as the animals age, and in older animals the same chronic stress treatment also produces neuron loss.[53, 63] These findings further underscore the idea that glucocorticoids compromise hippocampal function at several levels, and it appears that the severity and persistence of the effects of stress increase with age.

GLUCOCORTICOIDS AND SYNAPTIC PLASTICITY LEVELS

Partial deafferentation in selected regions of the adult brain results in a sequence of compensatory events involving both the remaining afferent fibers and denervated dendrites, culminating in a reorganization of the circuitry in the denervated area. These events include "sprouting" of the remaining afferent fibers, restoration of spine density and length, and replacement of vacated synaptic contacts. This phenomenon of reactive synaptogenesis has provided one of the most interesting models for the study of the local response to damage in the CNS. One of the best examples of such reactive synaptogenesis occurs in the dentate gyrus of the hippocampus in animals following entorhinal cortex lesions.[64–66] Entorhinal cortex lesions disrupt the perforant path, resulting in nearly 60% loss of synaptic input to the granule cells of the dentate gyrus. However, beginning as early as a few days after the lesion, new synapses are formed, and within 2 months virtually all the lost inputs are replaced.[65] Following fimbria–fornix lesions, synaptic replacement occurs as a result of the growth of fibers from the adrenergic neurons in the superior cervical ganglion into the denervated region.[67] In the case of entorhinal damage, the new synapses originate from the cholinergic cells of the medial septum, from glutaminergic commissural–associational pyramidal cells, and, to a lesser extent, from neurons of the contralateral entorhinal cortex.[68, 69] In the aged rat, lesion-induced compensatory sprouting in the hippocampus occurs more slowly[70] and, in some cases, less completely[71] than in younger animals. Glucocorticoids suppress hippocampal sprouting in young animals in a dose-related manner.[72] In young adult rats maintained on high glucocorticoid levels (mimicking those seen in the aged rats), hippocampal sprouting was significantly reduced following lesions of either the fimbria–fornix[24] or the entorhinal cortex.[72, 73] Thus, the elevated glucocorticoid levels apparent in the aged rat are associated with reduced synaptogenesis in response to local neuronal damage. These data raise the possibility that in addition to regulating the extent of tissue damage, glucocorticoids may actively dampen the compensatory response of surviving neurons in the presence of neuropathology, and this effect appears to be mediated by glucocorticoid effects on the expression of genes involved in synaptic repair.[74, 75]

The maintenance of hippocampal morphology is also adversely affected by exposure to elevated glucocorticoid levels. Thus, McEwen's group has shown that exposure to 7 days of chronic stress in young adult rats can produce significant atrophy of hippocampal pyramidal neurons. These effects are blocked by either adrenalectomy or administration of the glucocorticoid receptor antagonist, RU 38486. The effects are also blocked by Ca^{2+} channel blockers and suggest that the effects of stress are mediated by glucocorticoid effects on intracellular Ca^{2+} levels. Importantly, these effects are entirely reversible; although it is not clear if this is true for older animals.

CONSEQUENCES OF ELEVATED GLUCOCORTICOID LEVELS: NEURONAL SURVIVAL

The degree of glucocorticoid exposure appears to be a determinant of hippocampal pathology after neurological insult, as well as during the aging process. In humans, poststroke damage is positively correlated with circulating cortisol levels.[76] Briefly, Sapolsky's studies[77–83] have shown that glucocorticoids endanger hippocampal neurons by (1) limiting the availability of energy substrates, thus rendering neurons metabolically compromised, and (2) by increasing extracellular glutamate concentrations by reducing glutamate uptake by glial cells. Both corticosteroid actions are mediated through the glucocorticoid receptor.

Recent data from the laboratories of Slovitar and McEwen have illustrated another dimension to corticosteroid regulation of neuronal survival in the hippocampus. In these studies, long-term adrenalectomized animals showed a profound decrease in neuron density in the dentate gyrus, a hippocampal subregion.[84] In this case, the absence of glucocorticoid stimulation in adult animals resulted in neuron loss. Interestingly, the effects were observed uniquely in the granule cells of the dentate gyrus. Pyramidal cell density in Ammon's horn (CA1–CA4) was completely unaffected. In contrast, chronic exposure to elevated glucocorticoid levels results in the loss of pyramidal neurons in Ammon's horn, but the dentate is unaffected. Landfield et al.[6] reported that long-term adrenalectomized animals maintained under conditions of low corticosterone replacement showed no changes in neuron density in the dentate gyrus. The key here is the low level of corticosterone replacement. McEwen's group have found that low levels of corticosterone or aldosterone replacement are sufficient to block the effects of adrenalectomy on neuron density in the dentate gyrus[85]; both treatments selectively target the hippocampal mineralocorticoid receptor. In contrast, Sapolsky's group have shown that the loss of pyramidal neurons in Ammon's horn in the presence of elevated glucocorticoid levels is mediated by the glucocorticoid receptor. Thus, glucocorticoid potentiation of the neurotoxic effects of excitatory amino acids on hippocampal neurons is mimicked by RU 28362, a glucocorticoid receptor agonist, and blocked by RU 38486. In an interesting parallel, we have shown that during the early postnatal period, an active period of hippocampal development, there are high, adult-like levels of mineralocorticoid receptors and very low levels of glucocorticoid receptors.[86] In essence, low levels of corticosteroid stimulation is necessary for the maintenance of hippocampal granule neurons, and chronically elevated glucocorticoids are hazardous for the survival of pyramidal neurons. The ideal state is one of tight titration around low basal glucocorticoid levels, an important argument for efficient glucocorticoid negative feedback.

INDIVIDUAL DIFFERENCES IN HPA ACTIVITY AND BRAIN AGING

We have been attempting to understand the neurobiological mechanisms underlying individual differences in HPA function and the importance of such variation in determining vulnerability to pathology over the life span. The results of these studies have indicated that in the rat (1) changes in HPA function are not an inevitable consequence of aging and (2) that individual differences in HPA activity are closely related to hippocampal dysfunction and cognitive impairments.[2, 5, 30, 31, 35, 87]

We first examined the relationship between HPA function and hippocampal aging using the neonatal handling paradigm. Animals handled briefly once per day for the first weeks of life show increased hippocampal glucocorticoid receptor expression throughout life. Aging rats lose hippocampal corticosteroid receptors, become insensitive to glucocorticoid negative-feedback regulation,[88] and hypersecrete corticosterone under both basal and poststress conditions. However, handled (H) animals have significantly higher hippocampal glucocorticoid receptors than do nonhandled (NH) animals,[31, 32, 87] although both groups do show a loss of receptors. As expected on the basis of these receptor data, we found that the age-related increases in HPA activity were far less pronounced in the H rats.[31, 87] At each age tested, the H animals secreted less corticosterone during restraint stress and terminated corticosterone secretion following stress sooner than did the NH rats. Moreover, while basal levels of corticosterone rise by midlife in the NH rats, H animals show no increase in basal HPA activity even as late as 28 months of age.[31, 32, 87]

These findings suggested that in later life exposure to the highly catabolic glucocorticoids was greater in the NH animals, and as one might expect, the evidence for hippocampal damage was also greater. Aged NH rats showed increased hippocampal degeneration compared with old H rats.[31, 32, 87] As in humans, the hippocampus of the rat is known to be of considerable importance in learning and memory, and hippocampal injury profoundly disrupts cognition. These findings suggested that the older H rats, with attenuated cell loss in the hippocampus, should show less evidence of age-related cognitive impairments than older NH rats. Behavioral testing was performed using the Morris water maze,[89] a test of spatial memory. Spatial memory deficits emerged with age in the NH rats such that the 24-month-old NH animals took significantly longer to locate the platform (i.e., 3–4 times longer) than the 6-month-old animals on all but the first 3 of 18 trials.[31, 87] In contrast, among the H rats there were no statistically reliable age differences: The 24-month-old H rats performed as well as the 6-month-old animals. There were no differences in the performance of the H and NH animals at 6 months of age. Importantly, in subsequent testing H and NH rats of all ages performed similarly when the platform was made visible by raising it above the water level indicating that the H/NH differences were due to spatial rather than motor skills. These spatial memory deficits in the older NH animals are probably related to the hippocampal damage seen in these animals, as similar deficits are observed after damage to the dorsal hippocampus.[90]

Neuroendocrine dysfunction in the aged rat includes glucocorticoid hypersecretion and negative-feedback insensitivity, as well as hippocampal degeneration. These deficits form a complex and self-perpetuating cascade[34]; a consequence of glucocorticoid hypersecretion is accelerated neuron loss in the aging hippocampus (including corticosterone-concentrating neurons), and a consequence of hippocampal damage is adrenocortical negative-feedback insensitivity and glucocorticoid hypersecretion. The interaction of these abnormalities occurs with aging in the rat and is accelerated by conditions which further elevate glucocorticoid levels, such as stress. Neonatal handling increases glucocorticoid feedback sensitivity and reduces the HPA response to stress and, in later life, basal HPA activity. The critical features of this effect, that of reducing glucocorticoid concentrations under a variety of conditions, appear to prevent the degenerative "glucocorticoid cascade."

The diminished rate of hippocampal neuron loss in the aging H rats probably reflects the lower cumulative lifetime exposure to glucocorticoids. It should be noted that this outcome is the product of two apparently opposing trends. While the increased concentrations of hippocampal glucocorticoid receptors are related to the enhanced negative-feedback sensitivity and decreased glucocorticoid secretion in the H rats, the same increased receptor concentrations could conceivably sensitize the hippocampus to the endangering effects of glucocorticoids. In this instance the decreased secretion apparently outweighs the risk of increased target sensitivity, perhaps by ensuring that the prolonged glucocorticoid exposure necessary for the endangering effects does not occur. It should be noted that under normal resting conditions only about 10–15% of the hippocampal glucocorticoid receptor population is occupied by hormone.[30, 38, 41, 57] As glucocorticoid levels rise with age, it is the change in hormone levels rather than receptor density that determines the increase in the corticoid signal. It is important that H and NH animals do not differ in hippocampal neuron number and spatial memory ability at 6 months of age. Rather, the difference emerges over time as the function of an interaction with age.

These findings, together with the studies of Landfield, suggest that in a normal population of laboratory rats, individual differences in HPA activity should serve as a predictor of age-related hippocampal pathology. We[2] examined this question by screening a large sample of aged (22–28 months) rats using a test of spatial memory. If HPA dysfunction is associated with hippocampal pathology and not merely with advanced age, we would expect that a sample of aged, cognitively impaired (ACI) and aged, cognitively unimpaired (ACU) should differ considerably in HPA activity. We used the Morris water maze to screen more than 100 aged animals in order to select aged animals that were either cognitively impaired or unimpaired (>2 SD or <0.5 SD from the mean of 6-month-old animals, respectively). According to this criteria, about 30% of the animals were designated as ACI and a comparable percentage as ACU (underscoring the extreme variation in cognitive decline in aged rats; also see Gage et al.[91] and Gallagher and Pelleymounter[92]). Interestingly, both groups of aged animals showed a loss of hippocampal neurons; however, the decrease in neuron density was substantially greater in the ACI rats.

The ACI animals showed increased plasma ACTH and corticosterone levels under both basal and poststress conditions, whereas HPA activity in the ACU animals did not differ from 6-month-old control animals. As in the old NH animals, the increase in basal ACTH and corticosterone in the ACI rats was observed only in the PM phase of the cycle. During this period median eminence levels of both CRH and especially AVP were also significantly higher in the ACI rats.[93] Interestingly, while both aged groups showed a loss of hippocampal glucocorticoid receptors (with no change in hypothalamic or pituitary receptor density), the loss was significantly greater in the ACI animals. The ACI rats also showed a significant decrease in hippocampal mineralocorticoid receptor density. These findings demonstrate the overall loss of hippocampal corticosteroid receptors in the

ACI animals. These data were comparable to that of the previous handling studies, where the old H animals resembled the ACU rats in the Issa et al.[2] study. Taken together with the findings of the Landfield studies, these data strongly suggest that increased glucocorticoid levels are selectively associated with the occurrence of hippocampal pathology and impaired cognition in later life.

Thus, aged rats that show increased basal HPA activity also show increased evidence for hippocampal damage and spatial learning and memory impairments. In contrast, aged rats with normal basal HPA activity are indistinguishable from young rats on measures of hippocampal morphology and spatial learning and memory. Thus, amongst a population of aged rats basal glucocorticoid levels are highly correlated with performance on hippocampal-dependent forms of learning and memory.[3, 5]

GLUCOCORTICOIDS AND HUMAN AGING

Several studies have now shown that human aging is not generally associated with increased plasma cortisol levels.[17, 34, 42, 43] To the best of our knowledge there are no differences in plasma corticosteroid binding globulin (CBG) levels with age.[94] Likewise, there is no consistent evidence for glucocorticoid negative-feedback deficits in elderly human subjects.[45] Although, Weiner et al.[95] (1993) reported that postdexamethasone plasma cortisol levels were positively correlated with age, suggesting some negative-feedback insensitivity with age. Indeed, careful inspection of the data from the very thorough study of Waltman et al.[96] reveals an interesting trend. In this study the investigators examined both cortisol circadian rhythms and dexamethasone suppression of both ACTH and cortisol. In the aged patients there were no differences in the maximum (427 ± 140 versus 345 ± 55 nmol/liter) or minimum (51 ± 22 versus 28 ± 4 nmol/liter) cortisol levels. However, note that (1) the cortisol levels are generally higher in the older subjects and (2) the size of the standard deviation is substantially higher in the older subjects (sample size was the same for the two groups). The same trend was evident for the dexamethasone suppression test data. The increase in the standard deviation suggests greater variability in the older subjects. The nonsignificant increase in cortisol suggests that some aged humans do show elevations in cortisol levels. This is exactly the finding that emerged from the comprehensive study of Van Cauter et al.[97] examining changes in the circadian rhythm of cortisol with age. There was no change in the acrophase of the peak or nadir; however, the amplitude of the rhythm was decreased in older subjects due to an increase in overall cortisol levels in some aged subjects. What was striking was that the increased basal cortisol secretion occurred in some aged individuals and clearly not in others.

This is exactly the type of finding we initially saw in large groups of aged rats: A nonsignificant trend toward higher levels and increased variance about the mean in the older groups. We now know that this was due to the inclusion of animals that show no increase in HPA activity (ACU rats) and animals that show pronounced increases in HPA activity (ACI rats). These data suggest that there might be considerable variance in basal cortisol secretion among aged humans.

GLUCOCORTICOIDS AND AGE-RELATED NEUROPATHOLOGY IN HUMANS

Although increased HPA activity is not associated with aging in humans, as in the rat elevated glucocorticoid levels do seem to accompany age-related neuropathology. In studies with multiple samples (as is required in studies of hormones that show circadian rhythms), cortisol levels are significantly higher in Alzheimer's (SDAT) patients compared with control subjects. Davis et al.[98] reported significantly higher cortisol levels in SDAT patients. In our own studies, we[99] found that differences in cortisol levels were more or less consistent across the 24-hr cycle and that this was true for both males and females. Moreover, the magnitude of the difference in basal cortisol levels between SDAT patients and control subjects was substantial (+30 to 52% for males and +57 to 102% for females). These reports also indicated that the pattern of cortisol levels during the 24-hr cycle was very similar in SDAT patients and healthy control subjects. Thus despite changes in secretory patterns and some evidence of phase shifts, a circadian rhythm in hormone levels is preserved in later life. These findings suggest that while CNS factors involved in the drive over HPA activity may be altered in Alzheimer's disease, the signals from the circadian pacemakers that govern the pattern of hormone levels appear unchanged. Importantly, the mean 24-hr cortisol levels were inversely correlated with the memory test scores in SDAT patients[98]; although correlational, this finding does more directly link the endocrine and neurological states in SDAT patients. Finally, Raskind's group have provided evidence for decreased glucocorticoid feedback sensitivity in early-stage SDAT patients.[100]

The biological effect of steroid hormones is associated with the free component of the hormone (steroid not bound to plasma binders, about 5–15% of the steroid levels). CBG is the principle plasma binder for corticosteroids in mammals, and changes in plasma CBG levels can greatly affect free glucocorticoid levels. The brain uptake (as well as cerebrospinal fluid [CSF] levels) of cortisol is directly proportional to the free cortisol levels.[101] Interestingly, in preliminary studies we found a significant (~30%) decrease in plasma CBG levels and increased free cortisol concentrations in SDAT patients. Thus, measures of total cortisol levels may actually underestimate the magnitude of the elevated glucocorticoid signal in SDAT. In the rat CBG mRNA and CBG binding capacity is down regulated by elevated glucocorticoid levels. Thus, increases in cortisol levels can produce decreased CBG, perhaps explaining the effect observed in the SDAT patients. Moreover, cortisol levels of 20–25 µg/dl will saturate plasma CBG and result in dramatic increases in free hormone levels (basically most of the cortisol in circulation above this level is free). For these reasons, a measure of plasma CBG levels is essential in studies of HPA activity.

These findings have led to the hypothesis that increased glucocorticoid levels might promote hippocampal damage in later life in humans.[3, 6, 34] It is interesting that hippocampal neuron loss is a hallmark of Alzheimer's disease. Several years ago Sapolsky and Meaney[102] made a concerted effort to examine the glucocorticoid receptor binding in postmortem human brain tissue. We[103] have identified the receptor in biopsy samples of human temporal cortex; but unfortunately, it appears as though the glucocorticoid receptor decays quickly following death in the primate brain. We have recently examined this question using measures of changes in receptor mRNA expression in the postmortem human hippocampus, which shows high expression of both glucocorticoid receptor mRNA/neuron.[104] We found little evidence of any change in glucocorticoid receptor mRNA levels per neuron in brains affected by Alzheimer's disease. It should be recalled that there is considerable evidence for increased circulating cortisol levels in Alzheimer patients. These findings suggest that there is no effective, compensatory down regulation of glucocorticoid receptor expression in Alzheimer's disease. Thus, despite the presence of increased cortisol levels, and hippocampal neurons likely remain highly sensitive to the potentially damaging effects of glucocorticoids in Alzheimer patients.

Can increased glucocorticoid levels damage the human brain? To date, the meager evidence on the HPA axis and neuropathology in humans is, at best, consistent with the data from the rodent studies. Prolonged exposure to elevated glucocorticoid levels can damage the monkey hippocampus.[105] Also of considerable importance is a report by De Leon's group[106] that the degree of hippocampal atrophy in SDAT patients, as determined by positive emission tomography, was positively correlated with circulating cortisol levels. Finally Starkman et al.[107] reported that among elderly subjects cortisol levels were inversely correlated with hippocampal volume.

The relationship between cortisol hypersecretion and impaired cognitive function is not unique to Alzheimer's disease. Hypercorticoid states, including those derived from glucocorticoid therapies and Cushing's disease, are associated with evidence (obtained using computed tomography [CT] scans) of cerebral atrophy.[108–110] Both the impaired cognitive functioning and the cerebral atrophy are reversed following a decline in glucocorticoid levels. In depression, brain ventricular size has been found to be positively correlated with the degree of dexamethasone resistance and elevated cortisol levels.

GLUCOCORTICOIDS AND HUMAN COGNITION

Corticosteroid treatment has been known for some time to induce a reversible psychotic condition (so-called steroid psychosis) in certain individuals.[111] This condition also includes cognitive disturbances, including the loss of memory, attentional deficits, and problems with logical thinking. Varney et al.[112] described a "dementia-like" syndrome (including attentional deficits and memory impairments) in a group of patients given high doses of corticosteroids for disorders not related to the CNS and who did not show evidence of psychosis. Moreover, Wolkowitz et al.[113] found that subjects given a single 1-mg dose of dexamethasone showed memory and attentional deficits.

These findings are consistent with studies of Cushing's patients, where cortisol levels are profoundly elevated. Cushing's patients consistently show attentional deficits and memory impairments[107] and the magnitude of the impairments correlate with plasma cortisol levels.[107] Interestingly, the older subjects (> 45 years) were more seriously affected; these subjects showed significantly greater memory impairments than did younger subjects.[107] Importantly, these deficits were reversible; surgical intervention to correct the hypercortisolemic state resulted in normalized cortisol levels and improved cognitive performance.[107]

In addition, cognitive impairments associated are often associated with clinical depression and are correlated with increased cortisol levels. Several studies have documented the cognitive deficits often found in depressed patients, including attentional deficits and verbal and visual memory impairments.[113–118] Again, both the increase in plasma cortisol and the magnitude of the cognitive disturbances are more severe in elderly patients, often resulting in a condition of so-called "pseudo-dementia." Importantly, in hypercortisolemic patients (especially dexamethasone-resistent patients) the cognitive disturbances are more pronounced.[116]

The assessment of cognitive dysfunction in Cushing's and depressed patients is complicated by the disease state. However, it is intriguing that these findings are in the same direction as those few studies that have examined the effects of corticosteroid administration to healthy subjects (for a review, see Lupien and McEwen[119]). Dysfunction of the HPA axis is also associated with hormonal changes that extend beyond that of cortisol; both ACTH and CRH are known to affect neuronal function at several levels. However, it is interesting that in Cushing's disease cortisol and ACTH are elevated while corticotropin releasing factor (CRF) (as best as can be measured) is decreased. In depressed patients CRF appears to be, if anything, elevated. In normal subjects administered dexamethasone, both CRF and ACTH should be decreased. Thus, the only common feature among these conditions is the elevations in cortisol levels. This is hardly conclusive, but it is consistent with the idea that the active hormone in these conditions is cortisol.

A number of studies have provided evidence that exogenous administrations of corticosteroids impair cognitive function (for a complete review, see Lupien and McEwen[119]). Perhaps the best starting point for the present discussion is actually the recent study by Kirschbaum et al.[120] which showed that the oral administration of 10 mg of hydrocortisone leads to a significant decrease in memory performance as tested 60 min after hydrocortisone intake. In their study, they measured declarative and nondeclarative memory performance in order to assess the influence of corticosteroids on the hippocampal formation process. The logic for the inclusion of this mnesic dissociation is due to the fact that studies report that the hippocampus is essential for a specific kind of memory, notably declarative, while it is not essential for nondeclarative memory.[121] Declarative memory refers to conscious or voluntary recollection of previous information, whereas nondeclarative memory refers to the fact that experience changes the facility for recollection of previous information without affording conscious access to it (priming). Thus, this somewhat specialized role of the hippocampus serves as the basis for specific hypotheses regarding the effects of acute administration of corticosteroids on human cognition.[122, 123] The results showed that subjects who received hydrocortisone treatment presented an impaired performance in the declarative memory task but not in the nondeclarative memory task, thus suggesting that cortisol interacts with hippocampal neurons to induce cognitive deficits.

Although most of the effects of exogenous administrations of glucocorticoids upon human psychophysical and cognitive processing has been done using hydrocortisone as a compound, some other studies have used other compounds and reached different conclusions. In a series of experiments, Wolkowitz et al.[113, 124] have shown that poor performance (errors of commission; incorrectly identifying distractors as targets) on verbal memory tasks occurs in normal volunteers following the administration of prednisone (80 mg/day for 5 days). However, no such effects were observed with the administration of 1 mg of dexamethasone. The general cognitive deficit described in this study involved the relative inability to discriminate previously presented relevant information (target) from irrelevant information (distractors) in a test of recognition memory. The authors concluded that exogenous administration of corticosteroids may diminish the encoding of meaningful stimuli and impair selective attention (i.e., reducing the ability to discriminate relevant from irrelevant information).

Based on animal and human literature a recent model proposed by Lupien and McEwen[119] suggested that medium/high levels of circulating glucocorticoids may first affect the process of selective attention, thus impairing further explicit acquisition of information, whereas high/very high levels of circulating levels of corticosteroids may affect explicit memory. This suggestion has recently been confirmed by a dose–response hydrocortisone infusion study in young normal control subjects performed by Lupien et al.[125] In this study a 300 µg/kg/hr infusion of hydrocortisone impaired selective attention capacity, whereas a 600 µg/kg/hr infusion impaired both selective attention and explicit memory.

The results of these studies suggest that in humans, as in rodents, elevated glucocorticoids serve to compromise hippocampal function and promote cognitive impairments. We have been examining this hypothesis in studies with a population of elderly human subjects, examining cognitive function in relation to individual differences in HPA activity and hippocampal integrity.

HPA FUNCTION AND
HIPPOCAMPAL AGING IN HUMANS

We[126] reported results of a longitudinal study of healthy aged subjects that reflected the variability in HPA activity in later life. Seventeen female and 34 male subjects ranging from 60 to 90 years participated in this study (males = 73.7 years ± 6.6; females = 72.5 years ± 4.3). The status of the subjects was determined by a complete physical examination including ECG, EEG, CT scan, and a battery of laboratory tests for kidney, liver, and thyroid functions, hemogram, vitamin B_{12}, folate levels, as well as a neuropsychological assessment. In addition, subjects were excluded for the regular use of any medications. This served to ensure that the population was relatively free from any obvious disease at the time of recruitment into the study and annually thereafter.

A primary measure is that of circulating cortisol levels, which is examined once per year (but see the following). All subjects are sampled hourly for a 27-hr period using an indwelling forearm catheter kept patent with a 0.3% heparin saline solution. Throughout the course of sampling, illumination was maintained at 300 lux during the "daytime" (0700–2300 hr) and at 50 lux during the "nighttime" (2300–0700 hr).

SUBGROUPS OF ELDERLY SUBJECTS
WITH REGARD TO PATTERNS OF CORTISOL
SECRETION OVER YEARS

With the inclusion of young control subjects, we have observed no overall changes in plasma cortisol levels with age in our population. However, the longitudinal basal cortisol data revealed the existence of distinct subgroups of elderly subjects.[126] Simple regression analysis on plasma cortisol levels for each subject were conducted with year as the independent variable and cortisol concentration at each year as the dependent variable. The direction and amplitude of the slope of the regression line then served as the measure of the cortisol history per subject (thereafter "cortisol slope"). We found clear evidence for subgroups that show either (1) a progressive year-to-year increase in cortisol levels with currently high levels (PHC, $n = 12$), or (2) a progressive year-to-year increase in cortisol levels with currently moderate levels (PMC, $n = 29$), or (3) a progressive decrease in cortisol levels with currently low-to-moderate cortisol levels (NLC, $n = 10$). This profile provides two measures of interest. The first is the slope of the increase or decrease in basal cortisol over time as an indication of the dynamic change in adrenal activity. The second is the actual current basal cortisol level as an indication of the static state of adrenal activity in elderly human subjects. Both acute and chronic elevations of cortisol have been shown to induce cognitive deficits in human populations.[113, 119, 120, 122–124, 127, 128]

ENDOCRINE, CLINICAL, AND BEHAVIORAL
CORRELATES OF SUBGROUPS

PHC subjects show progressive increases in plasma cortisol levels over years. Moreover, the most recent plasma cortisol level is also significantly elevated in these subjects. However, there are no obvious differences in the circadian cortisol rhythm in these subjects in comparison to either PMC or NLC subjects (Fig. 2.1). In addition, there were no group differences in plasma CBG levels in either the AM or PM phase of the cycle. This was further supported by the strong correlation between total and free plasma cortisol levels (measured recently using salivary sampling) across subjects. Thus, plasma total cortisol levels predicted differences in free cortisol, an important finding since the biologically active form of the steroid is the free form.

In order to assess the validity of these subgroups of aged subjects, group differences were examined for endocrine, clinical, metabolic, and behavioral correlates using three endocrine measures (last 24-hr, averaged 24-hr, and cortisol slope). The groups did not differ with regard to weight, height, body mass index, pulse, diastolic blood pressure, or glucose levels. However, the PHC subjects showed significant elevations in systolic blood pressure, plasma total cholesterol, and triglycerides.[126] These results are in agreement with reports of glucocorticoid-induced states of hypertension, hyperlipidemia, and hypercholesterolemia.[8, 129, 130] Indeed, 70–80% of patients with Cushing's syndrome developed hypertension, and the majority of them showed remission of hypertension with successful treatment.[131] Other reports have shown that endogenous or exogenous glucocorticoid excess eliminates or reverses circadian blood pressure variation[132] and that the hypertension reported in these cases can be inhibited by a glu-

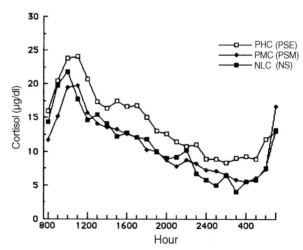

FIGURE 2.1   Mean plasma cortisol levels (μg/dl/hr) over a 24-hr period for the PHC (also called PSE), PMC (also called PSM), and NLC (also called NS) groups.

cocorticoid antagonist such as RU 486.[133, 134] The relation between systolic blood pressure and cortisol slope observed here suggests that glucocorticoid-induced hypertension may slowly develop in time in those aged individuals showing increases in cortisol levels with years. For our purposes, these data suggest that the increased basal cortisol levels of the PHC subjects are functionally significant.

NEUROPSYCHOLOGICAL CORRELATES
OF SUBGROUPS

The next question concerned the neuropsychological function in PHC, PMC, and NLC subjects. On the basis of the previous animal studies, we expected significant cognitive deficits in PHC subjects that were specific to tasks that are dependent upon hippocampal function. During 1992, eight subjects from the PHC group, six from the PMC group and five from the NLC group were tested using a neuropsychological assessment that evaluated declarative and nondeclarative memory, selective and divided attention, as well as verbal fluency and picture naming.[135] Correlational analyses performed between cognitive performance and cortisol slope showed that the slope of the change in cortisol levels over time predicted cognitive deficits in this elderly population. Aged subjects from the PHC group were impaired on tasks measuring declarative memory and selective attention when compared to aged subjects from the PHC and NLC groups. Although subjects from the PHC group showed deficits in declarative memory function, they showed a performance equivalent to that of the other groups on measures of nondeclarative memory. We further showed that subjects from the NLC group performed as well as young healthy subjects with regard to cognitive performance. Together, these findings suggest that elevated glucocorticoid levels can serve to directly influence cognitive function in later life in humans.

DEHYDROEPIANDROSTERONE LEVELS,
CORTISOL LEVELS, AND COGNITIVE FUNCTION IN
RELATION TO SUBGROUPS

It has been postulated that dehydroepiandrosterone (DHEA) and its sulfate (DHEA-S) are markers of successful aging and life expectancy, whereas cortisol levels may be associated with pathological brain aging.[136–138] We have retrospectively analyzed DHEA-S levels in this population of aged individuals for the years 1990–1991, 1991–1992 and 1992–1993 in order to determine if DHEA-S secretion is negatively associated with cortisol hypersecretion and cognitive deficits.[139] Subjects were retested in year 1992–1993 for 24-hr cortisol secretion as well as memory function, and we replicated the cortisol/memory impairment relationship previously reported. The results of the analysis of DHEA-S levels revealed decreasing DHEA-S levels in all subjects. Consistent with the hypothesis that DHEA-S may have a protective effect on the aging system, we have shown that subjects showing a significant increase in cortisol levels with years and memory impairment had the greatest decrease in DHEA-S levels. However, the reverse pattern was not observed in individuals showing decreasing cortisol levels and highly efficient cognitive function. These individuals also consistently presented lower DHEA-S levels than the other subjects. Finally, although cortisol levels were significantly correlated with cognitive function, DHEA-S levels were not significantly correlated with cognitive efficiency. These results suggest that cortisol secretion during aging may be a better biological marker of cognitive decline than DHEA-S levels and that changes in cognitive function are specifically related to cortisol and merely to changes in adrenal function.

ENVIRONMENTAL VALIDITY OF
THE LONGITUDINAL CORTISOL MEASURES
IN THIS ELDERLY POPULATION

Plasma sampling for cortisol is thorough, but rather invasive. Subjects are required to remain within the artificial environment of the clinical unit for 27 hr. The technique has allowed us to estimate parameters of the cortisol rhythm, but is less than ideal for in vivo measures of glucocorticoid levels. In order to obtain more representative samples we are now making extensive use of salivary sampling procedures. Salivary cortisol levels have been shown to be highly reliable measures of free cortisol levels.[140]

In order to determine whether plasma cortisol levels obtained in a 1-day laboratory testing environment provided a valid indication of differences in cortisol levels, we sampled a subgroup of subjects in their home environment for 24 hr obtaining salivary samples for cortisol radioimmunoassay. The subjects were asked to take one sample each hour for a 24-hr period, providing an approximation of the sampling routine used in the clinical unit. A Spearman's correlation coefficient was calculated between the 24-hr salivary free cortisol levels of the subjects and the 24-hr plasma samples of the same individuals. Please note that these samples were not taken at the same time as the plasma sampling was performed: Saliva and blood samplings were taken during the same year, no more than 1 month apart from each other. As expected, there was a highly significant correlation between plasma and salivary cortisol levels over the 24-hr period.

We then examined whether plasma cortisol levels obtained in a 1-day laboratory testing environment provided a valid indication of stable differences in cortisol. We obtained salivary samples for cortisol determinations from subjects in their home environment 4 times per day for each of 30 consecutive days. We also asked subjects to rate their affective state on a scale each time they took a saliva sample and to record their meal intake every day, for the 30-day period. The results showed clear differences in salivary cortisol levels between PHC and NLC subjects over the 1-month period (Fig. 2.2). Differences in free cortisol levels in salivary samples were significant throughout the day, with the greatest difference occurring at about the time of the diurnal peak (early AM samples), repli-

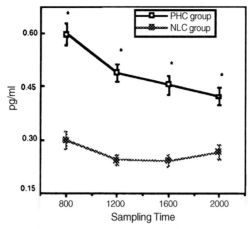

FIGURE 2.2   Mean (± SEM) cortisol levels (pg/dl) from salivary samples obtained from PHC and NLC subjects at various times during the day (800; 1200; 1600; and 2000). The individual datum points for each day are the means of cortisol levels for the PHC and NLC groups drawn from 30 consecutive days of sampling. The differences between groups were significant at each time point ($p < 0.05$).

cating a pattern previously observed in plasma samples.[126] Moreover, we found that those subjects showing the highest increase in cortisol levels (mainly those of the PHC group) had a higher intake of fat and sugar,[141] particularly in the morning, and also scored higher on feelings of stress, depression, and fatigue, again, particularly in the morning.[125]

STRESS RESPONSE AND ITS EFFECTS ON MEMORY FUNCTION IN THIS ELDERLY POPULATION

Elevated cortisol levels exert both short- and long-term effects on cognitive performance (see Lupien and McEwen[119] in appendix). The observation of acute and chronic effects of cortisol increase on cognitive function in humans leads to the question as to whether stress-induced increase in cortisol levels in elderly subjects may be related to the memory impairments frequently reported in these individuals. Indeed, aging has long been viewed in the gerontogeriatric literature as being either accelerated by stress factors or as reflecting a decreased adaptation to stress.[142, 143] In order to test whether stress responsiveness may play a role in memory impairments in the elderly individual, we examined cortisol reactivity and declarative/nondeclarative memory performance in response to a psychological stressor in elderly subjects. Declarative and nondeclarative memory, as well as salivary cortisol levels, were measured before and after a nonstressful (computer-generated detection task) and a stressful condition (public speaking task). The results showed that the stressful condition significantly decreased declarative memory performance while the nonstressful condition did not.[144]

Nondeclarative memory performance was not affected by either condition. Further analyses separating the subjects into "responders" (85% of them from the PHC group) and "nonresponders" in terms of stress-induced cortisol change revealed a very different pattern of cortisol secretion and declarative memory performance in both populations.

We showed that the responders started to present increased cortisol levels 80 minutes before the actual stressor and 35 minutes earlier than the nonresponders. Although the responders did not differ from the nonresponders on declarative memory performance before and after the nonstressful condition, they presented a lower declarative memory performance when retested before the actual stressor and still decreased this performance after the stressor.[145] The early increase in cortisol levels observed in the responder group suggests that the anticipation of the stress, rather than the actual stressor per se, may have played a more significant role in the stress-induced declarative memory deficits observed in this subgroup. It also suggests that the exposition of these individuals to high cortisol levels from the beginning of the experiment gave rise to the stress-induced declarative memory deficit observed before the stress, whereas the acute rise in cortisol levels observed after the stressor induced an additional negative effect on the consolidation and recall of this information. Together, these results suggest that the cortisol response to anticipation of stress and/or in response to stress in the elderly may specifically affect those memory functions that are dependent on hippocampal activity and further suggest that an altered cortisol responsivity to acute and/or chronic stress, with its detrimental effects on memory, could be an important factor explaining the genesis of memory deficits in aging.[145]

COGNITIVE DEFICITS RELATED TO HIPPOCAMPAL DAMAGE IN RELATION TO SUBGROUPS

The hippocampus has been implicated in performance on several cognitive tasks that are not essentially declarative by nature.[121] Several animal and human studies have shown that it is particularly sensitive to the time-limited, and spatial aspects of memory. Patients with amnesia due to hippocampal dysfunction show normal retention at short delays and impaired retention at longer delays.[146] They also present impaired spatial memory.[147] Thus, we examined PHC and NLC subjects using a test of immediate and delayed recall, and we measured spatial memory function in the same subjects using a human maze. Immediate and delayed memory was measured using visual presentation of 15 noncomplex line drawings of objects of everyday use. The subject was presented with the 15 line drawings for 3 seconds each and then asked to verbally recall as many line drawings as possible, immediately after the presentation or 24 hr later. The results showed that the PHC and NLC groups do not differ on tests of immediate memory, whereas the PHC subjects showed significant impairments on test of delayed recall.[148]

Spatial memory function was measured in the same individuals using a human maze. The surface area of the maze was 1500 ft^2 and the walls were 6 ft high, with no spatial cues either on the floor or on the ceiling.[144] The subject was shown a path by following the experimentor through the maze and was then required to redo it on his or her own. The subject had to learn a simple and a complex path. The complexity of the path was determined by the number of points of decision in the learned path. A point of decision is an intersection in the maze to which the subject must take a decision (turn left or right or go straight ahead). The simple path comprised three points of decision, and the complex path comprised five points of decision. All subjects learned the notion of a point of decision using a smaller maze of 500 ft^2. The time taken to find the correct path served as the measure of spatial memory function. Note that all subjects presented equivalent walking pace when measured on a pilot study, using the smaller human maze. The results showed that subjects from the PHC group took significantly longer to find their way when they had to learn a complex path when compared to subjects from the NLC group. The pattern emerging from these neuropsychological findings is consistent with the idea that cumulative exposure to elevated glucocorticoid levels compromises hippocampal integrity and thus performance on hippocampal-dependent cognitive tasks.

Studies of brain changes associated with dementia in later life have reported an anatomically specific relationship between hippocampal volume and memory performance. These observations have been extended to elderly populations showing mild cognitive impairments.[149] In order to examine the relationship between cortisol levels and hippocampal atrophy in our population, we examined hippocampal volume using magnetic resonance imaging (MRI) scans in the same subjects tested for neuropsychological performance.[148, 150] Two findings are noteworthy. First, the hippocampal volume of the elderly PHC subjects is reduced by 17% when compared to that of elderly subjects of the NLC group (Table 2.1). Second, this effect is unique to the hippocampus; we found no group differences in the volume of the parahippocampal and fusiform gyri, as well as in the other temporal lobe structures analyzed (see Table 2.1).

TABLE 2.1   MRI Volumes for Selected Regions (Right and Left Hemispheres Combined cm^3)

	Increasing/ high	Standard error	Decreasing/ moderate	Standard error
Hippocampus	4.00[a]	0.08	4.54[a]	0.13
Parahippocampal gyrus	5.50	0.44	5.42	0.47
Fusiform gyrus	8.57	1.11	8.24	0.69
Superior temporal gyrus	21.07	1.15	19.06	1.59
Middle and inferior temporal gyri	26.73	1.43	27.82	1.21
Head size	278.68	14.72	273.78	8.29

[a]Significant difference at p <.001.

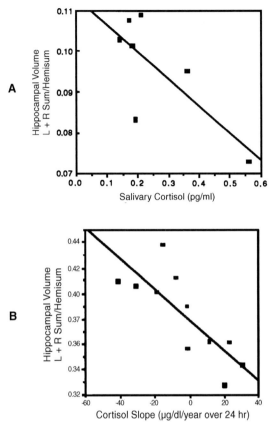

FIGURE 2.3   (A) Correlation between static salivary cortisol measure and hippocampal volume (Left + Right Sum/Hemisum) in seven elderly human subjects showing different patterns of cortisol secretion over the years. (B) Correlation between dynamic plasma cortisol measure and hippocampal volume (Left + Right Sum/Hemisum) in 11 elderly human subjects showing different patterns of cortisol secretion over the years.

Interestingly, the magnitude of the decrease in hippocampal volume in the PHC subjects is comparable to that previously reported for elderly subjects with age-related mild cognitive impairments (MCI; 14%;[149]). These brain changes in MCI subjects have been shown to be predictive of the development of dementia in later life.[149] Finally, there was a highly significant negative correlation between both the static (salivary; $r = -0.75$; $R2 = 0.56$; $p < 0.05$; Fig. 2.3) and dynamic (plasma cortisol slope; $r = -0.81$; $R2 = 0.65$; $p < 0.003$; see Fig. 2.3) measures of cortisol levels and hippocampal volume.

Taken together these studies provide support for the idea that individual differences in HPA activity serve to predict hippocampal dysfunction in later life. It is clear that a number of questions remain. We have little insight into the origins of the differences in basal HPA activity. To date, we have found that individuals with elevated HPA activity show increased scores on scales related to anxiety and altered patterns of fat consumption, which can be related to mood. Interestingly, the hippocampus has been strongly associated with the expression of anxiety,[151] and thus it is not clear whether these mood states lie at the base of the changes in neuroendocrine function or as a result of the hippocampal dysfunction. It is also not clear whether the changes in hippocampal volume are reversible at this time in life. Each of these issues lies at the heart of our current studies.

REFERENCES

1. Meaney, M. J., O'Donnell, D., Rowe, W., Tannenbaum, B., Steverman, A., Walker, D., Nair, N. P. V., & Lupien, S. J. (1995). Individual differences in hypothalamic-pituitary-adrenal activity in later life and hippocampal aging. *Exp. Gerontol., 30,* 229–251.
2. Issa, A., Gauthier, S., & Meaney, M. J. (1990). Hypothalamic-pituitary-adrenal activity in aged cognitively impaired and cognitively unimpaired aged rats. *J. Neurosci., 10,* 3247.
3. Landfield, P., Waymire, J., & Lynch, G. (1978). Hippocampal aging and adrenocorticoids: A quantitative correlation. *Science, 202,* 1098.
4. Sapolsky, R. M. (1992). Stress, the aging brain, and the mechanisms of neuron death (p. 423). Cambridge: MIT Press.
5. Yau, J. L. W., Olsson, T., Morris, R. G. M., Meaney, M. J., & Seckl, J. R. (1995). Chronic anti-depressant treatment improves spatial learning in young, but not aged rats: Correlation with corticosteroid receptor gene expression in the hippocampus. *Neuroscience, 66,* 571–581.
6. Landfield, P., Baskin, R. K., & Pitler, T. A. (1981). Brain-aging correlates: Retardation by hormonal-pharmacological treatments. *Science, 214,* 581.
7. Baxter, J.D., & Tyrell, J. B. (1987). The adrenal cortex. In P. Felig, J. D. Baxter, A. E. Broadus, & L. A. Frohman (Eds.), *Endocrinology and metabolism* (p. 385), New York: McGraw-Hill.
8. Brindley, D. N., & Rolland, Y. (1989). Possible connections between stress, diabetes, obesity, hypertension and altered lipoprotein metabolism that may result in atherosclerosis. *Clin. Sci., 77,* 453.
9. Antoni, F. A. (1986). Hypothalamic control of ACTH secretion: Advances since the discovery of 41-residue corticotropin-releasing factor. *Endocr. Rev., 7,* 351.
10. Antoni, F. A. (1993). Vasopressinergic control of pituitary adrenocorticotropin secretion comes of age. *Front. Neuroendocrinol., 14,* 76.
11. Rivier, C., & Plotsky, P. M. (1986). Mediation by corticotropin-releasing factor of adenohypophysial hormone secretion. *Ann. Rev. Physiol., 48,* 475.
12. Munck, A., Guyre, P. M., & Holbrook, N. J. (1984). Physiological functions of glucocorticoids in stress and their relations to pharmacological actions. *Endocr. Rev., 5,* 25.
13. Dallman, M. F., Akana, S., Cascio, C. S., Darlington, D. N., Jacobson, L., & Levin, N. (1987). Regulation of ACTH secretion: Variations on a theme of B. *Rec. Prog. Horm. Res., 43,* 113.
14. Dallman, M. F., Akana, S. F., Bradbury, M. J., Strack, A. M., Hanson, E. S., & Scribner, K. A. (1994). Regulation of the hypothalamo-pituitary-adrenal axis during stress: Feedback, facilitation and feeding. *Semin. Neurosci., 6,* 205.
15. de Kloet, E. R. (1991). Brain corticosteroid receptor balance and homeostatic control. *Front. Neuroendocrinol. 12,* 95.
16. Feldman, S., & Weidenfeld, J. (1994). Neural mechanisms involved in the corticosteroid feedback effects on the hypothalamo-pituitary-adrenocortical axis. *Prog. Neurobiol., 45,* 129.
17. Jacobson, L., & Sapolsky, R. M. (1991). The role of the hippocampus in feedback regulation of the hypothalamic-pituitary-adrenal axis. *Endocr. Rev., 12,* 118.
18. Keller-Wood, M., & Dallman, M. (1984). Corticosteroid inhibition of ACTH secretion. *Endocr. Rev., 5,* 1.

19. McEwen, B. S., DeKloet, E. R., & Rostene, W. H. (1986). Adrenal steroid receptors and actions in the nervous system. *Physiol. Rev., 66,* 1121.
20. Diorio, D., Viau, V., & Meaney, M. J. (1993). The role of the medial prefrontal cortex (cingulate gyrus) in the regulation of hypothalamic-pituitary-adrenal responses to stress. *J. Neurosci., 13,* 3839.
21. Moghaddam, B., Bolinao, M. L., Stein-Behrens, B., & Sapolsky, R. (1994). Glucocorticoids mediate the stress-induced extracellular accumulation of glutamate. *Brain Res., 655,* 251–254.
22. Meaney, M. J., & Aitken, D. H. (1985). The effects of early postnatal handling on the development of hippocampal glucocorticoid receptors: Temporal parameters. *Dev. Brain Res., 22,* 301.
23. Brett, L., Chong, G., Coyle, S., & Levine, S. (1983). The pituitary-adrenal response to novel stimulation and ether stress in young adult and aged rats. *Neurobiol. Aging, 4,* 133.
24. DeKosky, S., Scheff, S., & Cotman, C. (1984). Elevated corticosterone levels: A possible cause of reduced axon sprouting in aged animals. *Neuroendocrinology, 38,* 33.
25. Hess, G., & Riegle, G. (1970). Adrenocortical responsiveness to stress and ACTH in aging rats. *J. Gerontol., 25,* 354.
26. Heroux, J. A., Grigoriadis, D. E., & De Souza, E. B. (1991). Age-related decreases in corticotropin-releasing hormone (CRH) receptors in rat brain and anterior pituitary gland. *Brain Res., 542,* 155.
27. Sapolsky, R. M., Krey, L. C., & McEwen, B. S. (1983). Corticosterone receptors decline in a site-specific manner in the aged rat. *Brain Res., 289,* 235.
28. Sencar-Cupovic, L., & Milkovic, S. (1976). The development of sex differences in adrenal morphology and responsiveness to stress of rats from birth to end of life. *Mech. Ageing Dev., 5,* 1.
29. Tang, G., & Phillips, R. (1978). Some age-related changes in pituitary-adrenal function in the male laboratory rat. *J. Gerontol., 33,* 377.
30. Meaney, M. J., Viau, V., Bhatnagar, S., & Aitken, D. H. (1988). Occupancy and translocation of hippocampal glucocorticoid receptors during and following stress. *Brain Res., 445,* 198.
31. Meaney, M. J., Aitken, D. H., Bhatnagar, S., van Berkel, C., & Sapolsky, R. M. (1988). Postnatal handling attenuates neuroendocrine, anatomical, and cognitive impairments related to the aged hippocampus. *Science, 239,* 766.
32. Meaney, M. J., Aitken, D., Sharma, S., & Viau, V. (1992). Basal ACTH, corticosterone, and corticosterone-binding globulin levels over the diurnal cycle, and hippocampal type I and type II corticosteroid receptors in young and old, handled and nonhandled rats. *Neuroendocrinology, 55,* 204.
33. Sonntag, W. E., Goliszek, A. G., Brodish, A., & Eldridge, J. C. (1987). Diminished diurnal secretion of adrenocorticotropin (ACTH), but not corticosterone, in old male rats: Possible relation to increased adrenal sensitivity to ACTH *in vivo. Endocrinology, 120,* 2308.
34. Sapolsky, R. M., Krey, L. C., & McEwen, B. S. (1986). The neuroendocrinology of stress and aging: The glucocorticoid cascade hypothesis. *Endocr. Rev., 7,* 284.
35. Rowe, W., Steverman, A., Walker, M., Sharma, S., Barden N., Seckl, J. R., & Meaney, M. J. (1997). Antidepressants restore hypothalamic-pituitary-adrenal feedback function in aged, cognitively-impaired rats. *Neurobiol. Aging., 18,* 527–533.
36. Krieger, D. T. (1982). *Monographs in endocrinology* (Vol. 22). Berlin: Springer-Verlag.
37. Ritger, H., Veldhuis, H., & de Kloet, E. R. (1984). Spatial orientation of hippocampal corticosterone receptor systems of old rats: Effects of ACTH4-9 analogue ORG2766. *Brain Res., 309,* 393.
38. Reul, H. M., Tonnar, J., & DeKloet, E. R. (1987). Neurotrophic ACTH analogue promotes plasticity of type I corticosteroid receptors in brain of senescent male rats. *Neurobiol. Aging, 9,* 253.
39. Peiffer, A., Barden, N., Meaney, M. J., (1991). Age-related changes in glucocorticoid receptor binding and mRNA levels in the rat brain and pituitary. *Neurobiol. Aging, 12,* 475.
40. Meaney, M. J., O'Donnell, D., Viau, V., Bhatnagar, S., Sarrieau, A., Smythe, J. W., Shanks, N.,

& Walker, C.-D. (1994). Corticosteroid receptors in rat brain and pituitary during development and hypothalamic-pituitary-adrenal (HPA) function. In P. McLaughlin & I. Zagon (Eds.), *Receptors and the developing nervous system* (pp. 163–202). London: Chapman and Hall.

41. Reul, J. M. H. M., van den Bosch, F. R., DeKloet, E. R. (1987). Relative occupation of type-I and type-II corticosteroid receptors in rat brain following stress and dexamethasone treatment: Functional implications. *J. Endocrinol., 115,* 459.

42. Wilson, M., Greer, M., & Roberts, L. (1980). Hippocampal inhibition of pituitary-adrenocortical function in female rats. *Brain Res., 197,* 344.

43. Fischette, C., Komisurak, B., Ediner, H., Feder, H. H., & Seigel, A. (1980). Differential fornix ablations and the circadian rhythmicity of adrenal corticosterone secretion. *Brain Res., 195,* 73.

44. Sapolsky, R. M., Krey, L. C., & McEwen, B. S. (1984). Glucocorticoid-sensitive hippocampal neurons are involved in terminating the adrenocortical stress response. *Proc. Natl. Acad. Sci. USA, 81,* 6174.

45. Bradbury, M., & Dallman, M. F. (1989). Effects of type 1 and type 2 glucocorticoid receptor antagonists on ACTH levels in the PM. *Soc. Neurosci., 15,* 716.

46. Feldman, S., & Conforti, N. (1980). Participation of the dorsal hippocampus in the glucocorticoid negative feedback effect on adrenocortical activity. *Neuroendocrinology, 30,* 2.

47. Herman, J. P., Schafer, M., Young, E. A., Thompson, J., Douglass, J., Akil, H., & Watson, S. J. (1989). Evidence for hippocampal regulation of neuroendocrine neurons of the hypothalamo-pituitary-adrenocortical axis. *J. Neurosci., 9,* 3072.

48. Sapolsky, R. M., Armanini, M. P., Packan, D. R., Sutton, S. W., & Plotsky, P. M. (1990). Glucocorticoid feedback inhibition of adrenocorticotropic hormone secretagogue release: Relationship to corticosteroid receptor occupancy in various limbic sites. *Neuroendocrinology, 51,* 328.

49. Pfaff, D. W., Siva, M. T., & Weiss, J. M. (1971). Telemetered recording of hormone effects on hippocampal neurons. *Science, 172,* 394.

50. Joels, M., & De Kloet, E. R. (1989). Effect of glucocorticoids and norepinephrine on excitability in the hippocampus. *Science, 245,* 1502.

51. Kerr, D. S., Campbell, L. W., Hao, S.-Y., & Landfield, P. W. (1989). Corticosteroid modulation of hippocampal potentials: Increased effect with aging. *Science, 245,* 1505.

52. Landfield, P. W., & Pitler, T. A. (1984). Prolonged Ca-dependent afterhyperpolarizations in hippocampal neurons of the aged rat. *Science, 226,* 1089.

53. Kerr, D. S., Campbell, L. W., Applegate, M. D., Brodish, A., & Landfield, P. W. (1991). Chronic stress-induced acceleration of electrophysiologic and morphometric biomarkers of hippocampal aging. *J. Neurosci., 11,* 1316.

54. Foy, M. R., Stanton, M. E., Levin, S., & Thompson, R. F. (1987). Behavioral stress impairs long-term potentiation in rodent hippocampus. *Behav. Neural Biol., 48,* 138.

55. Bennet, M. C., Diamond, D. M., Fleshner, M., & Rose, G. M. (1991). Serum corticosterone level predicts the magnitude of hippocampal primed burst potentiation and depression in urethane-anesthetized rats. *Psychobiology, 19,* 301.

56. Diamond, D. M., Bennett, M. C., Fleshner, M., & Rose, G. M. (1992). Inverted-U relationship between the level of peripheral corticosterone and the magnitude of hippocampal primed burst potentiation. *Hippocampus, 2,* 421.

57. Reul, J. M. H. M., & DeKloet, E. R. (1985). Two receptor systems for corticosterone in rat brain: Microdistribution and differential occupation. *Endocrinology, 117,* 2505.

58. Diamond, D. M., Bennett, M. C., Stevens, K. E., Wilson, R. L., & Rose, G. M. (1990). Exposure to a novel environment interferes with the induction of hippocampal primed burst potentiation. *Psychobiology, 18,* 273.

59. Shors, T. J., Seib, T. B., Levine, S., & Thompson, R. F. (1989). Inescapable vs escapable shock modulates long-term potentiation in the rat hippocampus. *Science, 244,* 224.

60. Shors, T. J., Levine, S., & Thompson, R. F. (1990). Effect of adrenalectomy and demedullation on the stress-induced impairment of long-term potentiation. *Neuroendocrinology, 51,* 70.

61. Joels, M., & De Kloet, E. R. (1990). Mineralocorticoid receptor-mediated changes in membrane properties of rat CA1 pyramidal neurons in vitro. *Proc. Natl. Acad. Sci. USA, 87,* 4495.
62. Drakontides, A. B., Baker, T., & Riker, W. F. (1982). A morphological study of the effect of glucocorticoid treatment on delayed organophosphorus neuropathy. *Neurotoxicology, 3(4),* 165–177.
63. Bodnoff, S. R., Humphreys, A. G., Lehman, J. C., Diamond, D. M., Rose, G. M., & Meaney, M. J. (1995). Enduring effects of chronic corticosterone treatment on spatial learning, synaptic plasticity, and hippocampal neuropathology in young and mid-aged rats. *J. Neurosci., 15,* 61–69.
64. Lee, K. D., Stanford, E. J., Cotman, C. W., & Lynch, G. S. (1977). Ultrastructural evidence for bouton proliferation in partially deafferented dentate gyrus in the adult rat. *Exp. Brain. Res., 29,* 475.
65. Matthews, D. A., Cotman, C. W., & Lynch, G. (1976). An electron microscopic study of lesion-induced synaptogenesis in the dentate gyrus of the adult rat. *Brain Res., 115,* 1.
66. Scheff, S. W., Benardo, L. S., & Cotman, C. W. (1980). Decline in reactive fiber growth in the dentate gyrus of aged rats compared to young adult rats following entorhinal cortex removal. *Brain Res., 199,* 21.
67. Moore, R. Y., Bjorklund, A., & Stenevi, U. (1977). Plastic changes in the adrenergic innervation of the rat septal area in response to denervation. *Brain Res., 33,* 13.
68. Lynch, G., Matthews, D. A., Mosko, S., Parks, T., Cotman, C. W. (1972). Induced acetylcholinesterase rich layer in rat dentate gyrus following entorhinal lesions. *Brain Res., 42,* 311.
69. Steward, O., Cotman, C. W., & Lynch, G. (1976). Growth of a new fiber projection in the brain: Reinnervation of the dentate gyrus by contralateral entorhinal cortex following ipsilateral entorhinal cortex lesion. *Exp. Brain Res., 20,* 45.
70. Hoff, S. F., Scheff, S. W., Benardo, L. S., & Cotman, C. W. (1982). Lesions-induced synaptogenesis in the dentate gyrus of aged rats: I. Loss of reacquisition of normal synaptic density. *J. Comp. Neurol., 295,* 246.
71. Cotman, C. W., & Scheff, S. W. (1979). Compensatory synapse growth in aged animals after neuronal death. *Mech. Aging Develop., 9,* 103.
72. Scheff, S. W., Benardo, L. S., & Cotman, C. W. (1980). Hydrocortisone administration retards axon sprouting in rat dentate gyrus. *Exptl. Neurol., 68,* 195.
73. Scheff, S. W., & DeKosky, S. T. (1983). Steroid suppression of axon sprouting in the hippocampal dentate gyrus of the adult rat: Dose-response relationship. *Exptl. Neurol., 82,* 183.
74. Poirier, J., May, P. C., Osterburg, H. H., Geddes, J., Cotman, C., & Finch, C. E. (1989). Selective alterations of RNA in rat hippocampus after entorhinal cortex lesioning. *Proc. Natl. Acad. Sci. USA, 87,* 303.
75. Poirier, J., Baccichet, A., & Dea, D. (1991). Glucocorticoid modulation of apolipoprotein E and LDL receptor gene expression in the reinnervating hippocampus. *J. Neurochem., 57*(Suppl.), 590.
76. Foibel, J. P., Hardy, R., Campbell, M., & Goldstein, R. (1977). Joynt, plasma cortisol and catecholamine levels and survival of heart trauma. *JAMA, 238,* 1374.
77. Sapolsky, R. M. (1985). Glucocorticoid toxicity in the hippocampus: Temporal aspects of neuronal vulnerability. *Brain Res., 359,* 300.
78. Sapolsky, R. M. (1986). Glucocorticoid toxicity in the hippocampus: Temporal aspect of synergy with kainic acid. *Neuroendocrinology, 43,* 386.
79. Sapolsky, R. M. (1986). Glucocorticoid toxicity in the hippocampus: Reversal with supplementation with brain fuels. *J. Neurosci., 6,* 2240.
80. Sapolsky, R. M., Krey, L. C., & McEwen, B. S. (1985). Prolonged glucocorticoid exposure reduced hippocampal neuron number: Implications for aging. *J. Neurosci, 5,* 1221.
81. Sapolsky, R. M., & Pulsinelli, W. (1985). Glucocorticoids potentiate ischemic injury to neurons: Therapeutic implications. *Science, 229,* 1397.
82. Horner, H., Packan, D., & Sapolsky, R. M. (1990). Glucocorticoids inhibit glucose transport in cultured hippocampal neurons and glia. *Neuroendocrinology, 52,* 57.

83. Packan, D., & Sapolsky, R. M. (1990). Glucocorticoid endangerment of the hippocampus: Tissue, steroid, and receptor specificity. *Neuroendocrinology, 51,* 613.
84. Slovitar, R., Valiquette, G., Abrams, G., Ronk, E., Sollas, P., Paul, L., & Neubort, F. (1989). Selective loss of hippocampal granule cells in the mature rat brain after adrenalectomy. *Science, 243,* 535.
85. McEwen, B. S., & Gould, E. (1991). Adrenal steroid influences on the survival of hippocampal neurons. *Biochem. Pharmacol., 40,* 2393.
86. Meaney, M. J., Aitken, D. H., Bodnoff, S. R., Iny, L. J., & Sapolsky, R. M. (1985). The effects of postnatal handling on the development of the glucocorticoid receptor systems and stress recovery in the rat. *Prog. Neuropsychopharm. Biol. Psychiatry, 9,* 731.
87. Meaney, M. J., Aitken, D. H., & Sapolsky, R. M. (1991). Environmental regulation of the adrenocortical stress response in female rats and its implications for individual differences in aging. *Neurobiol. Aging, 21,* 323.
88. Sapolsky, R. M., Krey, L. C., & McEwen, B. S. (1986). The adrenocortical axis in the aged rat: Impaired sensitivity to both fast and delayed feedback inhibition. *Neurobiol. Aging, 7,* 331.
89. Morris, R. G. M. (1985). An attempt to dissociate "spatial-mapping" and "working-memory" theories of hippocampal function. In W. Seifert (Ed.), *Neurobiology of the hippocampus* (p. 405). New York: Academic Press.
90. Morris, R. G. M., Garrard, P., Rawlins, J. N. P., & O'Keefe, J. (1982). Place navigation impaired in rats with hippocampal lesions. *Nature, 297,* 681.
91. Gage, F. H., Kelly, P. A. T., & Bjorklund, A. (1984). Regional changes in brain glucose metabolism reflect cognitive impairments in aged rats. *J. Neurosci., 4,* 2856.
92. Gallagher, M., & Pelleymounter, M. (1988). Spatial learning deficits in old rats: A model for memory decline in the aged. *Neurobiol. Aging, 9,* 549.
93. Sarrieau, A., Rowe, W., O'Donnell, D., Levin, N., Seckl, J. R., & Meaney, M. J. (1992). Increased hypothalamic ACTH secretagogue synthesis in aged impaired vs. aged unimpaired rats. *Soc. Neurosci. Abstr., 19,* 567.
94. Meaney, M.S., & Sharma, S. (unpublished data).
95. Weiner, M. F., Vobach, S., Svetlik, D., & Risser, R. C. (1993). Cortisol secretion and Alzheimer's disease progression: A preliminary report. *Biol. Psychiatry, 34,* 158.
96. Waltman, C., Blackman, C. R., Chrousos, G. P., Riemann, C., & Harman, S. M. (1991). Spontaneous and glucocorticoid-induced adrenocorticotropic hormone and cortisol secretion are similar in healthy young and old men. *J. Clin. Endocrinol. Metabol., 73,* 495.
97. Van Cauter, E., Leproult, R., & Kupfer, D. J. (1996). Effects of gender and age on the levels and circadian rhythmicity of plasma cortisol. *J. Clin. Endocrinol. Metabol., 81,* 2468-2473.
98. Davis, K. L., Davis, B. M., Greenwald, B. S., Mohs, R. C., Mathé, A. A., Johns, C. A., & Horvath, T. B. (1986). Cortisol and Alzheimer's disease. I: Basal studies. *Am. J. Psychiatry, 143,* 300.
99. Nair, N. P. V., and Meaney, M. J., (unpublished data).
100. Raskind, M., Peskind, E., Rivard, M. F., Veith, R., & Barnes, R. (1982). Dexamethasone suppression test and cortisol circadian rhythm in primary degenerative dementia. *Am. J. Psychiatry, 139,* 1468.
101. Partridge, W. M., Sakiyama, R., & Judd, H. L. (1983). Protein-bound corticosterone in human serum is selectively transported into rat brain and liver in vivo. *J. Clin. Endocrinol. Metabol., 57,* 160.
102. Sapolsky, R. M., & Meaney, M. J. (1988). Postmortem decay in glucocorticoid binding in human and primate brain. *Brain Res., 448,* 182.
103. Sarrieau, A., Dussaillant, M., Sapolsky, R. M., Aitken, D. H., Olivier, O., Lal, S., Rostene, W. H., Quirion, R., & Meaney, M. J. (1988). Glucocorticoid binding sites in human temporal cortex. *Brain Res., 442,* 159.
104. Seckl, J. R., O'Donnell, D., Meaney, M. J., Yates, C., & Fink, G. (1994). Glucocorticoid and mineralocorticoid receptor mRNA expression in the hippocampus of Alzheimer's patients and age-matched controls. *Mol. Brain Res., 18,* 239–245.

48   SONIA J. LUPIEN AND MICHAEL J. MEANEY

105. Sapolsky, R., Zola-Morgan, S., & Squire, L. (1991). Inhibition of glucocorticoid secretion by the hippocampal formation in the primate. *J. Neurosci., 11,* 3695.
106. DeLeon, M. J., Golomb, J., George, A. E., Convit, A., Tarshish, C. Y., McRae, T., de Santi, S., Smith, G., Ferris, S. H., Noz, M., & Rusinek, H. (1993). The radiologic prediction of Alzheimer's disease: The atrophic hippocampal formation. *Am. J. Neuroradiol., 14,* 897.
107. Starkman, M. N., Gebarski, S. S., Berent, S., & Schteingart, D. E. (1992). Hippocampal formation volume, memory dysfunction, and cortisol levels in patients with Cushing's syndrome. *Biol. Psychiatry, 32,* 756.
108. Momose, K. J., Kjellberg, R. N., & Kliman, B. (1971). High incidence of cortical atrophy of the cerebral and cerebellar hemispheres. *Radiology, 99,* 341.
109. Bentson, J., Reza, M., Winter, J., & Wilson, G. (1978). Steroids and apparent cerebral atrophy on computed tomography scans. *J. Comput. Assist. Tomogr., 2,* 16–23.
110. Okuno, T., Ito, M., Yoshioka, M., & Nakano, Y. (1980). Cerebral atrophy following ACTH therapy. *J. Comput. Assist. Tomogr., 4,* 20.
111. Harris, G. W. (1972). Humours and hormones. *Proc. Soc. Endocrinol., 53,* i.
112. Varney, N. R., Alexander, B., & MacIndoe, J. H. (1984). Reversible steroid dementia in patients without steroid psychosis. *Am. J. Psychiatry, 141,* 369.
113. Wolkowitz, O. M., Reus, V. I., Weingartner, H., Thompson, K., Breier, A., Doran, A., Rubinow, D., & Pickar, D. (1990). Cognitive effects of corticosteroids. *Am. J. Psychiatry, 147,* 1297.
114. Weingartner, H., Cohen, R. M., & Martello, J. (1981). Cognitive processes in depression. *Arch. Gen. Psychiatry, 38,* 42.
115. Cohen, R. M., Weingartner, H., Smallberg, S., Pickar, D., & Murphy, D. L. (1982). Effort and cognition in depression. *Arch. Gen. Psychiatry., 39,* 593.
116. Rubinow, D., Post, R., Savard, R., & Gold, P. (1984). Cortisol hypersecretion and cognitive impairment in depression. *Arch., Gen. Psychiatry, 41,* 279.
117. Roy-Byrne, P. P., Weingartner, H., Bierer, L. M., Thompson, K., & Post, R. M. (1986). Effortful and automatic cognitive processes in depression. *Arch. Gen. Psychiatry, 43,* 265.
118. Wolkowitz, O. M., & Weingartner, H. (1988). Defining cognitive changes in depression and anxiety: A psychobiological analysis. *Psychiatr. Psychobiol., 3,* 1315.
119. Lupien, S., & McEwen, B. S. (1997). The acute effects of corticosteroids on cognition: Integration of animal and human model studies. *Brain Res. Rev., 24,* 1–27.
120. Kirschbaum, C., Wolf, O. T., May, M., Wippich, W., & Hellhammer, D. H. (1996). Stress and drug-induced elevation of cortisol levels impair explicit memory in healthy adults. *Life Sci., 58,* 1475.
121. Squire, L. R. (1992). Memory and the hippocampus: A synthesis from findings with rats, monkeys, and humans. *Psychol. Rev., 99,* 195.
122. Lupien, S., Gillin, C., Frakes, D., Soefje, S., & Hauger, R. L. (1995). Delayed but not immediate effects of a 100 minute hydrocortisone infusion on declarative memory performance in young normal adults. *Int. Soc. Psychoneuroendocrinol. Abstr.,* 25.
123. Newcomer, J. W., Craft, S., Hershey, T., Askins, K., & Bardgett, M. E. (1994). Glucocorticoid-induced impairment in declarative memory performance in adult humans. *J. Neurosci., 14,* 2047.
124. Wolkowitz, O. M., Weingartner, H., Rubinow, D. R., Jimerson, D., Kling, M., Berretini, W., Thompson, K., Breier, A., Doran, A., Reus, V. I., & Pickar, D. (1993). Steroid modulation of human memory: Biochemical correlates. *Biol. Psychiatry, 33,* 744.
125. Lupien, S., Tannebaum, E. M., Ohashi, E., & Meaney, M. J. (1996). Daily salivary cortisol levels in healthy elderlies correlate with feelings of fatigue, depression, and stress. *Int. Soc. Psychoneuroendocrinol. Abstr.,* 92.
126. Lupien, S., Lecours, A. R., Schwartz, G., Sharma, S., Hauger, R. L., Meaney, M. J., & Nair, N. P. V. (1995). Longitudinal study of basal cortisol levels in healthy elderly subjects: Evidence for sub-groups. *Neurobiol. Aging, 17,* 95.
127. Beckwith, B. E., Petros, T. V., Scaglione, C., & Nelson, J. (1986). Dose-dependent effects of hydrocortisone on memory in human males. *Physiol. Behav., 36,* 283.

128. Newcomer, J. W., Selke, G., Kelly, A. K., Paras, L., & Craft, S. (1995). Age-related differences in glucocorticoid effect on memory in human subjects. *Soc. Neurosci. Abstr., 21,* 61.

129. Heuser, I. J., Gotthardt, U., Schweiger, U., Schmider, J., Lammers, C. H., Dettling, M., & Holsboer, F. (1994). Age-associated changes of pituitary-adrenocortical hormone regulation in humans: Importance of gender. *Neurobiol. Aging, 15,* 227.

130. Miller, A. H., Sastry, G., Speranza, A. J., Lawlor, B. A., Mohs, R. C., Ryan, T. M., Gabriel, S. M., Serby, M., Schmeidler, J., & Davis, K. L. (1994). Lack of association between cortisol hypersecretion and nonsuppression on the DST in patients with Alzheimer's disease. *Am. J. Psychiatry, 151,* 267.

131. Mantero, F., & Boscaro, M. (1992). Glucocorticoid-dependent hypertension. *J. Steroid Biochem. Mol. Biol., 43,* 409.

132. Imai, Y., Abe, K., Sasaki, S., Minami, N., Nihei, M., Munakata, M., Murakami, O., Matsue, K., Sekino, H., Miura, Y., & Yoshinaga, K. (1988). Altered circadian blood pressure rhythm in patients with Cushing's syndrome. *Hypertension, 12,* 11.

133. Chrousos, G. P., Laue, L., Nieman, L. K., Kawai, S., Udelsman, R. U., Brandon, D. D., & Loriaux, D. L. (1988). Glucocorticoids and glucocorticoid antagonist: Lesson from RU 486. *Kidney Int., 34,* S18.

134. Whitworth, J. A. (1987). Mechanism of glucocorticoid-induced hypertension. *Kidney Int. 31,* 1213.

135. Lupien, S., Lecours, A. R., Lussier, I., Schwartz, G., Nair, N. P. V., & Meaney, M. J. (1994). Basal cortisol levels and cognitive deficits in human aging. *J. Neurosci., 14,* 2893.

136. Flood, J. F., Smith, G. E., & Roberts, E. (1988). Dehydroepiandrosterone and its sulfate enhance memory retention in mice. *Brain Res., 447,* 269.

137. Diamond, D. M., Branch, B. J., Fleshner, M., & Rose, G. M. (1995). Enhancement of hippocampal primed burst (PB) potentiation by dehydroepiandrosterone sulfate (DHEAS). *Soc. Neurosci. Abstr., 21,* 1686.

138. Morales, A. J., Nolan, J. J., Nelson, J. C., & Yen, S. S. C. (1994). Effects of replacement dose of dehydroepiandrosterone in men and women of advancing age. *J. Clin. Endocrinol. Metabol., 78,* 1360.

139. Lupien, S., Ngô, T., Rainville, C., Nair, N. P. V., Hauger, R. L., & Meaney, M. J. (1995). Spatial memory as measured by a human maze in aged subjects showing various patterns of cortisol secretion and memory function. *Soc. Neurosci., 21,* 1709.

140. Kirschbaum, C., & Hellhammer, D. H. (1994). Salivary cortisol in psychoneuroendocrine research: Recent developments and applications. *Psychoneuroendocrinology, 19,* 313.

141. Tannebaum, B. M., Lupien, S., Ohashi, E., & Meaney, M. J. (1996). Daily salivary cortisol levels in healthy elderlies: Relationship to dietary lifestyle. *Soc. Neurosci. Abstr., 22,* 1889.

142. Selye, H., & Tuchweber, B. (1976). Stress in relation to aging and disease. In A. V. Everitt & J. A. Burgess (Eds.), *Hypothalamus, pituitary, and aging* (p. 553). Springfield: Charles C. Thomas.

143. Costa, P. T., & McCrae, R. R. (1993). Psychological stress and coping in old age. In L. Goldberger & S. Bregnitz (Eds.), *Handbook of stress: Theoretical and clinical aspects* (pp. 402–412). New York: Free Press.

144. Lupien, S., Sharma, S., Arcand, J. F., Schwartz, G., Nair, N. P. V., Meaney, M. J., & Hauger, R. L. (1995). Dehydroepiandrosterone sulfate (DHEA-S) levels, cortisol levels and cognitive function in elderly human subjects. *Int. Soc. Psychoneuroendocrinology Abstr.,* 24.

145. Lupien, S., Gaudreau, S., Tchiteya, B. M., Maheu, F., Sharma, S., Nair, N. P. V., Hauger, R. L., & Meaney, M. J. (1997). Stress-induced declarative memory impairments in healthy elderly subjects: Relationship with cortisol reactivity. *J. Clin. Endocrinol. Metabol., 82,* 2070–2075.

146. Scolville, W. B., & Milner, B. (1956). Loss of recent memory after bilateral hippocampal lesions. *J. Neurol. Neurosurg. Psychiatry, 20,* 11.

147. O'Keefe, J., & Nadel, L. (1978). *The hippocampus as a cognitive map.* Oxford: Clarendon Press.

148. Lupien, S., DeLeon, M., DeSanti, S., Convit, A., Tannebaum, B. M., Nair, N. P. V., McEwen,

B. S., Hauger, R. L., & Meaney, M. J. (1996). Longitudinal increase in cortisol during human aging predicts hippocampal atrophy and memory deficits. *Soc. Neurosci. Abstr., 22,* 1889.

149. Convit, A., DeLeon, M. J., Tarshish, C., DeSanti, S., Kluger, A., Rusinek, H., & George, A. (1995). Hippocampal volume losses in minimally impaired elderly. *Lancet, 345,* 266.

150. Lupien, S., DeLeon, M., DeSanti, S., Convit, A., Nair, N. P. V., McEwen, B. S., Hauger, R. L., & Meaney, M. J. (in press). Longitudinal increase in cortisol during human aging predicts hippocampal atrophy and memory deficits. *Nat. Neurosci.*

151. Gray, J. A. (1985). Emotional behaviour and the limbic system. *Adv. Psychosom. Med., 13,* 1–25.

3

SPATIAL COGNITION AND FUNCTIONAL ALTERATIONS OF AGED RAT HIPPOCAMPUS

C. A. BARNES

Departments of Psychology and Neurology and
Arizona Research Laboratories Division of Neural Systems,
Memory and Aging, University of Arizona, Tucson, Arizona 85724

Most recent theories of hippocampal function[1–9] include some variant of the original suggestion of O'Keefe and Nadel[10] that the hippocampus functions as a "cognitive map." The implementation of new cell-specific lesion methods[11–13] and advances in the accuracy of imaging methods for localization of brain damage in humans[14–16] have greatly assisted this convergence of theoretical perspectives. Studies conducted in rats,[17, 18] monkeys,[19–24] and humans[25–27] clearly indicate that one common role of the hippocampus is learning and remembering spatial relationships or the spatial context of events. Interestingly old rats,[28–30] monkeys[31, 32] and humans[33, 34] each exhibit rather selective deficits in tasks whose performance is also selectively impaired by hippocampal lesions. This gives some credibility to the hypothesis that the common architecture of the hippocampus across these species gives rise to common functional properties and that these functions may be susceptible to similar age-related alterations.

Because of the relative ease of conducting neurophysiological studies in freely behaving rats (compared to other species), most of the available detailed data on brain–behavior relationships and hippocampal function during aging have been collected from these animals. Thus, the following review focuses upon the information processing characteristics of hippocampal cells recorded in young and old rats that are free to move and behave in extended environments. The striking selectivity of hippocampal pyramidal cell firing in relation to the animal's location led O'Keefe to call these cells "place cells."[35] Place-specific firing is now known to exist in all three major subfields of the hippocampus.[36, 37] The region of an environment over which a given place cell fires is referred to as the cell's "place field," and the integration of all the place fields of individual cells is presumed to constitute the animal's cognitive map of its environment.[10]

TECHNICAL ADVANCES IN SINGLE CELL RECORDING STUDIES

There is little question that advances over the past two decades in the technology of behavioral neurophysiology have played a key role in advancing our understanding of the normal operation of the hippocampus in spatial navigation and the deterioration of this function during aging. A major first step was the implementation of quantitative methods for monitoring behavior simultaneously with spikes from single cells.[38–40] Another advance was the stereo recording concept[41] in which triangulation principles were applied to the isolation and identification of spikes from two- or four-pronged electrode probes ("stereotrodes" or "tetrodes"). The yield of single cells (sometimes 15–20 per probe with current methods) and reliability of single-cell isolation in densely packed cortical tissue[42] was improved substantially with this technique. The most recent technical advance in single-cell recording methodology was the development of a mechanical device ("hyperdrive") that enables the independent movement of 12 tetrode probes and a 48-channel signal processing and data acquisition system necessary to acquire and analyze large sets of real-time behavior and spike data.[43, 44] This technique enables the collection of data simultaneously from large numbers of cells (typically 30–50, occasionally up to 150) and greatly facilitates the experimental distinction between dynamic changes in activity patterns that occur independently in single cells versus those that can only be understood in terms of the nonlinear interactions of large pools of synaptically coupled neurons within a network.

DO OLD RATS HAVE BROADER, LESS RELIABLE PLACE FIELDS?

The first study of the effects of aging on firing characteristics of hippocampal neurons in unrestrained, awake animals was conducted by Barnes et al.[45] (Table 3.1), who trained young and old rats to traverse a radial 8-arm maze for food reward (at arm ends). They used a "forced choice" procedure in which access to the maze arms was controlled such that a rat was required to visit each of the 8 arms once in random sequence on any trial. Quantitative video tracking was used

TABLE 3.1 Experiments in Which Hippocampal Single Cell Activity Is Compared in Freely Behaving Young and Old Rats

	Group	Age[a]	Strain	Behavioral status	Electrodes	#	# Rats	# CS cells	n for analysis
Barnes et al.[45]	Old	28 months	F344	Unknown	Single wire	1	5	27	Cells
	Young	15 months	F344	Unknown	Single wire	1	5	27	
Markus et al.[53]	Old	26 months	F344	Unknown	Stereotrode	4	7	127	Rats
	Young	11 months	F344	Unknown	Stereotrode	4	5	108	
Mizumori et al.[56]	Old	26 months	F344	Known	Stereotrode	4	13	197	Cells
	Young	11 months	F344	Known	Stereotrode	4	13	223	
Tanila et al.[58]	Old	27 months	Long-Evans	Impaired	Tetrode	1	4	111	Cells /rats
	Old	27 months	Long-Evans	Unimpaired	Tetrode	1	3	104	
	Young	5 months	Long-Evans	Known	Tetrode	1	7	146	
Tanila et al.[59]	Old	27 months	Long-Evans	Impaired	Tetrode	1	2	8	Cells
	Old	27 months	Long-Evans	Unimpaired	Tetrode	1	2	14	
	Young	5 months	Long-Evans	Known	Tetrode	1	3	13	
Barnes et al.[62]	Old	28 months	F344	Known	Tetrode	12	6	987	Rats
	Young	12 months	F344	Known	Tetrode	12	6	942	
Shen et al.[57]	Old	27 months	F344	Known	Tetrode	12	6	590	Rats
	Young	11 months	F344	Known	Tetrode	12	6	847	

[a]Approximate mean age during cell recording.

to relate firing rates of individual cells to the location and behavior of the animal. As predicted from previous in vitro studies of the basic biophysics of hippocampal cells,[46] it was found that the average firing rates of CA1 pyramidal cells during active behavior was the same in young and old rats; however, the total area of the maze over which a given place cell fired (specificity) and the consistency with which the cell fired in the same position on the maze during multiple traversals (reliability) were both reduced in the old rats. Because cells were recorded only one at a time, this phenomenon was interpreted at the cellular level, as indicating that the spatial information conveyed by each cell was simply less precise. As discussed presently, however, recent parallel recording studies suggest a different interpretation, one that necessitates a system level description.

In addition to the lack of multicell recording, two other drawbacks of the original Barnes et al.[45] study serve as illustrations to guide future work in this field. First, as was standard practice in electrophysiological studies at the time, the n used in the Barnes et al.[45] study for purposes of statistical analysis was the number of cells, rather than number of animals. This practice is inappropriate for age comparisons, as cells within an animal are not independent. The appropriate experimental design for age comparisons is to obtain data from as many cells as possible from each individual and to use the within-animal average as the representative value for statistical comparison (i.e., n should be the number of animals). The necessity of this experimental design is becoming increasingly widely recognized by investigators in the field, and its application has been greatly facilitated by advances in parallel recording methods. Second, although it is likely that the old rats as a group were spatial memory impaired compared with the young rats [particularly considering the old ages that were used in the Barnes et al.[45] study], direct assessment of their memory performance was not made. Behavioral screening is now widely accepted as a mandatory component of studies of the neurobiological mechanisms of behavioral alterations in aging.

EFFECTS OF CUE AND TASK DEMAND MANIPULATIONS ON PLACE CELL FIRING PROPERTIES OF YOUNG AND OLD RATS

In addition to technical and methodological advances, over the past decade a better understanding of the factors that control firing patterns of hippocampal cells has emerged. For example, it is now known that when environments are changed, one of two distinct phenomena can occur: Place fields can either shift or rotate as an ensemble, preserving the spatial relationships of the fields *relative to each other,* or the entire distribution of fields is replaced by an uncorrelated, new distribution with similar cohesiveness.[37, 47–50] Furthermore, whereas place cells can maintain their general place field firing patterns in darkness,[36, 51, 52] the specificity and reliability of place cell firing on a maze in the dark is affected by whether the rat can see the environment before traversing the maze in the dark.[53]

The simple arrangement of objects in an environment can also determine the extent to which the same physical landmarks influence place cell firing.[54] In addition, simply changing the demands of a behavioral task, in an identical environment, can cause complete rearrangements of place field firing.[55]

Comparisons of the firing patterns of young and old place cells during controlled manipulations of environmental variables have revealed some properties that appear to be stable with age and some that are altered. Markus et al.[53] (see Table 3.1), for example, compared the specificity and reliability of hippocampal place cells in young and old rats using a forced-choice procedure on an 8-arm radial maze that manipulated the lighting in the room. The room lights were alternated on and off in blocks of 8 arm choices (i.e., 8 correct arm choices with the lights on, then 8 correct choices with the lights off, etc.). The effect of eliminating visual cues external to the maze in the "lights off" condition was to reduce the specificity and reliability of place field firing, to the same extent, in both age groups. Furthermore, regardless of what environmental manipulations are made, it appears that the basic temporal firing characteristics of place cells are not different between young and old rats, including mean and maximal firing rates,[45, 53, 56-58] interspike interval distributions,[45] and the phase of the theta rhythm at which place cells fire in the place field[57] (see Table 3.1).

On the other hand, some experimental manipulations do distinguish between age groups. For example, Tanila et al.[58] (see Table 3.1) found that old, memory-impaired rats had place fields that were more strongly under the control of distal visual cues when both distal and proximal cue rearrangements were made. Although the young and memory-intact old rat cells did show some instances of distal cue control of cell firing, more cells were controlled by cues local to the maze and those constant in the room (i.e., air ventilation sound, corridor noise, etc.). These data suggest that the old memory-impaired rats were most strongly influenced by the distal visual landmarks and less influenced by internal maze sources. This might be interpreted as an inability of the old rats to integrate a broad spectrum of available landmarks.[58] Alternatively, the results may indicate that the old memory-impaired rats show the strongest binding to the distant landmarks, which may be particularly prominent and in real-life situations more stable than local odors or other proximal cues. Visual and odor discrimination learning was not different between age groups, suggesting that the present result was not due to altered sensory factors.[58]

Age differences also emerge, in a hippocampal region-specific manner, when young and old rats are required to perform two procedural versions of the 8-arm radial maze that have different memory requirements (i.e., forced choice versus memory version). When the specificity, reliability, and directional firing indices were compared under these two behavioral conditions, Mizumori et al.[56] (see Table 3.1) found no age differences in the firing properties of CA1 or CA3 pyramidal cells in the forced-choice (no memory) version of the radial maze task; however, in the memory version of the task, the CA1 place cells of old rats showed greater place field specificity and directionality than those of young rats.

On the other hand, old hippocampal cells recorded from the CA3 region in the memory task showed reduced specificity, reliability, and directional firing properties compared to young cells recorded there. These data indicate that the demands of the behavioral task can influence firing characteristics of the cells and that these properties change in an age-dependent fashion. Furthermore, Tanila et al.[59] (see Table 3.1) reported that compared to young and old memory-intact rats, two old memory-impaired rats showed diminished selectivity of place cell firing when exposed to completely novel environments and fewer place field rearrangements in the new situations.

Overall, these data tend to suggest that some aspects of hippocampal cell firing are intact in old rats, although age differences can be observed when tasks with different mnemonic loads are given, cue manipulations are implemented, or new environments are introduced. Although some of these data are internally consistent (such as no overall firing rate differences between age groups), several apparent inconsistencies arise, some of which may be explained by more recent findings and some of which continue to be unexplained, as discussed below.

DIFFERENCES BETWEEN STUDIES THAT REMAIN UNEXPLAINED

Although some of the major discrepancies (discussed in the following sections) between experimental outcomes can be resolved in the light of newer parallel recording studies, some cannot easily be explained. There are at least three outstanding findings in this category. They occur in relation to data from one laboratory[58, 59] in which place fields were defined slightly differently from the criteria of Muller et al.,[40] the criteria used in the manuscripts from other laboratories. This factor may not explain the differences observed, but it is worth considering. The first discrepancy involves the number of place fields observed per cell: Although Tanila et al.[58] conclude that old rats show more place fields per cell, neither Mizumori et al.[56] nor Shen et al.[57] observed an age difference in this variable. It is unlikely that this discrepancy can be explained by procedural differences, because the data were obtained under rather similar conditions in each laboratory. Because the Tanila et al.[58] finding of more place fields in old rats was not significant when statistics were conducted by subject rather than by cell, it is possible that this result is spurious. Another discrepant result involves the report of fewer "silent" cells in old rats.[59] Shen et al.[57] found that about half of old and young place cells were silent during the behavioral task, with no difference in this proportion between age groups. Because Shen et al.[57] had sample sizes of 847 young and 590 old cells versus 13 young and 22 old cells in the Tanila et al.[59] study, this may reduce to a sampling issue. Whether the different task demands in the two experiments contributed to the differences in proportions of silent cells is an open question.

The other finding that is at variance is the significant "firing onset delay" difference reported by Tanila et al.,[58] which was statistically significant when *n* was the number of cells or subjects. This variable was defined as the "number of trials completed before firing appears," which on average was 1 or 2 laps for old memory-impaired rats and 2 or 3 laps for old memory-intact rats and young rats. This implies that on average, place cells in this experiment did not fire on the first trial and that the young and memory-intact rats showed accentuated delay in firing for several trials. Shen et al.[57] also analyzed their data lap by lap. In their experiment, place cells fired more than one spike in the place field on the first lap, in both young and old rats, at least 90% of the time. Certainly, as will be discussed below, firing does become more robust with repeated traversals of a maze during a session[53, 60]; however, complete absence of firing on initial trials appears in other studies to be a relatively rare phenomenon.[43, 57, 60, 61] One additional difference in the Tanila et al.[58] experiment is the fact that medial forebrain bundle stimulation was used as reinforcement rather than food reward. Whether this factor contributed to the delay in firing observed in the Tanila et al.[58] situation will be extremely interesting to investigate.

SPATIAL SELECTIVITY: INTERACTION WITH EXPERIENCE DURING A RECORDING SESSION

A single answer to the question of whether place field specificity is altered with aging is not possible, as it appears to depend on hippocampal region and behavioral task demands. In fact, every category of possible outcome has been observed: decreases in old rat place field specificity,[45, 56, 58, 59] no change,[53, 56–58, 62] and increases in place field specificity of old rats.[56, 57] With the application of parallel recording methods to age comparisons of place field characteristics, partial resolution to these apparently discrepant results can be offered. These suggestions derive from the results of a recent experiment conducted while rats made repeated traversals around a rectangular track apparatus,[60] described in the following.

Theoretical arguments have been made that the hippocampus might store sequence or route information through Hebbian asymmetrical strengthening of synapses between cells with overlapping place fields on a route.[63–67] A prediction of these theoretical models is that asymmetrical synaptic enhancement of hippocampal connections would cause an expansion of place fields over repeated traversals of a route. Because place cells would be activated in the same temporal sequence every time the rat follows the same route, this would lead to synaptic strengthening between cells with place fields on that route. The postsynaptic neurons would thus tend to fire earlier and more robustly after repeated traversals. This should lead to place field expansion, and this expansion should be observed in a direction opposite to the direction of the route.

Recently Mehta et al.[60] have observed exactly this predicted asymmetrical place field expansion in CA1 pyramidal cells of young rats. The increase in size and backward shift of the place field relative to the direction the rat is moving occurs within a few traversals of the route. If a rat is placed onto a new route after the expansion has occurred on a familiar route, the new fields again start out small and expand with experience, consistent with the idea that asymmetrical Hebbian synaptic strengthening may be at work in the registration of sequences of spatial experience. In an age comparison of this experience-dependent place field expansion effect in CA1 pyramidal cells, Shen et al.[57] found that old rats failed to exhibit the expansion of place fields that normally occurs in young rats during the first few traversals of a familiar route on a given day. On lap 1 of a rectangular track there was no significant difference between place field sizes and specificity between age groups, but the young rats exhibited significantly larger place fields on laps 5, 10, and 15 (Fig. 3.1).

These data provide a potential explanation for some of the seemingly contradictory results in more recent age comparisons of place field specificity. First of all, it is clear from the experiments discussed above that specificity measures are dependent on the immediate past experience of a rat, even in a relatively familiar environment. The observed increased specificity (smaller place cell size) of CA1 place cells observed in old rats[56] may be due to the fact that place field specificity measures are derived from place fields averaged over numerous trials. Because the old rats do not show the experience-dependent place field expansion effect[57] (see Fig. 3.1), on average their fields would be smaller than those of young rats (even though on the first trial they would be the same size in the two

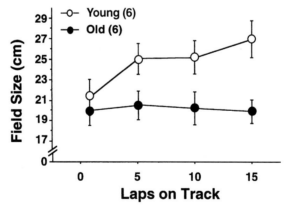

FIGURE 3.1   Data after Shen et al.[57] showing the effect of age on experience-dependent place field expansion while the rat traversed a rectangular track maze. Shown are the mean (±SEM) place field sizes for laps 1, 5, 10, and 15 for six young and six old rats. The place field sizes expanded significantly from lap 1 to lap 5 for the young rats, but the old rats did not exhibit the same expansion effect. The lack of place field expansion in the old rats provides indirect evidence for a failure of asymmetrical LTP-like processes in the hippocampus during aging. Reprinted with permission from *The Journal of Neuroscience*, 1997.

age groups). Although not statistically significant, the old, memory-impaired rats in the Tanila et al.[58] experiment also tended to have smaller place field sizes (2.6 versus 1.7 pixels for young and old, respectively). This serves to emphasize the importance of equating the "experience" of the rats on a maze on a given day, and possibly across days as well, when comparing their place field specificity.

There are a number of possible reasons for the lack of place field expansion during experience in old rats. One of the strongest possibilities is reduced long-term potentiation (LTP), which may result from fewer functional contacts made by CA3 Schaffer collateral axons to CA1 pyramidal cells and therefore reduced net synaptic input.[68, 69] There is also evidence for reduced temporal summation and total depolarization during bursts of high frequency afferent activation[70] and reduced frequency potentiation.[71] Any of these changes might lead to a reduced cooperativity for a given input, and hence less LTP; however, it must be recognized that it is difficult to equate electrical stimulation-induced LTP with what might be going on naturally during behavior. Nevertheless, such an LTP deficit in old rats has been observed under certain experimental conditions at the Schaffer collateral–CA1 synapse.[72, 73] Whatever the underlying mechanism of the experience-dependent place field expansion, the expected functional consequence of its reduction in old rats is somewhat counterintuitive (i.e., it might be assumed that smaller field sizes would mean more precision). When considering how information is conveyed by populations of neurons, however, transmission actually can be improved if each cell is broadly, rather than narrowly, tuned.[74–76] The expansion of place field sizes increases the number of neurons whose place fields overlap at any location. Information theory suggests that the population code for location conveys more bits of information as the place field expands, and thus place field broadening in young rats (i.e., reduction of "specificity") paradoxically increases the potential transmission of spatial information. The lack of field broadening observed in old rats might therefore be expected to lead to a loss of precision in the information transmitted as a consequence of experience. It may also indicate a reduced ability in old rats to remember routes.

MULTISTABILITY OF PLACE CELL MAPS IN OLD RATS

In contrast to the "increased specificity" observed in place fields of old rats in recent studies, the original study of Barnes et al.[45] indicated both a reduced specificity and a reduced reliability over multiple trials on the 8-arm maze. Recent data provide an explanation of this discrepancy[62] (see Table 3.1). Rats were trained to run around a track maze in a very familiar environment. Two recording sessions were conducted per day, and three types of manipulations were conducted between the first and second recording session within a day. The first manipulation involved ~20 min of exploration of a previously unvisited portion of the track. The second manipulation type was free exploration of 6 novel environments for 10 min each. The third was to return to their home cage for 1 hr. There were no statistically significant differences between the effects of these three treatments, thus the data are pooled in the discussion below. To determine the consistency ("reliability") of place cell firing between the two recording sessions of the day, spatial correlations were computed from the firing-rate maps in both sessions (Fig. 3.2). Correlations of 1 would reflect perfect correspondence in spatial firing patterns between sessions (retrieval of the same map), and a score of 0 would indicate that the fields in the second session were randomly related to those in the first (complete "remapping"). In young rats the place field maps were highly correlated between the first and second recording episodes, as were the maps for old rats in about 2/3 of the sessions. In the remaining third of sessions, however, old rats exhibited complete rearrangement of the place fields, as though they had "chosen" the wrong map.[62] Within a session, old rats showed stable maps. That is, once a map was chosen, the old rats retained the same map throughout the session. In other words, if the rat is not removed from the apparatus, the fields of old rats are just as reliable as those of the young.

The multistability of mapping for a given environment in old rats provides an explanation for the apparent discrepancy between more recent experiments[53, 56–58] and the results of the Barnes et al.[45] study. In the latter study, when place fields were averaged over trials, old rats exhibited place fields that covered more

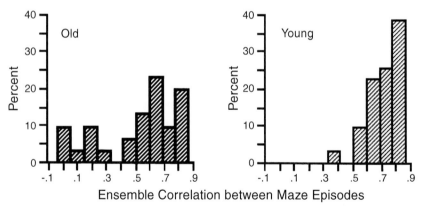

FIGURE 3.2   Data after Barnes et al.[62] showing the frequency distributions of the average place field correlations between the first and second recording episodes in a familiar room. The rats were removed from the room between these recording episodes. Notice that for the young rats the distribution tends to be unimodal with overall high correlations (i.e., the map was essentially the same on the first and second recording episodes during the day). For the old rats, on the other hand, the distribution tended to be bimodal. On about a third of the days, the correlations were essentially zero (i.e., the map chosen on the second episode was different from that chosen on the first), whereas on the remaining days the between-episode correlations were about as high as those of young rats. This multistability of hippocampal maps in old rats suggests an explanation for deficits in place recognition observed in all mammals tested on spatial tasks to date. Reprinted with permission from *Nature, 388,* 272–275, 1997, Macmillan Magazines, Limited.

arms of the maze and were less reliable than young rats (i.e., if a cell fired in one arm on one trial, it might not fire on that arm on the next trial). It was impossible to determine, recording one cell at a time in that experiment, whether this was a function of single cells showing instability or whether ensembles of cells were changing in a systematic fashion. The Barnes et al.[62] experiment suggests that the key procedural difference between the forced choice 8-arm maze task in the 1983 experiment and the Markus et al.[53] and Mizumori et al.[56] experiments is the fact that the rats were removed from the maze to disentangle the recording cable (a commutator to prevent cable twisting, now in routine use, was not then available). Rats were not disturbed in the latter two experiments. The implication for interpretation of the Barnes et al.[45] experiment is the possibility that old rats may have remapped between trials, when they were removed from the maze. Because the place fields were averaged over eight trials for the analysis, the increased place field size and unreliability may well have resulted from old animals exhibiting one map on some trials and another map on other trials.

One way in which to understand the implications of this multistability of maps in old animals is to place it into the context of theoretical models that suggest how the hippocampus represents information relevant to space. One theory[7, 77] proposes that there is a prewired matrix of connections that define a large but limited set of possible configurations of place fields that can serve as coordinate frameworks. These blank grids or "charts" form the basis of maps of specific environments through associative learning. During learning, landmarks or events (distal and proximal visual cues, odors, sounds, reward sites) become bound through Hebbian synaptic strengthening (such as LTP) to a particular configuration of place fields. The binding of these landmarks to the map serve to establish spatial relationships among arbitrary features and events in the environment and to ensure the correct recall of a map upon reentry to an environment. This theoretical view predicts that disruption of LTP during aging could result in failures of initial binding of information to the map and in retrieval of the correct map. This would occur in spite of the fact that place cell firing properties would appear relatively intact.

In support of the above theoretical ideas, there is, in fact, evidence for reduced persistence of experimentally induced LTP of hippocampal synapses[28] in memory-impaired old rats. In addition, indirect evidence[68] suggests that there is a loss of synaptic connectivity among hippocampal pyramidal cells in aged rats, which may lead, under some conditions, to a reduced probability or magnitude of LTP induction.[72, 73] Thus, it is possible that age-related failure of induction or persistence of LTP at hippocampal synapses plays a role in the map multistability observed in old rats. If external stimuli normally become associated with locations on a map via LTP-like mechanisms, old rats would be disadvantaged both in establishing spatial relationships among features and events in an environment and in selecting the correct map upon entry into a familiar environment. Even though old animals apparently have intact frameworks for representing spatial relationships, they may be unable to use them effectively because they cannot bind information about the external world to coordinates in these frameworks.

The multistability or bimodal behavior of ensembles of hippocampal place cells prompted a rather specific prediction concerning the behavior of old rats in spatial tasks. On trials or days in which the old rat "chose the correct map," behavioral performance might be at quite high levels, whereas on those trials in which the wrong map was chosen, performance might look somewhat random. To assess whether such bimodal performance occurs during the learning of the spatial version of the Morris swim task, data from about 100 young and 100 old rats tested under identical conditions were reexamined in the light of this hypothesis.[62] Early in training, both groups exhibited a pronounced bimodal distribution of performance scores, sometimes exhibiting short swim paths and sometimes long ones, before escape onto the hidden platform. By the last day of training, however, the young rats showed essentially unimodal, accurate paths to the target. On the other hand, the old rats continued to show the strong bimodality of search path lengths on the last day of training. These behavioral results are thus consistent with the neurophysiological data suggesting that place field "maps" reflect preexisting structure in the synaptic matrix of the hippocampal formation and that what is changed in the aging process is the ability to maintain a binding between these internal structures and external landmarks and events. Given the abundance of data suggesting a central role of the hippocampus in orchestrating the memory consolidation process,[1, 4] these data may have implications for this function, as well as a possible explanation for the greater tendency of older organisms to get lost.

SUMMARY AND CONCLUSIONS

The ability of convergent inputs to cooperate in the induction of LTP[78] ful-fills a major tenet of Hebb's[79] neurophysiological postulate for a plausible mechanism of associative learning. This has made LTP a prime candidate for a mechanism for associative memory and also a prime suspect in the etiology of age-related memory dysfunction.[71, 80–83] Because it has proven exceedingly difficult to observe LTP-like changes in hippocampal synapses during behavior,[84, 85] it is not yet possible to determine how or even whether the age-related changes observed under artificial stimulation-induced LTP conditions are translated into functional changes in spatial memory capacity. Nevertheless, educated guesses can be made about how such changes might impact upon the interactions among populations of hippocampal neurons. The advent of methods for recording simultaneously from large numbers of hippocampal neurons in rats has provided a new window for observing such ensemble interactions.[43] The data resulting from age comparisons using these new methods, reviewed above, suggest an LTP deficit in old rats (lack of place field expansion[57] and weaker binding of features to maps in old rats (perhaps because of less robust LTP[62]). This, presumably, results in inappropriate map selection by old rats even in familiar environments and behavior that suggests a greater tendency for the animal to become lost. We may now be on the trail of the underlying mechanisms responsible for some forms of episodic memory dysfunction during the process of normal aging.

ACKNOWLEDGMENTS

The recent work described in this chapter from Barnes' laboratory was supported by AG12609 and MH01227. I am grateful to D. S. Snyder and E. Wang for organizing the stimulating and productive Workshop on Brain Aging which prompted this chapter, and to B. L. McNaughton, E. Ellsworth, K. Gothard, N. Insel, and J. Shen for critical comments on this manuscript.

REFERENCES

1. Squire, L. R. (1992). Memory and the hippocampus: A synthesis from findings with rats, monkeys, and humans. *Psychol. Rev., 99,* 195–231.
2. Gluck, M. A., & Myers, C. E. (1993). Hippocampal mediation of stimulus representation: A computational theory. *Hippocampus, 3,* 491–516.
3. Cohen, N. J., & Eichenbaum, H. (1993). Memory, amnesia, and the hippocampal system. Cambridge: MIT Press.
4. McClelland, J. L., McNaughton, B. L., & O'Reilly, R. C. (1995). Why are there complementary learning systems in hippocampus and neocortex: Insights from the successes and failures of connectionist models of learning and memory. *Psychol. Rev., 5,* 245–286.
5. Rudy, J. W., & Sutherland, R. J. (1995). Configural association theory and the hippocampal formation: An appraisal and reconfiguration. *Hippocampus, 5,* 375–389.
6. Levy, W. B. (1996). A sequence predicting CA3 is a flexible associator that learns and uses context to solve hippocampal-like tasks. *Hippocampus, 6,* 579–590.
7. McNaughton, B. L., Barnes, C. A., Gothard, K. M., Jung, M. W., Knierim, J. J., Kudrimoti, H. S., Qin, Y.-L., Skaggs, W. E., Gerrard, J. L., Suster, M., & Weaver, K. L. (1996). Deciphering the hippocampal polyglot: The hippocampus as a path integration system. *J. Exp. Biol., 199,* 173–185.
8. Redish, D. A., & Touretzky, D. S. (1997). Cognitive maps beyond the hippocampus. *Hippocampus, 7,* 15–35.
9. Rolls, E. T. (1996). A theory of hippocampal function in memory. *Hippocampus, 6,* 601–620.
10. O'Keefe, J., & Nadel, L. (1978). *The hippocampus as a cognitive map.* Oxford: Clarendon Press.
11. Jarrard, L. E. (1993). On the role of the hippocampus in learning and memory in the rat. *Behav. Neural Biol., 60,* 9–26.
12. Alvarez, P., Zola-Morgan, S., & Squire, L. R. (1995). Damage limited to the hippocampal region produces long-lasting memory impairment in monkeys. *J. Neurosci., 15,* 3796–3807.
13. Murray, E. A., & Mishkin, M. (1996). 40-minute visual recognition memory in rhesus monkeys with hippocampal lesions. *Soc. Neurosci. Abstr., 22,* 281.
14. Cabeza, R., Grady, C. L., Nyber, L., McIntosh, A. R., Tulving, E., Kapus, S., Jennings, J., Houle, S., & Craik, F. I. M. (1997). Age-related differences in neural activity during memory encoding and retrieval: A positron emission tomography study. *J. Neurosci., 17,* 391–400.
15. Corkin, S., Amaral, D. G., Gonzalez, R. G., Johnson, K. A., & Hyman, B. T. (1997). H. M.'s medial temporal lobe lesion: Findings from magnetic resonance imaging. *J. Neurosci., 17,* 3964–3979.
16. Paus, T., Jech, R., Thompson, C. J., Comeau, R., Peters, T., & Evans, A. C. (1997). Transcranial magnetic stimulation during positron emission tomography: A new method for studying connectivity of the human cerebral cortex. *J. Neurosci., 17,* 3178–3184.
17. O'Keefe, J., Nadel, L., Keightley, S., & Kill, D. (1975). Fornix lesions selectively abolish place learning in the rat. *Exp. Neurol., 48,* 152–166.

18. Morris, R. G. M., Garrud, P., Rawlins, J. N. P., & O'Keefe, J. (1982). Place navigation impaired in rats with hippocampal lesions. *Nature, 297,* 681–683.
19. Mahut, H. (1972). A selective spatial deficit in monkeys after transection of the fornix. *Neuropsychologia, 10,* 65–74.
20. Parkinson, J. K., Murray, E. A., & Mishkin, M. (1988). A selective mnemonic role of the hippocampus in monkeys: Memory for location of objects. *J. Neurosci., 8,* 4159–4167.
21. Angeli, S. J., Murray, E. A., & Mishkin, M. (1993). Hippocampectomized monkeys can remember one place but not two. *Neuropsychologia, 31,* 1021–1030.
22. Eifuku, S., Nishijo, H., Kita, T., & Ono, T. (1995). Neuronal activity in the primate hippocampal formation during a conditional association task based on the subject's location. *J. Neurosci., 15,* 4952–4969.
23. Rolls, E. T., & O'Mara, S. M. (1995). View-responsive neurons in the primate hippocampal complex. *Hippocampus, 5,* 409–424.
24. Gaffan, D., & Parker, A. (1996). Interaction of perirhinal cortex with the fornix-fimbria: Memory for objects and "object-in-place" memory. *J. Neurosci., 16,* 5864–5869.
25. Corkin, S. (1965). Tactually-guided maze-learning in man: Effects of bilateral hippocampal, bilateral frontal, and unilateral cerebral lesions. *Neuropsychologia, 3,* 339–351.
26. Smith, M. L., & Milner, B. (1981). The role of the right hippocampus in the recall of spatial location. *Neuropsychologia, 19,* 781–793.
27. Pigott, S., & Milner, B. (1993). Memory for different aspects of complex visual scenes after unilateral temporal- or frontal-lobe resection. *Neuropsychologia, 31,* 1–15.
28. Barnes, C. A. (1979). Memory deficits associated with senescence: A neurophysiological and behavioral study in the rat. *J. Comp. Physiol. Psych. 931,* 74–104.
29. Markowska, A. L., Stone, W. S., Ingram, D. K., Reynolds, J., Gold, P. E., Conti, L. H., Pontecorvo, M. J., Wenk, G., & Olton, D. S. (1989). Individual differences in aging: Behavioral and neurobiological correlates. *Neurobiol. Aging, 10,* 31–43.
30. Gallagher, M., & Rapp, P. R. (1997). The use of animal models to study the effects of aging on cognition. *Ann. Rev. Psychol., 48,* 339–370.
31. Lai, Z. C., Moss, M. B., Killiany, R. J., Rosene, D. L., & Herndon, J. G. (1995). Executive system dysfunction in the aged monkey: Spatial and object reversal learning. *Neurobiol. Aging, 16,* 947–954.
32. Rapp, P. R., Kansky, M. T., & Roberts, J. A. (1997). Impaired spatial information processing in aged moneys with preserved recognition memory. *Neuroreport, 8,* 1923–1928.
33. Uttl, B., & Graf, P. (1993). Episodic spatial memory in adulthood. *Psychol. Aging, 8,* 257–273.
34. Wilkniss, S. M., Jones, M. G., Korol, D. L., Gold, P. E., & Manning, C. A. (1997). Age-related differences in an ecologically-based study of route learning. *Psychol. Aging, 12,* 372–375.
35. O'Keefe, J., & Dostrovsky, J. (1971). The hippocampus as a spatial map. Preliminary evidence from unit activity in the freely-moving rat. *Brain Res., 34,* 71–75.
36. O'Keefe, J. (1976). Place units in the hippocampus of the freely moving rat. *Exp. Neurol., 51,* 8–109.
37. Jung, M. W., & McNaughton, B. L. (1993). Spatial selectivity of unit activity in the hippocampal granular layer. *Hippocampus, 3,* 65–182.
38. O'Keefe, J. (1983). Spatial memory within and without the hippocampal system. In W. Seifert (Ed.), *Neurobiology of the hippocampus* (pp. 375–403). New York: Academic Press.
39. McNaughton, B. L., Barnes, C. A., & O'Keefe, J. (1983). The contributions of position, direction, and velocity to single unit activity in the hippocampus of freely-moving rats. *Exp. Brain Res., 52,* 41–49.
40. Muller, R. U., Kubie, J. L., & Ranck, J. B., Jr. (1987). Spatial firing patterns of hippocampal complex-spike cells in a fixed environment. *J. Neurosci., 77,* 1935–1950.
41. McNaughton, B. L., O'Keefe, J., & Barnes, C. A. (1983). The stereotrode: A new technique for simultaneous isolation of several single units in the central nervous system from multiple unit records. *J. Neurosci. Meth., 8,* 391–397.

42. Gray, C. M., Maldonado, P. E., Wilson, M. A., & McNaughton, B. L. (1995). Tetrodes markedly improve the reliability and yield of multiple single-unit isolation from multi-unit recordings on cat striate cortex. *J. Neurosci. Meth., 63,* 43–54.

43. Wilson, M. A., & McNaughton, B. L. (1993). Dynamics of the hippocampal ensemble code for space. *Science, 261,* 1055–1058.

44. Gothard, K. M., Skaggs, W. E., Moore, K. M., & McNaughton, B.L. (1996). Binding of hippocampal CA1 neural activity to multiple reference frames in a landmark-based navigation task. *J. Neurosci., 16,* 823–835.

45. Barnes, C. A., McNaughton, B. L., & O'Keefe, J. (1983). Loss of place specificity in hippocampal complex spike cells of senescent rat. *Neurobiol. Aging 4,* 113–119.

46. Barnes, C. A., & McNaughton, B. L. (1980). Physiological compensation for loss of afferent synapses in rat hippocampal granule cells during senescence. *J. Phys., 309,* 473–485.

47. O'Keefe, J., & Speakman, A. (1987). Single unit activity in the rat hippocampus during a spatial memory task. *Exp. Brain Res., 68,* 1–27.

48. Sharp, P. E., Kubie, J. L., & Muller, R. U. (1990). Firing properties of hippocampal neurons in a visually symmetrical environment: Contributions of multiple sensory cues and mnemonic properties. *J. Neurosci., 10,* 3093–3105.

49. Bostock, E., Muller, R. U., & Kubie, J. L. (1991). Experience-dependent modifications of hippocampal place cell firing. *Hippocampus, 1,* 193–206.

50. Knierim, J. J., Kudrimoti, H. S., & McNaughton, B. L. (1995). Place cells, head direction cells, and the learning of landmark stability. *J. Neurosci., 15,* 1648–1659.

51. McNaughton, B. L., Leonard, B., & Chen, L. (1989). Cortical-hippocampal interactions and cognitive mapping: A hypothesis based on reintegration of the parietal and inferotemporal pathways for visual processing. *Psychobiology, 17,* 230–235.

52. Quirk, G. J., Muller, R. U., & Kubie, J. L. (1990). The firing of hippocampal place cells in the dark depends on the rat's recent experience. *J. Neurosci., 10,* 2008–2017.

53. Markus, E. J., Barnes, C. A., McNaughton, B. L., Gladden, V. L., & Skaggs, W. E. (1994). Spatial information content and reliability of hippocampal CA1 neurons: Effects of visual input. *Hippocampus, 4,* 410–421.

54. Cressant, A., Muller, R. U., & Poucet, B. (1997). Failure of centrally placed objects to control the firing fields of hippocampal place cells. *J. Neurosci., 17,* 2531–2542.

55. Markus, E. J., Qin, Y.-L., Leonard, B., Skaggs, W. E., McNaughton, B. L., & Barnes, C. A. (1995). Interactions between location and task affect the spatial and directional firing of hippocampal neurons. *J. Neurosci., 15,* 7079–7094.

56. Mizumori, S. J. Y., Lavoie, A. M., & Kalyani, A. (1996). Redistribution of spatial representation in the hippocampus of aged rats performing a spatial memory task. *Behav. Neurosci., 110,* 1006–1016.

57. Shen, J., Barnes, C. A., McNaughton, B. L., Skaggs, W. E., & Weaver, K. L. (1997). The effect of aging on experience-dependent plasticity of hippocampal place cells. *J. Neurosci., 17,* 6769–6782.

58. Tanila, H., Shapiro, M., Gallagher, M., & Eichenbaum, H. (1997). Brain aging: Changes in the nature of information coding by the hippocampus. *J. Neurosci., 17,* 5155–5166.

59. Tanila, H., Sipila, P., Shapiro, M., & Eichenbaum, H. (1997). Brain aging: Impaired coding of novel environmental cues. *J. Neurosci., 17,* 5167–5174.

60. Mehta, M. R., Barnes, C. A., & McNaughton, B. L. (1997). Asymmetric expansion of hippocampal place fields: Evidence for Hebbian sequence learning. *Proc. Natl. Acad. Sci., 94,* 8918–8921.

61. Hill, A. J. (1978). First occurrence of hippocampal spatial firing in a new environment. *Exp. Neurol., 62,* 282–297.

62. Barnes, C. A., Suster, M. S., Shen, J., & McNaughton, B. L. (1997). Cognitive map multistability in aged rat hippocampus. *Nature, 388,* 272–275.

63. McNaughton, B. L., & Morris, R. G. M. (1987). Hippocampal synaptic enhancement and information storage within a distributed memory system. *Trends Neurosci., 10,* 408–415.

64. Levy, W. B. (1989). A computational approach to hippocampal function. In R. D. Hawkins & G. H. Bower (Eds.), *Computational models of learning in simple neural systems* (pp. 243–305). New York: Academic Press.

65. Blum, K. I., & Abbott, L. F. (1996). A model of spatial map formation in the hippocampus of the rat. *Neural Computing, 8,* 85–93.

66. Skaggs, W. E., McNaughton, B. L., Wilson, M. A., & Barnes, C. A. (1996). Theta phase precession in hippocampal neuronal populations and the compression of temporal sequences. *Hippocampus, 6,* 149–172.

67. Tsodyks, M. V., Skaggs, W. E., Sejnowski, T. J., & McNaughton, B. L. (1996). Population dynamics and theta rhythm phase precession of hippocampal place cell firing: A spiking neuron model. *Hippocampus, 6,* 271–280.

68. Barnes, C. A., Rao, G., Foster, T. C., & McNaughton, B. L. (1992). Region specific age effects on AMPA sensitivity: Electrophysiological evidence for loss of synaptic contacts in hippocampal field CA1. *Hippocampus, 2,* 457–468.

69. Barnes, C. A., Rao, G., & McNaughton, B. L. (1996). Functional integrity of NMDA-dependent LTP induction mechanisms across the lifespan of F-344 rats. *Learning & Memory, 3,* 124–137.

70. Rosenzweig, E. S., Rao, G., McNaughton, B. L., & Barnes, C. A. (1997). The role of temporal summation in age-related LTP-induction deficits. *Hippocampus, 7,* 549–558.

71. Landfield, P. W. (1988). Hippocampal neurobiological mechanisms of age-related memory dysfunction. *Neurobiol. Aging, 9,* 571–579.

72. Deupree, D. L., Turner, D. A., & Watters, C. L. (1991). Spatial performance correlates with in vitro potentiation in young and aged Fischer 344 rats. *Brain Res., 554,* 1–9.

73. Moore, C. I., Browning, M. D., & Rose, G. M. (1993). Hippocampal plasticity induced by primed burst, but not long-term potentiation, stimulation is impaired in area CA1 of aged Fischer 344 rats. *Hippocampus, 3,* 57–66.

74. Lehky, S. R., & Sejnowski, T. J. (1990). Neural model of stereoacuity and depth interpolation based on a distributed representation of stereo disparity. *J. Neurosci., 10,* 2281–2299.

75. Treves, A., Barnes, C. A., & Rolls, E. T. (1996). Quantitative analysis of network models and of hippocampal data. In T. Ono, B. L. McNaughton, S. Molotchnikoff, E. T. Rolls, & H. Nishijo (Eds.), *Perception, memory and emotion: Frontier in neuroscience.* (pp. 567–579). Oxford: Pergamon Press.

76. Treves, A., Skaggs, W. E., & Barnes, C. A. (1996). How much of the hippocampus can be explained by functional constraints? *Hippocampus, 6,* 666–674.

77. Samsonovich, A., & McNaughton, B. L. (1997). Path integration and cognitive mapping in a continuous attractor model of the hippocampus. *J. Neurosci., 17,* 5900–5920.

78. McNaughton, B. L., Douglas, R. M., & Goddard, G. V. (1978). Synaptic enhancement in fascia dentata: Cooperativity among coactive afferents. *Brain Res., 157,* 277–293.

79. Hebb, D. O. (1949). *The organization of behavior.* New York: Wiley.

80. Landfield, P. W., Waymire, J. C., & Lynch, G. (1978). Hippocampal aging and adrenocorticoids: Quantitative correlations. *Science, 202,* 1098–1102.

81. Barnes, C. A., & McNaughton, B. L. (1980). Spatial memory and hippocampal synaptic plasticity in middle-aged and senescent rats. In D. Stein (Ed.), *Psychobiology of aging: Problems and perspectives* (pp. 253–272). New York: Elsevier Press/North-Holland.

82. de Toledo-Morrell, L., Geinisman, Y., & Morrell, F. (1988). Age-dependent alterations in hippocampal synaptic plasticity: Relation to memory disorders. *Neurobiol. Aging, 9,* 581–590.

83. Barnes, C. A., Treves, A., Rao, G., & Shen, J. (1994). Electrophysiological markers of cognitive aging: Region specificity and computational consequences. *Semin. Neurosci., 6,* 359–367.

84. Barnes, C. A. (1995). Involvement of LTP in memory: Are we "searching under the street light?" *Neuron, 15,* 751–754.

85. Martinez, J. L., Jr., & Derrick, B. E. (1996). Long-term potentiation and learning. *Ann. Rev. Psychol., 47,* 173–203.

4

IDENTIFICATION OF MOLECULAR AND CELLULAR MECHANISMS OF LEARNING AND MEMORY

THE IMPACT OF GENE TARGETING

ALCINO J. SILVA, KARL PETER GIESE, AND PAUL W. FRANKLAND

Cold Spring Harbor Laboratory
Cold Spring Harbor, New York 11724

Numerous studies demonstrate that aging can be associated with a marked cognitive decline. Understanding the mechanisms of learning and memory will be critical for unraveling the causes of this cognitive decline.[1, 2] Finding "magic bullets" against these problems will depend on a detailed knowledge of the molecular and cellular mechanisms responsible for the pathophysiology involved. Recently, transgenic and gene targeting techniques have been introduced to the study of learning and memory.[3-5] It is now possible to add, delete, and modify any cloned gene; in the near future, it will be possible to have complete control over when and where these mutations occur in the brain. Genetically modified mice can have very specific changes in biochemistry, electrophysiology, neuroanatomy, circuit function, and behavior. Therefore with these mice it is possible to study the molecular, cellular, neuroanatomical, and behavioral impact of mutations in single genes. In this way, genetic studies are attempting to forge connections between phenomena from different levels of biological complexity. This integrative approach will have a crucial role not only in uncovering mechanisms of learning and memory (L&M), but also in furthering the understanding of the pathophysiological states underlying cognitive disorders, such as those associated with aging.

Most genetic studies of L&M have explored hypotheses that had previously been studied with other approaches. For example, the possible connection between long-term potentiation (LTP) and learning has been investigated with behavioral pharmacology and in vivo recordings. There is an extensive literature on the possible role of LTP in L&M; therefore further experiments, including genetic studies, are easy to frame within that rich experimental tradition. In contrast, studies examining the possible role of other cellular mechanisms in L&M face greater problems. Although there are numerous experimental results demonstrating that LTP could not possibly be the sole cellular mechanism used by the brain to process, store, and recall information, comparatively little has been done to explore the potential roles of other mechanisms.

One of the many intrinsic problems of this type of research is that there is no single experiment that can unambiguously connect any molecular or cellular mechanism with L&M. Instead, several different types of studies are required to determine whether any one mechanism is involved in L&M. These include at least (1) the development of biologically based learning models that propose an explanation for the role of a given mechanism in L&M, (2) lesion experiments with different techniques and in different species, and (3) direct observations during learning with different techniques and in different species. Thus, the task of unraveling the molecular and cellular basis of L&M is daunting and painstakingly slow, requiring the collaborative work of many groups over decades. Despite the magnitude of the task, significant progress has been made in several fronts in recent years. Here, we review studies illustrating the role of genetics in unraveling the molecular and cellular basis of L&M. These studies implicate LTP, short-term plasticity, and changes in spike properties in L&M. Importantly, these mechanisms are known to change during aging and may be involved in the cognitive decline often observed during aging.[1, 2]

LTP AS A CELLULAR MECHANISM FOR L&M

MODELS

In determining whether a given molecular or cellular mechanism is involved in L&M, it is important to develop rigorous models that explain how this specific mechanism functions in L&M. Biologically based models can strengthen the connection between a given mechanism and L&M, since they provide an explanation of how the two are connected. Without a clear hypothesis it is difficult to both interpret the results of L&M experiments and plan future studies.

Most rodent studies investigating the cellular mechanisms of memory have focused on long-lasting changes in synaptic efficacy (e.g., long-term potentiation and long-term depression). These studies are based on the hypothesis that memories can be stored in neural circuits by adjusting synaptic strengths in that circuit.[6] In 1949 Hebb proposed that when one neuron excites another repeatedly a biochemical or structural change takes place that has a long-lasting effect on the efficacy with which the two neurons communicate.[6] This idea has had a lasting impact in theoretical neuroscience, and there are numerous elaborate L&M models implementing Hebbian ideas.[7] Thus, the hypothesis implicating LTP in L&M is comfortably entrenched within the rich neuroconnectionist modeling tradition.

OBSERVATIONS

Brain regions such as the hippocampus, with an acknowledged participation in L&M, express various forms of LTP and long-term depression (LTD). These electrophysiological phenomena have the very properties thought to be crucial for memory formation (associativity, specificity, reversibility, and stability).[8, 9] Additionally, direct observation during learning found that hippocampal neurons fire in patterns that are ideal for the induction of LTP.[10, 11] Interestingly, aged rats with spatial learning abnormalities are thought to have deficits in both the induction and maintenance of LTP.[1] Thus, these findings demonstrate that LTP-like synaptic phenomena may occur during learning. However, they do not show that LTP is required for learning.

LESIONS

The induction of some forms of LTP involves the activation of N-methyl-D-aspartate receptors (NMDAR), and a subsequent postsynaptic increase in $[Ca^{2+}]$.[8, 12] This increase in $[Ca^{2+}]$ is known to activate a number of protein kinases and other enzymes that are thought to trigger a long-lasting enhancement of synaptic transmission.[8, 12] Pharmacological blockers of LTP in rats (e.g., 2-aminophosphovaleric acid [APV], an NMDAR blocker) were shown to also disrupt memory formation (see for example Davis et al.[13], and Kim et al.[14]), suggesting that a block of LTP impairs learning.

In addition to the experiments with pharmacological lesions, there are also a number of genetic lesion experiments that have explored the role of LTP in L&M. For example, genetic deletions of the α-calcium-calmodulin kinase II (αCaMKII),[15] Fyn tyrosine kinase,[16] type I adenylate cyclase,[17] NMDAR ε1 subunit,[18] and metabotropic glutamate receptor 5 (mGluR5)[19] affect LTP in the hippocampal CA1 region and hippocampal-dependent L&M, such as performance in the hidden-platform version of the Morris water maze.[20]

PROBLEMS, CONTRADICTIONS

The evidence described above shows that (1) modeling studies, (2) observations during learning, and (3) two types of lesion experiments (pharmacological and genetic) in two different species (mice and rats) implicated LTP in learning. However, not all experiments carried out with either pharmacological or genetic lesions supported the hypothesis that LTP underlies L&M. Mutants lacking either the β1 isoform of the protein kinase A catalytic subunit (Cβ1⁻)[21] or the β isoform of the inhibitory subunit (RIβ⁻) of this kinase[22] have deficient CA3 LTP, CA1 LTD, and CA1 LTP depotentiation.[21–23] However, both mutants performed normally in an extensive battery of learning tests, which included the visible and hidden-platform tests of the water maze, as well as cued and contextual conditioning.[23]

Genetic lesions are not alone in questioning the model that long-term changes in glutamatergic synaptic transmission (LTP) are essential for learning. Recent pharmacological experiments suggest that the effects of NMDAR blockers on learning tests, such as the water maze,[13] may have been confounded by performance impairments caused by the drugs used.[24, 25] For example, rats treated with the NMDAR antagonist APV show thigmotaxic swimming, difficulty in climbing onto and staying on the water maze escape platform, and hyperactivity. These abnormal behaviors seem unrelated to learning and could have been responsible for the inability of the rats to learn the spatial location of the platform in the water maze. In contrast, rats treated with APV, but pretrained on the behavioral demands of the water maze, appeared to perform normally, indicating that NMDAR function (and by consequence NMDAR-dependent LTP) is not essential for spatial learning.[24–26]

For each of the genetic and pharmacological experiments just mentioned, however, there are important considerations that question whether the results really dissociate LTP from learning. For example, it is known that there are various forms of LTP in the hippocampus and that only some of those are dependent on the function of NMDARs.[12] Therefore, application of APV does not block all LTP in the hippocampus. Importantly, NMDAR-dependent LTP only lasts a few hours, and the later phases of LTP are not sensitive to NMDAR blockers.[12] It is also possible that the tasks used are not sensitive enough to detect the learning deficits. For example, a mutation of the gene for the cAMP responsive element binding protein (CREB) does not affect performance in the water maze when the mice are trained with 1-hr intertrial intervals,[27] but it does impair LTP.[28] Training with shorter intervals (i.e., 1 min), however, uncovered a spatial learning deficit in these mutants.[27] Clearly, there are many possible confounds in experiments that appear to dissociate LTP from learning. Nevertheless, similar uncertainties also exist in experiments (some of which were reviewed here) that argue for a role of LTP in learning. The very complexity that makes unraveling mechanisms of L&M extremely difficult also hinders the falsification of established models.

OTHER MECHANISMS

One of the problems with determining the role of LTP in L&M is that most of the effort has been focused on a single cellular mechanism (LTP), and comparatively very little attention has been given to other concurrent processes that may interact, complement, compensate for, and modify the role of LTP in L&M. It is impossible to determine where and how a single piece fits in a large puzzle if all the other pieces are ignored.

LTP increases the probability that a neuron will fire an action potential, and therefore this mechanism can have an immediate effect on spike dynamics. Since information in the brain is encoded by the number and pattern of action potentials, LTP should have a pronounced and long-lasting impact on neuronal networks. However, NMDAR-dependent LTP is not the only cellular mechanism capable of changing spike dynamics. For example, changes in the number or activity of K^+, Na^+ or Cl^- channels can also affect the dynamics of spike trains. Consequently, modulation of these channels could also have a role in learning.

Similarly, short-term plasticity is also known to have a pronounced (although brief) effect on spike dynamics. For example, paired-pulse facilitation (PPF) refers to the potentiation observed for the second of a pair of synaptic stimuli. Due to the facilitation of synaptic transmission, in general the second of a pair of stimuli (20–500 ms apart) is more effective at inducing a spike than a single isolated stimulus.[29] Following synaptic stimulation by a short spike train, neurotransmitter release is increased for seconds to minutes (augmentation and posttetanic potentiation).[30, 31] Therefore PPF, augmentation, and posttetanic potentiation can have an immediate impact on the pattern and frequencies of spike trains. These are only a few examples of cellular mechanisms that together with LTP may be responsible for the processing, storage, and recall of information in neuronal circuits. Interestingly, studies with aged animals have identified a number of neurophysiological changes that may be correlated with their learning deficits, reinforcing the idea that L&M engages a number of molecular and cellular mechanisms.[1] The studies that follow provide evidence that in mammals, mechanisms other than LTP, have a direct effect on L&M. Understanding how these mechanisms affect L&M will be crucial for defining the role of LTP.

SHORT-TERM PLASTICITY AND LEARNING

MODELS

Short-term plasticity (STP) could serve as a temporary memory buffer.[32] Just as LTP is thought to be required for long-term memory, short-term changes in synaptic strength may store information for brief periods of time. For example, in the water maze, animals use distal cues to locate the hidden platform. In any trial the animal has to continuously remember the places it just visited in order to organize its searches for the platform. Despite its critical value during each search for the platform, this information (working memory) may have no long-term value, and therefore it is probably discarded as quickly as it is acquired. Working memory may require synaptic changes (short-term plasticity) that can be easily induced and quickly erased. Long-term storage of this type of information could overwhelm the capacity of hippocampal networks with superfluous information.

STP may also be critical for encoding the timing or sequence of events in neuronal networks.[33] For example, the relative sequence of spatial information that the animals are exposed to, as they search for the hidden platform, may be an important component of the cognitive processes involved in learning the spatial relationships of spatial cues (objects) in the room. Since the animals cannot be exposed simultaneously to all spatial cues that surround the pool, the specific sequence of visual information acquired during their searches for the platform could be critical for determining the spatial relationships between the objects seen from the pool.

The different mechanisms of STP (PPF, posttetanic potentiation, augmentation, paired-pulse depression, etc.) could be used by circuits to process, modify, filter, and integrate information. STP seems specifically adept for this function because its properties are highly responsive to a neuron's recent history of activation, as well as to the architecture of circuits and to brain-wide modulatory mechanisms.[30, 31]

OBSERVATIONS

Little is known about the function of short-term (milliseconds to seconds) changes in synaptic strength.[30, 31] Nevertheless, studies with a variety of organisms suggest that STP could have a role in learning.[30, 31] Short-term changes in synaptic efficacy seem to underlie habituation of withdrawal responses in *Aplysia*[30, 34] and the habituation of escape responses in vertebrate and invertebrate species.[35–37] For example, softly touching the tail of *Aplysia* triggers an inhibition of its siphon withdrawal reflex. Importantly, a short-term potentiation in synaptic release in L30 inhibitory neurons appears to have a crucial role in this habituation response. STP in L30 neurons (caused by touching the tail) inhibits neurons that activate siphon motor neurons. In contrast, tail shock, which triggers sensitization of this response, reduces L30 STP (particularly augmentation) and consequently leads to an enhanced siphon withdrawal response. This is an elegant demonstration of how STP could have a profound impact on the function of a neuronetwork, allowing it to adapt quickly to changing environments.[30] Interestingly, a number of studies demonstrated changes in STP during aging. For example, frequency potentiation and posttetanic potentiation are decreased in old rats.[1] These deficits could contribute to the learning deficits characterized in aging animals.[2]

GENETIC LESIONS

To determine whether STP in the central nervous system (CNS) is required for mammalian learning, we studied 4 different lines of mutant mice lacking key presynaptic proteins that are known to affect the regulation of neurotransmitter release.[38] Mice heterozygous for the α-calcium-calmodulin kinase II (αCaMKII$^{+/-}$) have lower PPF and increased augmentation (Aug).[39] Mice lacking synapsin II (SyII$^{-/-}$), as well as synapsin I and II mutants (SyI/II$^{-/-}$), show normal PPF, but lower Aug. In contrast, mice lacking synapsin I (SyI$^{-/-}$) show increased PPF, but normal Aug.[40, 41] It is noteworthy that hippocampal CA1 LTP seemed unaffected in all of these mutants.[38, 40, 41] The loss of synapsin II also does not affect LTP in the mossy fiber pathway of the hippocampus.[42] Importantly, Sy II$^{-/-}$, SyI/II$^{-/-}$, and CaMKII$^{+/-}$ mutants with a decrease in either Aug or PPF, show learning deficits, but the increase in PPF does not appear to disrupt learning in Sy I$^{-/-}$ mutants.[38]

Two different learning tasks were used to test the mutants: water maze and fear conditioning. The mutations studied appeared to have a larger impact on the hippocampal-dependent versions of these tasks. For example, the CaMKII$^{+/-}$ mutants performed poorly in the spatial version of the water maze, a task which is sensitive to hippocampal lesions. In contrast, their performance was normal in the visible-platform version of the water maze, a task which is unaffected by hippocampal lesions. These results show that their spatial learning deficits are not due to abnormalities in vision, motivation, and motor coordination.[38]

In addition, electrophysiolgical studies in the neuromuscular junction of *Drosophila* learning mutants revealed impairments in STP.[43, 44] It is conceivable that these mutants also have similar impairments in STP in the brain, and that these deficits are responsible for their L&M abnormalities.[45]

The work summarized above supports the idea that STP is critical for learning. First, modeling results provide a framework for the role of STP in learning. Second, direct electrophysiological observation of invertebrate circuits demonstrated that short-term plasticity has a key role in habituation in invertebrates and in the modulation of motor responses in vertebrate and invertebrate species.[30] Third, our genetic lesion experiments reviewed here provide several lines of evidence (different mutants and different tasks) for the hypothesis that a decrease in either Aug or PPF leads to learning deficits. Furthermore, the observation that mice with an increase in PPF do not show learning impairments demonstrated that only certain disruptions of presynaptic plasticity have an impact on learning. Altogether these results indicate that STP has a role in learning.

SPIKING AND LEARNING

MODELS

Information in the brain is thought to be encoded by the number and/or pattern of action potentials (spikes).[46] Consequently, mechanisms that affect the generation of spikes or spike trains could have a role in learning. For example, the slow afterhyperpolarization (sAHP) is a key determinant of spike train characteristics. Large sAHPs suppress neuronal firing, while neurons spike readily under conditions that reduce the sAHP. Importantly, a variety of biologically based models have proposed a key role for the sAHP in L&M.[47, 48] For example, acetylcholine is known to suppress synaptic transmission and decrease the sAHP. This binary function has been proposed to serve as a filtering device that prevents the contamination of new memories with old memories inadvertantly reactivated during learning.[47] The reduction of the sAHP may also serve to stabilize memory.[49] Increased excitability in neurons encoding the learned information may ensure their future co-activation, thus stabilizing the memory.

Based on recent findings in our laboratory (see the following), we propose that during learning the transient A-type K^+ currents are modulated so that they loose their fast inactivation and behave like delayed rectifiers. This modulation should result in a decrease in the width of spikes in a spike train.[50] The decrease in spike width then would result in decreased calcium influx during the spike train, thus reducing the calcium-dependent sAHP.[51] In this section we will review evidence that supports the model that A-type K^+ channel modulation of the sAHP may be required for learning.

OBSERVATIONS

A-type K^+ channels were proposed to have a role in invertebrate learning, such as in the conditioning of the phototactic behavior of *Hermissenda*. This marine mollusk is positively phototatic and therefore it is possible to use light as a conditioned stimulus. *Hermissenda* is also very sensitive to motion, and high-speed rotation has been used as an unconditioned stimulus. Pairing these two stimuli causes a long-term suppression of the normal positive phototatic response. Interestingly, this conditioning procedure reduces A-type K^+ channel currents in cells (B photoreceptors) that have a key role in mediating phototatic conditioning. In addition, there was also an increase in excitability in these cells. These results implicate A-type K^+ channel currents in associative learning in *Hermissenda*.[52]

Remarkably, single-electrode voltage clamp studies in the motor cortex (layers III and V) of cats revealed a decrease in an early outward current (A-type K^+ current) induced by depolarization in cells that developed a conditioning response, but not in cells that did not.[53] The properties of this decreased current were similar to those of the A-type K^+ currents. The decrease in A-type K^+ currents may be related to the increase in spike discharges also observed in conditioned cells of the motor cortex.[53] Thus, these elegant observations indicate that, just as in *Hermissenda,* Pavlovian conditioning in the motor cortex of cats results in a decrease of A-type K^+ currents and in an increase in spike discharges.[53]

Perhaps learning triggers changes in A-type K^+ currents because these currents can affect spike dynamics. According to our proposal, in both the *Hermissenda* and in the motor cortex studies, the postlearning decreases in fast inactivating currents (A-like) may be due to their conversion (e.g., by phosphorylation) into slowly inactivating currents, resulting in decreased sAHP. Consistent with this hypothesis, previous studies in rabbits showed that trace eyeblink conditioning, a hippocampal-dependent form of learning, results in a reduction of the sAHP in the CA1 region of the hippocampus that lasted several days.[54] Reductions in the sAHP lead to the same increases in spike discharges observed in conditioned cells of the motor cortex of cats.[53] We propose that changes in A-type K^+ currents are responsible for these reductions in the sAHP.

Interestingly, aging leads to an increase in the sAHP which is associated with cognitive deficits.[55] Remarkably, these cognitive deficits can be treated with nimodipine, a drug that blocks L-type calcium channels and reduces the sAHP.[54] These findings provide evidence for the hypothesis that the increase in sAHP observed with aging is one of the factors contributing to cognitive decline.

LESIONS

Studies in *Drosophila* showed that the mutation of a *Shaker* A-type K^+ channel impairs olfactory conditioning.[56] After repeated exposures of an odor paired with electric shock, flies learn to avoid the conditioned odor in a T-maze. Interestingly, *Shaker* mutants are defective in both their rate and maximum level of acquisition of this conditioned response. In the first 30 min following training, memory decays faster in *Shaker* mutants than in control flies. Curiously, the decay rates beyond the initial 30-min period following training appear to be normal in these flies. These results suggest that *Shaker* (A-type) K^+ channel function may be critical for short-term memory in *Drosophila*.[56] Furthermore, antisense inhibition of a *Shaker*-related K^+ channel subunit (Kv1.1) impaired L&M in mice and rats, indicating that K^+ channel function is also important for mammalian learning.[57] Kv1.1 is a delayed rectifier K^+ channel subunit (not A-type).

To determine whether disruptions of A-type K^+ channels affected mammalian L&M, our laboratory derived a mouse mutant for an auxiliary β-subunit (Kvβ1.1)[58] that confers A-type inactivation on otherwise noninactivating *Shaker*-related K^+ channels.[59] Consistent with this hypothesis, our results suggest that the the loss of Kvβ1.1 transformed some A-type K^+ channels into noninactivating, delayed rectifier-type K^+ channels. The Kvβ1.1-mutation, however, did not eliminate all A-type K^+ current in the CA1 cells of the hippocampus, since the deletion of Kvβ1.1 did not affect the α-subunits which are independently capable of A-type fast inactivation.[60]

Importantly, the loss of Kvβ1.1 in the mutants led to a reduction in both spike broadening during repetitive firing and the sAHP.[58] Since the cumulative inactivation of A-type K^+ channels during a spike train reduces the repolarizing current in later spikes (and therefore increases their duration),[50] the reduction in the inactivation of A-type K^+ currents in the mutants may be directly responsible for their decrease in spike broadening. Spike broadening can control calcium-influx during a spike train[61]; therefore the decrease in sAHP observed in the mutants may result from reduced spike broadening.

Consistent with the hypothesis that the sAHP in CA1 cells is required for hippocampal learning, the Kvβ1.1 mutant mice showed impairments in two learning tasks: the Morris water maze and the social transmission of food preference.[58] Kvβ1.1 mutant mice were able to initially learn normally the Morris water maze. However, they were impaired in their ability to learn a new location for the hidden platform. Importantly, the mutation did not affect other behaviors tested. For example, both fear conditionig and exploratory behavior in an open field appeared normal. The analysis of the Kvβ1.1 mutants indicated that A-type K+ channels affect the sAHP, and it indicates that the decrease in sAHP impaired learning that requires changes in recently learned information.

How could the lower sAHP of the mutants result in their specific learning impairments? The key finding is that while "naive" mutants learned the location of a hidden platform in the water maze test, once trained the mutants were impaired in their ability to learn a new platform location. Modeling studies suggested that reductions in the sAHP after learning stabilize learned information.[49] Since the mutants already have a low sAHP, further reductions during learning[54] may prevent any additional learning. Perhaps, in the mutants the increase in excitability is such that the networks coding for the old information are inescapably activated by the related new information, compromising any changes during relearning.

The results summarized above support the idea that the modulation of A-type K+ channels is critical for learning. First, a large number of modeling and electrophysiological studies demonstrate that information in the brain is encoded by the number and/or pattern of spike trains. A-type K+ channels have a key role in the modulation of spike trains, since they affect both spike broadening during a spike train and the sAHP (hyperpolarization inhibits neuronal firing). Therefore it is reasonable to propose that the modulation of A-type K+ channels may be critical for learning. Second, direct electrophysiological observations in vertebrate and invertebrate preparations documented decreases in A-type K+ currents after learning. Third, genetic lesions of A-type K+ channels in *Drosophila* and mice impaired learning.

The findings reviewed here also support the idea that learning triggers decreases in A-type K+ currents. Perhaps, phosphorylation slows down the inactivation of these currents, so that they resemble delayed rectifiers. This change in inactivation kinetics then reduces spike width during spike trains and decreases the sAHP. The resulting increased excitability may stabilize learned information by ensuring the co-activation of those cells any time the same circuit receives related information. The stabilizing influence of the increased excitability may only be needed temporarily until structural changes cement the synaptic modifications induced during learning.

FUTURE OF GENETIC TECHNIQUES

Most genetic studies of L&M have used techniques that do not allow control over the time and the place of the mutations. It is often difficult to determine whether the effects of mutations are due to alterations in development or to changes in the adult function of the altered or deleted proteins. Similarly, since mutations tend to affect several brain regions, traditional gene targeting techniques cannot pinpoint the exact brain structures responsible for the behavioral deficits of the mutants. Recent developments in transgenic technology promise to circumvent these limitations. For example, a newly developed technique allows the restriction of a mutation to specific regions of the brain.[62] This technique takes advantage of promoters with a restricted expression in the brain.[62] For example, the αCaMKII promoter is mostly active in specific regions of the postnatal forebrain.[63] In transgenic mice, this promoter can be used to control the expression of the Cre recombinase,[64] an enzyme that deletes genes flanked by loxP sites (small DNA sequences recognized by Cre).[62] Thus, deletion of the gene of interest (flanked by loxP sites) takes place only in tissues of the transgenic mouse where the Cre recombinase is expressed.[62]

Another new genetic strategy combines a tissue specific promoter (e.g., CaMKII) with the tetracycline transactivator system.[65] With this system it is possible to have some control over the time (and the place) when a given mutant protein is expressed in mice.[66] This technique allows the direct comparison of the same transgenic mice with or without the mutant protein. Therefore with this technique it may be possible to separate the developmental effects of a genetic lesion from its effects on function in adult animals.

The ability to target specific domains within a protein will be also crucial for genetic studies of L&M. Most of the initial studies used mutants where entire proteins were deleted. Consequently, with these mutants it was impossible to determine which aspects of protein function were critical for L&M. For example, the complete deletion of the αCaMK II in mice (αCaMKII^{null}) disrupts hippocampal LTP and spatial learning.[67, 68] Beside its Ca^{2+}-calmodulin (CaM)–dependent activity, αCaMKII can undergo autophosphorylation, resulting in CaM-independent activity. The complete deletion of the αCaMKII protein in αCaMKII^{null} mutants disrupts both the CaM-dependent and independent activity of this kinase. To determine whether the CaM-independent activity of the αCaMKII was critical for learning, we used a novel gene targeting procedure (pointlox procedure) to introduce a point mutation into the αCaMKII gene.[69] This mutation (a substitution of threonine 286 for alanine or T286A) blocked the autophosphorylation of this kinase and its CaM-independent activity, without affecting its CaM-dependent activity. Despite normal CaM-dependent activity, the αCaMKII^{T286A} mutants showed no LTP and no spatial learning in the Morris water maze, demonstrating that the autophosphorylation of αCaMKII is required for LTP and learning.[69]

The ability to derive animals with specific amino acid substitutions in genes of interest may also be critical to fine-tune hypothesis of L&M. For example, besides their deficit in LTP, the αCaMKII^{null} mutants revealed a dramatic increase in augmentation. Thus, without further experiments it was impossible to determine whether the deficit in LTP or the increase in augmentation was responsible for the learning impairments. This problem, however, was addressed by the analysis of the αCaMKII^{T286A} mutants: These mutants are deficient in LTP and learning, but have normal augmentation (unpublished observations). Thus, just as inducible mutants will provide a better indication of when a protein is critical for L&M, the study of mice with point mutations will help to determine which biochemical properties of a given protein are critical for L&M.

CONCLUSIONS

The studies just reviewed indicate that there is no single experiment that can unambiguously connect any cellular or molecular mechanism with L&M. Instead, to determine whether any one mechanism is required for L&M requires several different types of studies including (1) the development of biologically based learning models, (2) lesion experiments with different techniques (including genetics), and (3) direct observations during learning. In the future it may also be possible to artificially induce learning by manipulating candidate mechanisms. Here, we reviewed studies that implicate LTP, STP, and changes in A-currents in L&M. The strength of these studies is that they combined evidence from each of the first three categories mentioned above. For example, pharmacological and genetic lesions studies, direct observations during learning in several species, and modeling experiments support the hypothesis that L&M involves changes in A-type K$^+$ currents. Individually, each of these experiments is difficult to interpret because there are many possible explanations for the results. The interpretation of these experiments, however, is far more constrained when all of the results are considered together. The convergence of data from experiments with multiple species and techniques strengthens arguments for the involvement of a given mechanism in L&M. Without this convergence of data, it is difficult to rule out alternative hypotheses and other explanations based on unique attributes of any one given behavioral system.

Since it is likely that LTP, STP, and A-type K$^+$ currents are involved in learning, understanding the interplay between each of these physiological factors will be important in defining the molecular and cellular events leading to learning. For example, it is possible that the inconsistencies reported in the LTP–L&M literature are due to complex interactions between LTP and other processes; for example, an increase in excitability may be capable of compensating for lower levels of LTP. Despite the daunting problems facing the search for the molecular and cellular mechanisms of L&M, there is good reason for optimism. First, with genetic techniques it is possible to modify the function of any protein in mice. Second, revolutionary imaging techniques and improvements in electrophysiological recordings allow unprecedented abilities to observe functioning brains during L&M. Third, faster computers capable of handling large amounts of information will be able to handle increasingly more realistic models of L&M. These models will be extremely useful for exploring specific ideas about mechanisms of L&M. All together, these advances promise to open a new and exciting chapter in the study of the molecular and cellular mechanisms of L&M. The breakthroughs on the horizon will also have an enormous impact on our understanding of the mechanisms of cognitive disorders, such as those associated with aging.

ACKNOWLEDGMENTS

We are grateful to Stephen Snyder and Eugenia Wang for discussions that led to the writing of this chapter. This work was possible because of grants from the Whitehall Foundation, Beckman Foundation, Klingenstein Foundation, McKnight Foundation, and the NIH (R01 AG13622–01) to AJS, and the German Research Council (DFG) to KPG.

REFERENCES

1. Barnes, C. A. (1994). Normal aging: Regionally specific changes in hippocampal synaptic transmission. *Trends Neurosci., 17,* 13–18.
2. Gallagher, M., & Rapp, P. R. (1997). The use of animal models to study the effects of aging on cognition. *Annu. Rev. Psychol., 48,* 339–370.
3. Grant, S. N., & Silva, A. J. (1994). Targeting learning. *Trends Neurosci., 17,* 71–75.
4. Mayford, M., Abel, T., & Kandel, E. R. (1995). Transgenic approaches to cognition. *Curr. Opin. Neurobiol., 5,* 141–148.
5. Tonegawa, S. (1995). Mammalian learning and memory studied by gene targeting. *Ann. N. Y. Acad. Sci., 758,* 213–217.
6. Hebb, D. O. (1949). *The organization of behavior.* New York: Wiley.
7. Churchland, P. S., & Sejnowski, T. J. (1992). *The computational brain.* Cambridge: MIT Press.
8. Bliss, T. V. P., & Collingridge, G. L. (1993). A synaptic model of memory: Long-term-potentiation. *Nature, 361,* 31–39.
9. Malenka, R. C. (1994). Synaptic plasticity in the hippocampus: LTP and LTD. *Cell, 78,* 535–538.
10. Larson, J., Wong, D., & Lynch, G. (1986). Patterned stimulation at the theta frequency is optimal for the induction of hippocampal long-term potentiation. *Brain Res., 368,* 347–350.
11. Otto, T., Eichenbaum, H., Wiener, S. I., & Wible, C. G. (1991). Learning-related patterns of CA1 spike trains parallel stimulation parameters optimal for inducing hippocampal long-term potentiation. *Hippocampus, 1,* 181–192.
12. Huang, Y. Y., Nguyen, P. V., Abel, T., & Kandel, E. R. (1996). Long lasting forms of synaptic potentiation in the mammalian hippocampus. *Learn. Mem., 3,* 74–85.
13. Davis, S., Butcher, S. P., & Morris, R. G. M. (1992). The NMDA receptor antagonist D-2-amino-5-phosphonopentanoate (D-AP5) impairs spatial learning and LTP *in vivo* at intracerebral concentrations comparable to those that block LTP *in vitro. J. Neurosci., 12,* 21–34.
14. Kim, J. J., Fanselow, M. S., DeCola, J. P., & Fernadez, J.-L. (1992). Selective impairment of long-term but not short-term conditional fear by the *N*-methyl-D-aspartate antagonist APV. *Behav. Neurosci., 106,* 591–596.
15. Silva, A. J., Paylor, R., Wehner, J. M., & Tonegawa, S. (1992). Impaired spatial learning in alpha-calcium-calmodulin kinase II mutant mice. *Science, 257,* 206–211.
16. Grant, S. G. N., O'Dell, J., Karl, K. A., Stein, P. L., Soriano, P., & Kandel, E. R. (1992). Impaired long-term potentiation, spatial learning, and hippocampal development in Fyn mutant mice. *Science, 258,* 1903–1910.
17. Wu, Z. L., Thomas, S. A., Villacres, E. C., Xia, Z., Simmons, M. L., Chavkin, C., Palmiter, R. D., & Storm, D. R. (1995). Altered behavior and long-term potentiation in type I adenylyl cyclase mutant mice. *Proc. Natl. Acad. Sci. USA, 92,* 220–224.
18. Sakimura, K., Kutsuwada, T., Ito, I., Manabe, T., Takayama, C., Kushiya, E., Yagi, T., Aizawa, S.,

Inoue, Y., Sugiyama, H., & Mishina, M. (1995). Reduced hippocampal LTP and spatial learning in mice lacking NMDA receptor epsilon 1 subunit. *Nature, 373,* 151–155.

19. Lu, Y. M., Jia, Z., Janus, C., Henderson, J. T., Gerlai, R., Wojtowicz, J. M., & Roder, J. C. (1997). Mice lacking metabotropic glutamate receptor 5 show impaired learning and reduced CA1 long-term potentiation (LTP) but normal CA3 LTP. *J. Neurosci., 17,* 5196–5206.

20. Morris, R. G. M., Garrud, P., Rawlins, J. N. P., & O'Keefe, J. (1982). Place navigation impaired in rats with hippocampal lesions. *Nature, 297,* 681–683.

21. Qi, M., Zhuo, M., Skalhegg, B. S., Brandon, E. P., Kandel, E. R., McKnight, G. S., & Idzerda, R. L. (1996). Impaired hippocampal plasticity in mice lacking the Cbeta1 catalytic subunit of cAMP-dependent protein kinase. *Proc. Natl. Acad. Sci. USA, 93,* 1571–1576.

22. Brandon, E. P., Zhuo, M., Huang, Y. Y., Qi, M., Gerhold, K. A., Burton, K. A., Kandel, E. R., McKnight, G. S., & Idzerda, R. L. (1995). Hippocampal long-term depression and depotentiation are defective in mice carrying a targeted disruption of the gene encoding the RI beta subunit of cAMP-dependent protein kinase. *Proc. Natl. Acad. Sci. USA, 92,* 8851–8855.

23. Huang, Y. Y., Kandel, E. R., Varshavsky, L., Brandon, E. P., Qi, M., Idzerda, R. L., McKnight, G. S., & Bourtchouladze, R. (1995). A genetic test of the effects of mutations in PKA on mossy fiber LTP and its relation to spatial and contextual learning. *Cell, 83,* 1211–1222.

24. Cain, D. P., Saucier, D., Hall, J., Hargreaves, E. L., & Boon, F. (1996). Detailed behavioral analysis of water maze acquisition under APV or CNQX: Contribution of sensorimotor disturbances to drug-induced acquisition deficits. *Behav. Neurosci., 110,* 86–102.

25. Saucier, D., & Cain, D. P. (1995). Spatial learning without NMDA receptor-dependent long-term potentiation. *Nature, 378,* 186–189.

26. Bannerman, D. M., Good, M. A., Butcher, S. P., Ramsay, M., & Morris, R. G. (1995). Distinct components of spatial learning revealed by prior training and NMDA receptor blockade. *Nature, 378,* 182–186.

27. Kogan, J., Frankland, P. W., Blendy, J. B., Coblentz, J., Marowitz, Z., Schutz, G., & Silva, A. J. (1997). Spaced training induces normal long-term memory in CREB mutant mice. *Curr. Biol., 7,* 1–11.

28. Bourtchuladze, R., Frenguelli, B., Blendy, J., Cioffi, D., Schutz, G., & Silva, A. J. (1994). Deficient long-term memory in mice with a targeted mutation of the cAMP-responsive element binding protein. *Cell, 79,* 59–68.

29. Bolshakov, V. Y., & Siegelbaum, S. A. (1995). Regulation of hippocampal transmitter release during development and long-term potentiation. *Science, 269,* 1730–1734.

30. Fisher, S. A., Fisher, T. M., & Carew, T. J. (1997). Multiple overlapping processes underlying short-term plasticity. *Trends Neurosci., 20,* 170–177.

31. Zucker, R. S. (1989). Short-term synaptic plasticity. *Ann. Rev. Neurosci., 12,* 13–31.

32. Little, W. A., & Shaw, G. L. (1975). A statistical theory of short and long-term memory. *Behav. Biol., 14,* 115–133.

33. Buonomano, D. V., & Merzenick, M. M. (1995). Transformation of temporal information into a spatial code by a neural network based on realistic neuronal properties. *Science, 267,* 1028–1031.

34. Castelluci, V., Pinsker, H., Kupfermann, I., & Kandel, E. R. (1970). Neuronal mechanisms of habituation and dishabituation of the gill-withdrawal reflex in *Aplysia. Science 167,* 1745–1748.

35. Auerbach, A. A., & Bennet, M. V. L. (1969). Chemically mediated transmission at a giant fiber synapse in the central nervous system of a vertebrate. *J. Gen. Physiol., 53,* 183–210.

36. Zilber-Gachelin, N. F., & Chartier, M. P. (1973). Modification of motor reflex responses due to repetition of the peripheral stimulus in the cockroach. I. Habituation at the level of an isolated abdominal ganglion. *J. Exp. Biol. 59,* 359–382.

37. Zucker, R. S. (1972). Crayfish escape behavior and central synapses. II. Physiological mechanisms underlying behavioral habituation. *J. Neurophysiol., 35,* 621–637.

38. Silva, A. J., Rosahl, T. W., Chapman, P. F., Marowitz, Z., Cioffi, D., Sudhof, T. C., & Bourtchuladze, R. (1996). Impaired learning in mice with abnormal short-term plasticity. *Curr. Biol., 6,* 1509–1518.

39. Chapman, P. F., Frenguelli, B., Smith, A., Chen, C.-M., & Silva, A. J. (1995). The α-calcium-calmodulin kinase II: A bi-directional modulator of pre-synaptic plasticity. *Neuron, 14,* 591–597.

40. Rosahl, T. W., Geppert, M., Spillane, D., Herz, J., Hammer, R. E., Malenka, R. C., & Sudhof, T. C. (1993). Short-term synaptic plasticity is altered in mice lacking synapsin I. *Cell, 75,* 661–670.

41. Rosahl, T. W., Spillane, D., Missler, M., Herz, J., Selig, D. K., Wolff, J. R., Hammer, R. E., Malenka, R. C., & Sudhof, T. C. (1995). Essential functions of synapsins I and II in synaptic vesicle regulation. *Nature, 375,* 488–493.

42. Spillane, D. M., Rosahl, T. W., Sudhof, T. C., & Malenka, R. C. (1995). Long term potentiation in mice lacking synapsins. *Neuropharmacology, 34,* 1573–1579.

43. Broadie, K., Rushton, E., Skoulakis, E., & Davis, R. (1997). Leonardo, a *Drosophila* 14-3-3 protein involved in learning, regulates presynaptic function. *Neuron, 19,* 391–402.

44. Zhong, Y., & Wu, C. F. (1991). Altered synaptic plasticity in *Drosophila* memory mutants with a defective cyclic AMP cascade. *Science, 251,* 198–201.

45. Tully, T. (1991). Genetic dissection of learning and memory in *Drosophila melanogaster.* In J. Madden IV (Ed.), *Neurobiology of learning, emotion and affect* (pp. 30–66). New York: Raven Press.

46. Rieke, F., Warland, D., van Steveninck, R., & Bialek, W. (1997). *Spikes: Exploring the neural code.* Cambridge: MIT Press.

47. Hasselmo, M. E., & Bower, J. (1993). Acetylcholine and memory. *Trends Neurosci., 16,* 218–222.

48. Lisman, J. E., & Idiart, M. A. P. (1995). Storage of 7 ± 2 short-term memories in oscillatory subcycles. *Science, 267,* 1512–1515.

49. Berner, J. (1991). Functional roles for the slow afterhyperpolarization: Inferences from incorporation of features of the slow gK+ (Ca) into densely connected artificial neural networks. *Network, 2,* 1–6.

50. Ma, M., & Koester, J. (1996). The role of K+ currents in frequency-dependent spike broadening in *Aplysia* R20 neurons: A dynamic-clamp analysis. *J. Neurosci., 16,* 4089–4101.

51. Storm, J. F. (1990). Potassium currents in hippocampal pyramidal cells. *Prog. Brain Res., 83,* 161–187.

52. Alkon, D. (1984). Calcium-mediated reduction of ionic currents: A biophysical memory trace. *Science, 226,* 1037–1045.

53. Woody, C. D., Gruen, E., & Birt, D. (1991). Changes in membrane currents during Pavlovian conditioning of single cortical neurons. *Brain Res., 539,* 76–84.

54. Disterhoft, J. F., Thompson, L. T., Moyer, J. R., Jr., & Mogul, D. J. (1996). Calcium-dependent afterhyperpolarization and learning in young and aging hippocampus. *Life Sci., 59,* 413–420.

55. Landfield, P. W., & Pitler, T. A. (1984). Prolonged Ca^{2+}-dependent afterhyperpolarizations in hippocampal neurons of aged rats. *Science, 226,* 1089–1092.

56. Cowan, T. M., & Siegel, R. W. (1986). *Drosophila* mutations that alter ionic conduction disrupt acquisition and retention of a conditioned odor avoidance response. *J. Neurogenet., 3,* 187–201.

57. Meiri, N., Ghelardini, C., Tesco, G., Galeotti, N., Dahl, D., Tomsic, D., Cavallaro, S., Quattrone, A., Capaccioli, S., Bartolini, A., & Alkon, D. L. (1997). Reversible antisense inhibition of Shaker-like Kv1.1 potassium channel expression impairs associative memory in mouse and rat. *Proc. Natl. Acad. Sci. USA, 94,* 4430–4434.

58. Giese, K. P., Storm, J. F., Reuter, D., Fedorov, N. B., Shao, L.-R., Leicher, T., Pongs, O., & Silva, A. J. (submitted). Reduced K+ channel inactivation, spike broadening and afterhyperpolarization in Kvβ1.1-deficient mice with impaired learning.

59. Rettig, J., Heinemann, S. H., Wunder, F., Lorra, C., Parcej, D. N., Dolly, J. O., & Pongs, O. (1994). Inactivation properties of voltage-gated K+ channels altered by presence of β-subunit. *Nature, 369,* 289–294.

60. Sheng, M., Tsaur, M.-L., Jan, Y. N., & Jan, L. Y. (1992). Subcellular segregation of two A-type K+ channel proteins in rat central neurons. *Neuron, 9,* 271–284.

61. Jackson, M. B., Konnerth, A., & Augustine, G. J. (1991). Action potential broadening and frequency dependent facilitation of calcium signals in pituitary nerve terminals. *Proc. Natl. Acad. Sci. USA, 88,* 380–384.

62. Tsien, J. Z., Chen, D. F., Mercer, E. H., Anderson, D. J., Mayford, M., Kandel, E. R., & Tonegawa, S. (1996). Subregion- and cell type-restricted gene knockout in mouse brain. *Cell, 87,* 1317–1326.

63. Mayford, M., Wang, J., Kandel, E. R., & O'Dell, T. J. (1995). CaMKII regulates the frequency-response function of hippocampal synapses for the production of both LTD and LTP. *Cell, 81,* 891–904.

64. Sauer, B. (1993). Manipulation of transgenes by site-specific recombination: Use of Cre recombinase. *Meth. Enzymol., 225,* 890–900.

65. Furth, P. A., St. Onge, L., Boger, H., Gruss, P., Gossen, M., Kistner, A., Bujard, H., & Hennighausen, L. (1994). Temporal control of gene expression in transgeic mice by a tetracycline-responsive promoter. *Proc. Natl. Acad. Sci. USA, 91,* 9302–9306.

66. Mayford, M., Bach, M. E., Huang, Y. Y., Wang, L., Hawkins, R. D., & Kandel, E. R. (1996). Control of memory formation through regulated expression of a CaMKII transgene. *Science, 274,* 1678–1683.

67. Silva, A. J., & Chapman, P. F. (1994). The alpha-calcium calmodulin kinase II and the plasticity of neurons, circuits and behavior. *The Neurosciences, 6,* 53–58.

68. Silva, A. J., Wang, Y., Paylor, R., Wehner, J. M., Stevens, C. F., & Tonegawa, S. (1992). Alpha calcium/calmodulin kinase II mutant mice: Deficient long-term potentiation and impaired spatial learning. *Cold Spring Harb. Symp. Quant. Biol., 57,* 527–539.

69. Giese, K. P., Fedorov, N. B., Filipkowski, R. K., & Silva, A. J. (1997). Autophosphorylation at thr 286 of the α calcium-calmodulin-kinase II in LTP and learning. *Science, 279,* 870–873.

5

NORMAL AGING AND ALZHEIMER'S DISEASE

BRADLEY T. HYMAN AND
TERESA GÓMEZ-ISLA

Neurology Department
Massachusetts General Hospital
Boston, Massachusetts 02114

One of the most difficult issues for researchers and clinicians focused on the study of the human aging process is to define what "normal" is. The majority of neurodegenerative processes that cause impaired cognition have an increased prevalence with age, adding hardship to the task of setting criteria for "normal" or healthy aging. The lack of universal standards is well reflected by the nosological and diagnostic problem that for several decades has attempted to distinguish Alzheimer's disease (AD) from "normal" or healthy aging.[1-3] AD represents the most prevalent cause of dementia in elderly individuals. One of its earliest and foremost clinical manifestations is memory impairment, and impairment of acquisition of new information remains its hallmark feature throughout the course. According to many, the idea that mild memory loss is also a common concomitant of aging raises a theoretical model whereby AD and aging lie on a continuous spectrum.[4,5] According to this idea, AD would represent exaggerated aging rather than a true disease.[6] On the other hand, mild memory difficulties are so commonplace among the elderly population that there is a natural reluctance to assign what appears to be a modest and benign decrement in abilities that occurs with advanced age to a disease process.

From a neuropathological point of view, there is also controversy about whether aging and AD represent a continuum or are dichotomous.[7-10] The two main neuropathological hallmarks of AD, neurofibrillary tangles (NFTs) and senile plaques (SPs), occur frequently in aging brains.[3, 11–15] NFTs are nearly universal in the hippocampal formation in nondemented individuals who reach the seventh decade of life, and their number increases with age.[16, 17] Amyloid deposits in the form of SPs may also be present in many nondemented individuals.[18] More than that, the topographical distribution of NFTs and SPs in the brains of elderly persons matches the hierarchical pattern of vulnerability seen in AD.[14, 16, 17, 19–22]

An alternative point of view is that AD and normal aging are two well-differentiated processes clearly demarcated from both clinical and pathological perspectives. Recent studies suggest that in truly healthy aging cognition holds over time, with many examples of individuals who "successfully" reach their elder years without a significant decline in their cognitive performance. According to these observations, it has been proposed that a sustained decline of even modest proportions should be an alert about an underlying pathological condition rather than representing benign senescence.[3, 23–26]

We can rephrase these issues as a series of questions that can be addressed through clinical–pathological studies of carefully selected populations. Is cognitive impairment inevitable with advanced age? Are the neuropathological hallmarks of AD (NFTs and SPs) always present in brains after a certain age (i.e., in some part are they a "natural consequence" of aging, or do they represent distinct processes)? Are we losing neurons with increasing age, or is neuronal loss a sign of a pathophysiological process?

Taking advantage of the knowledge that the NFTs, SPs, and loss of neurons and synapses in AD are in some way related to the pathophysiology of the disease, we have tried to clarify the relationship between them and the clinical manifestations of AD. We have approached the problem of understanding the clinical consequences of specific AD pathological changes in an attempt to define what lesions are specific for AD and more importantly what lesions differentiate AD from "normal" aging.

Neuropathological study of the AD brain demonstrates that NFTs and SPs follow remarkably consistent patterns of distribution.[20, 27] One of the earliest and more severe alterations in the AD brain is the development of NFTs in a subset of neurons in the medial temporal lobe.[13, 14, 20, 27–30] Neurons in layers II and IV of entorhinal cortex, and to a lesser extent also the perirhinal cortex, CA1/subiculum of the hippocampus, inferior temporal gyrus, the amygdala and posterior parahippocampal gyri, cholinergic basal forebrain, and dorsal raphe,[14] are among the first to be affected, likely compromising the integrity of crucial neural systems related to memory.[31] Later in the disease process, as the cognitive deficits become much broader, NFTs are present in increasing amounts in feedforward and feedback projection neurons in layers II, III, and V in high-order association cortices,[20, 32, 33] followed by first-order association cortices, and ultimately other subcortical structures and primary sensory and motor cortices. Primary motor and sensory cortices remain nearly unaffected by NFTs even late in the disease process.[14, 16, 20]

In contrast to NFTs, SPs have a more varied and broader distribution. Although they also affect association cortices, they appear in relatively smaller numbers in critical components of memory-related neural systems as the hippocampal formation, whereas they may be present in large amounts in regions, such as primary motor and sensory cortices, that remain clinically silent, even at the very late stages of the disease.[14, 20]

When we used quantitative methods to assess the number of NFTs and Aβ deposits in SPs in individuals with various durations or severities of disease, we discovered that NFTs parallel the severity of clinical illness, whereas SPs do not correlate with clinical symptoms or duration of illness.[17, 30, 34–36] Braak and Braak[27] have taken advantage of this consistent pattern of hierarchical vulnerability to propose a "staging" scheme for AD that is entirely dependent on tangle formation and does not rely on plaque formation.

NFTs and SPs can be directly assessed as "positive phenomena." By contrast, neuronal and synaptic losses are negative phenomena that can only be calculated from normative data on age-matched control populations. Neuronal and synaptic counts represent an attempt to quantitatively evaluate a final common pathway of neuronal dysfunction. In our own and others' experience, loss of neurons and synapses, has a fairly strong correlation with the degree of cognitive impairment,[36, 37] as would be expected for direct measures of neural system integrity. However, because losses of neurons and synapses are not uniform throughout the brain, an appreciation of the complexity of the brain architecture is critical.

Studies on neuronal loss in different neurodegenerative diseases and the aging process first began a 100 years ago.[38, 39] More recently, studies have been directed at assessing the occurrence, extent, topographical distribution, and role of neuronal loss in AD and normal aging.[8, 9, 13, 40–42] Data from these studies are frequently contradictory and difficult to compare because the population selection, sampling schemes, and counting techniques used varied widely. There is overall consensus, however, that neuronal loss occurs selectively in vulnerable areas of AD brains compared to age-matched normal specimens.[8, 43–47] There is more controversy than consensus regarding the issue of neuronal depletion as a common concomitant of aging not associated with specific diseases.[9, 40, 41, 48–51]

In collaboration with investigators in the AD research units at Washington University and the Mayo Clinic, we have had the opportunity to compare quantitative neuropathological features of individuals known to be cognitively normal prior to death, individuals with the earliest clinically detectable signs of dementia of the Alzheimer type, and individuals with well-established AD. We have applied stereological unbiased techniques[52, 53] to analyze the structural integrity of two regions of interest in AD that are known to be consistently affected by both NFTs and SPs[14, 20, 31, 54]: (1) the entorhinal cortex (EC) as a critical component of the medial temporal lobe memory-related neural system and (2) the superior temporal sulcus region (STS) as a representative of high-order association cortex; the STS is one of only three association areas in the nonhuman primate brain that receives input from each sensory modality.[29, 55, 56] We reasoned that neuropathological changes and neuronal loss in these two brain regions might contribute to the cognitive impairment in AD and, if absent in brains from age-matched nondemented individuals, could help to differentiate even the very mild stages of AD from normal aging.

We studied 20 individuals evaluated clinically at Washington University.[57] From the clinical assessments the 20 subjects were subdivided into two groups. The AD group included 10 clinically demented individuals (mean age ± SD = 84.2 ± 9.9 years; range 67–95 years) rated according to the Clinical Dementia Rating (CDR) scale. Four of the AD cases were rated CDR = 0.5 (very mild or questionable impairment), 1 was rated CDR = 1 (mild), and 5 were rated CDR = 3 (severe cognitive impairment). All 10 of them had a subsequent neuropathological diagnosis of definite AD by standard neuropathological criteria.[58, 59]

The control group included 10 individuals (mean age ± SD = 75 ± 9 years; range 60–89 years) who clinically were rated cognitively normal (CDR = 0). None of the control group met criteria for AD by neuropathological examination.[58, 59] Lamina-specific neuronal counts were carried out following a systematically random sampling scheme on Nissl-stained sections representing the entire EC.[53]

In the cognitively normal (CDR = 0) individuals (control subjects), there were about 650,000 neurons in layer II, 1 million neurons in layer IV, and 7 million neurons in the entire EC. In the control group, no change in the number of neurons was observed between 60 and 90 years old. The AD group with the mildest clinically detectable dementia (CDR = 0.5) had 32% fewer EC neurons than control subjects, with the number of neurons in layer II decreasing by 60% and in layer IV by 40% compared to control subjects. The group with severe AD (CDR = 3) had 69% fewer EC neurons than control subjects, and the number of neurons in layers II and IV decreased by about 90 and 70%, respectively, compared to control subjects.

These observations suggest that no significant neuronal loss takes place in the EC in cognitively intact elderly individuals, at least in the age range of about 60 to about 90. Assuming that individuals who ultimately develop AD start with about the same number of neurons as those who do not, our analysis of neuronal counts in EC of AD brains reveals that there is a very severe neuronal loss in the EC even in very mild AD cases that are at the threshold for clinical detection of dementia. This neuronal loss in AD parallels the known susceptibility of layers II and IV of EC for NFT formation, and it is so marked that it may have started well before onset of clinically detectable changes in performance on simple neuropsychometric tests.

We also studied the STS in brains of 45 individuals with definite AD who had

TABLE 5.1 Neuronal Number Remains Stable in the Superior Temporal Sulcus Region during Normal Aging

Age (years)	n	$X \pm SD \ (\times 10^3)$
<60	5	94.28 ± 7.0
61–70	6	94.36 ± 11.6
71–80	10	92.17 ± 8.8
81–90	7	96.36 ± 9.2
>90	4	84.87 ± 8.3

been examined and followed in the clinical units of the Massachusetts General Hospital, Washington University, or Mayo Clinic (average duration of illness 8.2 ± 5.2 years, range 0.3–21 years) and 28 nondemented control subjects (average age 71.8 ± 14.8 years, range from the 6th decade to the 10th decade)[60]; we present here updated results from our ongoing studies. Using similar statistically unbiased stereological counting techniques, we counted the number of STS neurons per 50-μm-thick cross section. This study allowed us to assess the structural integrity of the association cortices.

Again there was no change in the number of STS neurons per 50-μm-thick section in nondemented control subjects from the sixth to the ninth decades of life (average ± SD, 9.6 ± 1.0 × 10⁴). (Table 5.1). In AD, more than 50% of the neurons were lost (4.8 ± 2.3 × 10⁴). Neuronal loss increases in parallel with the duration and severity of illness (Table 5.2). A floor effect of neuronal loss is seen in STS after about 10 years; in other words, about a third of neurons are simply not vulnerable to AD degeneration regardless of duration of illness. Of note, the degree of loss paralleled but was much higher than the amount of NFTs found, perhaps by as much as two orders of magnitude. No statistical relationship could be uncovered between neuronal loss and the amount of Aβ deposits or SPs.

TABLE 5.2 Neuronal Loss in the Superior Temporal Sulcus Region in Alzheimer's Disease Reflects Duration of Symptoms

	Duration (years)	N	Number of neurons $X \pm SD \ (\times 10^3)$
Control subjects (ages 60–90)	—	28	96.0 ± 10.0
Alzheimer's disease	0–3	10	77.7 ± 23.5
	4–6	13	54.7 ± 16.1
	7–9	12	38.4 ± 7.7
	10–12	6	34.1 ± 7.7
	13–15	8	33.3 ± 7.8
	>15	4	29.7 ± 6.8

We were especially interested in the changes that occur in the STS early in AD. Certainly because the STS is a high-order association cortex, we expected some changes to occur when the disease process involved anything more than memory difficulties. We therefore examined the group of carefully defined CDR = 0.5 cases in whom we had already examined the EC from the perspective of STS neuron counts. In contrast with the marked loss of neurons in EC in patients with very mild dementia, the study of the STS in the same CDR = 0.5 patients shows no statistically significant difference in STS neuron number compared to nondemented control subjects, suggesting that the STS does not undergo neuronal loss until a much later point in the disease process, when clinical symptoms of a more advanced dementia are present.

The fact that neuronal loss does not occur in these two regions during the normal aging process is consistent with the hypothesis that AD and aging are not part of a continuous spectrum and suggests that normal aging and AD can be differentiated from a neuropathological perspective.

The contribution of NFT and Aβ deposits to neuronal loss and to the clinical evolution of AD remains controversial. Some early studies emphasized a prime role of amyloid accumulation in the progression and severity of dementia.[61–64] Other studies, including our own, found a closer correlation between NFT accumulation and clinical evolution of the disease than with amyloid accumulation.[17, 30, 34, 35, 65, 66] We had anticipated that the correlation between the magnitude of neuronal loss and the degree of neuropathological changes in AD may be better for NFTs than for SPs. The data confirmed our prediction: in both, the EC and the STS, we observed a significant negative correlation between the number of neurons and the amount of NFT and neuritic plaques, but no significant correlation was found between either of these parameters and the number of total SPs. Despite the presence of a few NFTs in the EC and some scattered SPs in the medial temporal lobe structures in some of the nondemented individuals included in these series, no significant neuronal loss could be documented in brain regions selectively targeted by AD pathology. These individuals where able to reach their elder years successfully, without clinically detectable symptoms of cognitive impairment.

In summary, we have observed that neuronal loss does not occur in the EC or the STS in cognitively intact elderly persons within the four decades assessed. By contrast, there is a selective and very dramatic loss of neurons in the EC even at the mildest stages of dementia in AD, likely even at a presymptomatic point. This alteration almost certainly contributes to the memory impairment that characterizes the earliest clinical phases of this type of dementia. At later points of the disease, when the illness is perhaps more clinically apparent, the STS association cortex also undergoes neuronal loss as clinical symptoms of a more advanced cognitive impairment become prominent. These findings support a model where AD and normal aging are not part of a continuum, and can be differentiated both anatomically and clinically.

ACKNOWLEDGMENTS

This work was supported by NIA AG08487, the Massachusetts and Washington University Alzheimer Disease Research Centers, and the Alzheimer Disease Registry at the Mayo Clinic.

REFERENCES

1. Drachman, D. A. (1983). How normal aging relates to dementia: A critique and classification. In D. Samuel, S. Algeri, & S. Gershon (Eds.), *Aging of the brain: Vol. 22* (pp. 19–31). New York: Raven.
2. Berg, L. (1988). The Clinical Dementia Rating (CDR) of Washington University. *Psychopharmacol. Bull., 24,* 637–639.
3. Morris, J. C., McKeel, D. W., & Storandt, M. (1991). Very mild Alzheimer's disease: Informant-based clinical, psychometric, and pathologic distinction from normal aging. *Neurology, 41,* 469–478.
4. Bartus, R. T., Dean, R. Ld., Beer, B., & Lippa A. S. (1982). The cholinergic hypothesis of geriatric memory dysfunction. *Science, 217(4558),* 408–414.
5. Brayne, C., & Calloway, P. (1988). Normal ageing, impaired cognitive function and senile dementia of the Alzheimer's type: A continuum? *Lancet, 1,* 1265–1267.
6. Drachman, D. A. (1994). If we live long enough, will be all demented? *Neurology, 44,* 1563–1565.
7. Coleman, P. D., & Flood, D. G. (1987). Neuron numbers and dendritic extent in normal aging and Alzheimer's disease. *Neurobiol. Aging, 8,* 521–545.
8. Terry, R. D., Peck, A., DeTeresa, R., Schechter, R., & Horoupian, D. S. (1981). Some morphometric aspects of the brain in senile dementia of the Alzheimer type. *Ann. Neurol., 10,* 184–192.
9. Terry, R. D., DeTeresa, R., & Hansen, L. A. (1987). Neocortical cell counts in normal human adult aging. *Ann. Neurol., 21,* 530–539.
10. West, M., Coleman, P., Flood, D., & Troncoso, J. C. (1994). Differences in the pattern of hippocampal neuronal loss in normal ageing and Alzheimer's disease. *Lancet, 344,* 769–772.
11. Tomlinson, B. E., Blessed, G., & Roth, M. (1968). Observations on the brains of nondemented old people. *J. Neurol. Sci., 7,* 331–356.
12. Dayan, A. D. (1970). Quantitative histological studies in the aged human brain. I. Senile plaques and neurofibrillary tangles in normal patients. *Acta. Neuropathol. (Berl.), 16,* 85–94.
13. Ball, M. (1977). Neuron loss, neurofibrillary tangles, and granulovacuolar degeneration in the hippocampus with ageing and dementia. A quantitative study. *Acta. Neuropathol. (Berl.), 37,* 111–118.
14. Arriagada, P. V., Marzloff, K., & Hyman, B. T. (1992). Distribution of Alzheimer-type pathological changes in nondemented elderly matches the pattern in Alzheimer's disease. *Neurology, 42,* 1681–1688.
15. Bouras, C., Hof, P. R., & Morrison, J. H. (1993). Neurofibrillary tangle densities in the hippocampal formation in a nondemented population define subgroups of patients with differential early pathological changes. *Neurosci. Lett., 153,* 131–135.
16. Price, D., Davis, P., Morris, J., & White, D. (1991). The distribution of tangles, plaques and related immunohistochemical markers in healthy aging and Alzheimer's disease. *Neurobiol. Aging, 12,* 295–312.
17. Arriagada, P. V., Growdon, H. H., Hedley-Whyte, E. T., & Hyman, B. T. (1992). Neurofibrillary tangles but not senile plaques parallel duration and severity of Alzheimer disease. *Neurology, 42,* 631–639.
18. Delaere, P., Duyckaerts, C., Masters, C., Beyreuther, F., Piette, F., & Hauw J.-J. (1990). Large

amounts of neocortical β A4 deposits without neuritic plaques nor tangles in a psychometrically assessed, non-demented person. *Neurosci. Lett., 116,* 87–93.

19. Mann, D. M. A., Tucker, C. M., & Yates, P. O. (1987). The topographic distribution of senile plaques and neurofibrillary tangles in the brains of non-demented persons of different ages. *Neuropathol. Appl. Neurobiol., 13,* 123–139.

20. Arnold, S. E., Hyman, B. T., Flory, J., Damasio, A. R., & Van Hoesen, G. W. (1991). The topographical and neuroanatomical distribution of neurofibrillary tangles and neuritic plaques in cerebral cortex of patients with Alzheimer's disease. *Cerebral Cortex, 1,* 103–116.

21. Hof, P. R., Bouras, C., Perl, D. P., Sparks, D. L., Mehta, N., & Morrison, J. H. (1995). Age-related distribution of neuropathological changes in the cerebral cortex of patients with Down's syndrome. Quantitative regional analysis and comparison with Alzheimer's disease. *Arch. Neurol., 52(4),* 379–391.

22. Hyman, B. T., West, H. L., Gómez-Isla, T., & Mui, S. (1995). Quantitative neuropathology in Alzheimer's disease: Neuronal loss in high-order association cortex parallels dementia. In K. Iqbal, J. A. Mortimer, B. Winblad, & H. M. Wisniewski (Eds.), *Research advances in Alzheimer's disease and related disorders* (pp. 453–460). New York: Wiley.

23. Roth, M. (1986). The association of clinical and neurological findings and its bearing on the classification and aetiology of Alzheimer's disease. *Br. Med. Bull., 42,* 42–45.

24. Morris, J. C., & Fulling, K. (1988). Early Alzheimer's disease. Diagnostic considerations. *Arch. Neurol., 45,* 345–349.

25. Linn, R. T., Wolf, P. A., Bachman, D. L., Knoefel, J. E., Cobb, J. L., Belanger, A. J., Kaplan E. F., & D'Agostino, R. B. (1995). The "preclinical phase" of probable Alzheimer's disease. A 13-year prospective study of the Framingham cohort. *Arch. Neurol., 52(5),* 485–490.

26. Morris, J. C., Storandt, M., McKeel, Jr., D. W., Rubin, E. H., Price, J. L., Grant, E. A., & Berg L. (1996). Cerebral amyloid deposition and diffuse plaques in "normal" aging: Evidence for presymptomatic and very mild Alzheimer's disease. *Neurology, 46(3),* 707–719.

27. Braak, H., & Braak, E. (1991). Neuropathological staging of Alzheimer related changes. *Acta. Neuropathol., 82,* 239–259.

28. Hyman, B. T., Van Hoesen, G. W., Damasio, A. R., & Barnes, C. L. (1984). Alzheimer's disease: Cell specific pathology isolates the hippocampal formation in Alzheimer's disease. *Science, 225,* 1168–1170.

29. Barnes, C. L., & Pandya, D. N. (1992). Efferent cortical connections of multimodal cortex of the superior temporal sulcus in the rhesus monkey. *J. Comp. Neurol., 318,* 222–244.

30. Berg, L., McKeel, D. W., Miller, J. P., Baty, J., & Morris, J. C. (1993). Neuropathological indexes of Alzheimer's disease in demented and nondemented persons aged 80 and older. *Arch. Neurol., 50,* 349–358.

31. Hyman, B. T., Van Hoesen, G. W., & Damasio, A. R. (1990). Memory-related neural systems in Alzheimer's disease: An anatomical study. *Neurology, 40,* 1721–1730.

32. Lewis, D., Campbell, M., Terry, R., & Morrison, J. (1987). Laminar and regional distributions of neurofibrillary tangles and neuritic plaques in Alzheimer's disease: A quantitative study of visual and auditory cortices. *J. Neurosci., 7,* 1799–1808.

33. Hof, P. R., Cox, K., & Morrison, J. H. (1990). Quantitative analysis of a vulnerable subset of pyramidal neurons in Alzheimer's disease. I. Superior frontal and inferior temporal cortex. *J. Comp. Neurol., 301,* 44–54.

34. Hyman, B. T., Marzloff, K., & Arriagada, P. V. (1993). The lack of accumulation of senile plaques or amyloid burden in Alzheimer's disease suggests a dynamic balance between amyloid deposition and resolution. *J. Neuropathol. Exp. Neurol., 53,* 594–600.

35. Nagy, Z., Esiri, M., Jobst, K., Morris, J., King, E.-F., McDonald, B., Litchfield, S., Smith, A., Barnetson, L., & Smith A. (1995). Relative roles of plaques and tangles in the dementia of Alzheimer's disease: Correlations using three sets of neuropathological criteria. *Dementia, 6,* 21–31.

36. Gómez-Isla, T., West, H. L., Rebeck, G. W., Harr, S. D., Growdon, J. H., Locasio, J. T., Perls, T.

T., Lipsitz, L. A., & Hyman, B. T. (1996). Clinical and pathological correlates of apolipoprotein E e4 in Alzheimer's disease. *Ann. Neurol., 39,* 62–70.

37. Terry, R. D., Masliah, E., & Salmon, D. P. (1991). Physical basis of cognitive alterations in Alzheimer's disease: Synapse loss is the major correlate of cognitive impairment. *Ann. Neurol., 41,* 572–580.
38. Hammarberg. (1895). *Studien über klinik und pathologie der idiotie.* E. Berling: Upsala.
39. Thompson, H. (1899). The total number of functional cells in the cerebral cortex of man, and the percentage of the total volume of the cortex composed of nerve cell bodies. *J. Comp. Neurol., 9,* 115–140.
40. Brody, H. (1955). Organization of the cerebral cortex . III. A study of aging in the human cerebral cortex. *J. Comp. Neurol., 102,* 511–556.
41. Colon, E. (1972). The elderly brain, a quantitative analysis in the cerebral cortex of two cases. *Psychiatr. Neurol. Neurochir. (Amst)., 75,* 261-270.
42. Shefer, V. (1973). Absolute number of neurons and thickness of the cerbral cortex during aging, senile and vascular dementia and Pick's and Alzheimer's diseases. *Neurosci. Behav. Physiol., 6,* 319–324.
43. Doucette, R., Fishman, M., Hachinski, V., & Mersky, H. (1986). Cell loss from the nucleus basalis of Meynert in Alzheimer's disease. *Can. J. Neurol. Sci., 13,* 435–440.
44. Etienne, P., Robitaille, Y., Wood, P., Guathier, S., Nair, N., & Quirion, R. (1986). Nucleus basalis neuronal loss, neuritic plaques and choline acetyltransferase activity in advanced Alzheimer's disease. *Neuroscience, 19,* 1279–1291.
45. Curcio, C., Kemper, T. (1984). Nucleus raphe dorsalis in dementia of the Alzheimer type: Neurofibrillary changes and neuronal packing density. *J. Neuropathol. Exp. Neurol., 43,* 359–368.
46. Yamamoto, T., & Hirano, A. (1985). Nucleus raphe dorsalis in Alzheimer's disease: Neurofibrillary tangles and loss of large neurons. *Ann. Neurol., 17,* 573–577.
47. Scott, S., DeKosky, S., Sparks, D., Knox, C., Scheff, S. (1992). Amygdala cell loss and atrophy in Alzheimer's disease. *Ann. Neurol., 32,* 555–563.
48. Henderson, G., Tomlinson, B., & Gibson, P. (1980). Cell counts in human cerebral cortex in normal adults throughout life using an image analysis computer. *J. Neurol. Sci., 46,* 113–136.
49. Anderson, J., Hubbard, B., Coghill, G., & Slidders, W. (1983). The effect of advanced old age on the neurone content of the cerebral cortex. Observations with an automatic image analyser point counting method. *J. Neurol. Sci., 58,* 235–246.
50. Chui, H., Bondareff, W., Zarow, C., & Slager, U. (1983). Stability of neuronal number in the human nucleus basalis of Meynert with age. *Neurobiol. Aging, 5,* 83–88.
51. Haug, H. (1985). Are neurons of the human cerebral cortex really lost during aging? A morphometric examination. In J. Traber, & W. Gispen (Eds.), *Senile dementia of the Alzheimer type.* Berlin/Heidelberg: Springer-Verlag.
52. Coggeshall, R. E. (1992). A consideration of neural counting methods. *Trends Neurosci., 15,* 9–13.
53. West, M. (1993). New stereological methods for counting neurons. *Neurobiol. Aging,*
54. Hyman, B. T., Van Hoesen, G. W., Kromer, L. J., & Damasio, A. R. (1986). Perforant pathway changes and the memory impairment of Alzheimer's disease. *Ann. Neurol., 20,* 473–482.
55. Pandya, D. N., & Yeterian, E. H. (1985). Architecture and connections of cortical association areas. In A. Peters & E. G. Jones (Eds.), *Cerebral cortex,* (pp. 3–61). New York: Plenum Press.
56. Van Hoesen, G. W. (1993). The modern concept of association cortex. *Curr. Opin. Neurobiol., 3,* 150–154.
57. Gómez-Isla, T., Price, J. L., McKeel, D. W., Morris, J. C., Growdon, J. H., & Hyman, B. T. (1996). Profound loss of layer II entorhinal cortex neurons in very mild Alzheimer's disease. *J. Neurosci., 16,* 4491–4500.
58. Khachaturian, Z. S. (1985). Diagnosis of Alzheimer's disease. *Arch. Neurol., 42,* 1097–1105.
59. Mirra, S. S., Heyman, A., & McKeel, D. (1991). The consortium to establish a registry for Alzheimer's disease (CERAD). II. Standardization of the neuropathologic assessment of Alzheimer's disease. *Neurology, 41,* 479–486.

60. Gómez-Isla, T., Hollister, R., West, H., Mui, S., Growdon, J. H., Petersen, R. C., Parisi, J. E., & Hyman, B. T. (1997). Neuronal loss correlates with but exceeds neurofibrillary tangles in Alzheimer's disease. *Ann. Neurol., 41,* 17–24.

61. Blessed, G., Tomlinson, B. E., & Roth, M. The association between quantitative measures of dementia and of senile change in the cerebral grey matter of elderly subjects. *Br. J. Psychiatry, 114,* 797–811.

62. Perry, E., Tomlinson, B., Blessed, G., Bergmann, K., Gibson, P., & Perry R. (1978). Correlation of cholinergic abnormalities with senile plaques and mental test scores in senile dementia. *Br. Med. J., 2,* 1457–1459.

63. Delaere, P., Duyckaerts, C., Brion, J., Poulain, V., & Hauw, J. (1989). Tau, paired helical filaments and amyloid in the neocortex: A morphometric study of 15 cases with graded intellectual status in ageing and senile dementia of Alzheimer type. *Acta Neuropathol., 77,* 645–653.

64. Duyckaerts, C., Hauw, J., Basternaire, F., Piette F., Poulain C., Rainsard, V., Javoy-Agid, F., & Berthoux, P. (1986). Laminar distribution of neocortical senile plaques in senile dementia of the Alzheimer type. *Acta Neuropathol., 70,* 249–256.

65. Wilcock, G. K., & Esiri, M. M. Plaques, tangles and dementia. A quantitative study. *J. Neurol. Sci., 56,* 343–356.

66. McKee, A., Kosik, K., & Kowall, N. (1991). Neuritic pathology and dementia in Alzheimer's disease. *Ann. Neurol., 30,* 156–165.

6

TRANSGENIC MOUSE MODELS WITH NEUROFILAMENT-INDUCED PATHOLOGIES

JEAN-PIERRE JULIEN

Centre for Research in Neuroscience, McGill University
The Montréal General Hospital Research Institute
Montréal, Québec, Canada H3G 1A4

Neurofilaments represent a major cytoskeletal component in large neurons. There is growing evidence that perturbations in the normal metabolism of neurofilaments are associated with diseases. The abnormal accumulations of neurofilaments within neurons, often called spheroids or Lewy bodies, are frequently observed in human neurodegenerative disorders, including amyotrophic lateral sclerosis (ALS), Parkinson's disease, and Alzheimer's disease. Moreover, dramatic declines in the levels of neurofilament mRNAs have been reported in these diseases and during aging. I will discuss here the evidence, based primarily on transgenic mouse studies, that neurofilament disorganization can contribute to pathogenesis.

NEUROFILAMENT STRUCTURE AND FUNCTION

Neurofilaments are made up by the copolymerization of three intermediate filament proteins, NF-L (61 kDa), NF-M (90 kDa), and NF-H (110 kDa).[1] Neurofilaments are present in most populations of neurons in the nervous system but they are particularly abundant in large myelinated axons of the peripheral nervous system that originate from motor and sensory neurons. In addition to conferring a mechanical support to the cell, neurofilaments play a role in mediating the caliber of large myelinated axons. This is an important function because the axonal caliber is a major determinant of conduction velocity. The unequivocal proof of neurofilament involvement in the control of axonal caliber was recently provided from the analysis of animals lacking axonal neurofilaments including a quivering quail mutant deficient in NF-L protein,[2] a transgenic mouse expressing a NF-H/lacZ fusion construct,[3] and an NF-L knock-out mouse.[4] All these animals showed a dramatic hypotrophy of axons due to the scarcity of neurofilaments.

The three neurofilament subunits share with other members of the intermediate filament family a central domain of approximately 310 amino acids, which is involved in the formation of coiled-coil structures. The current model of a 10-nm filament structure is that two coiled-coil dimers of protein subunit line up in a staggered fashion to form an antiparallel tetramer.[5] Eight tetramers packed together in helical array are required to make a ropelike 10-nm filament. Neurofilaments are obligate heteropolymers requiring NF-L with either NF-M or NF-H for polymer formation.[6, 7] There is evidence for the existence of two types of heterotetramers in neurofilaments, one containing NF-L and NF-M and the other containing NF-L and NF-H. A remarkable feature of the NF-M and NF-H proteins is their long carboxy terminal tail domain that forms side-arm projections at the periphery of the filament.[8] The tail domain of NF-H has a high content in charged amino acids and with multiple repeats of Lys-Ser-Pro (KSP) that account for the unusual high content of phosphoserine residues in this protein.[9, 10, 11] It is believed that local changes in phosphorylation of NF-H could regulate the spacing between neurofilaments and thereby the axonal caliber. Evidence for this has come from studies on *Trembler* mice and hypomyelinating transgenic mice in which reduced levels of NF-H phosphorylation resulted in decreases in axon caliber.[12, 13]

PERTURBATIONS OF NEUROFILAMENT METABOLISM IN NEURODEGENERATIVE DISEASES AND IN AGING

There is evidence of perturbations in the normal metabolism of neurofilaments in neurodegenerative diseases. The abnormal accumulation of neurofilaments is frequently observed within degenerating neurons in ALS, Alzheimer's disease, and Parkinson's disease (Table 6.1). In addition, dramatic declines in the levels of neurofilament mRNAs have been detected in patients with neurodegenerative diseases as compared to age-matched controls (see Table 6.1). In motor neurons of ALS patients, Bergeron et al.[14] reported a 60% decrease in the levels of NF-L mRNA. In Parkinson's disease, the levels of NF-L and NF-H mRNAs were found to be selectively reduced in substantia nigra neurons by 30% and 70%, respectively.[15] A decline of 70% in NF-L mRNA level has also been detect-

TABLE 6.1   Human Diseases with Abnormal Neurofilament Accumulations

Disease	Abnormalities	References
ALS (70% cases)	NF depositions in motor neurons	53, 19
	Decline of 60% in NF-L mRNA	14
Parkinson's disease (100% cases)	Lewy bodies in substantia nigra and locus coeruleus	54
	Declines of 30% NF-L mRNA and 70% NF-H mRNA	15
Alzheimer's disease (20% cases)	Cortical Lewy bodies	55
	Decline of 70% in NF-L mRNA	16
Lewy body dementia	Cortical Lewy bodies	54
Guam-parkinsonism (100% cases)	NF depositions in motor neurons	56
Giant axonal neuropathy	NF accumulations in peripheral axons	57
Lhermitte-Duclos disease	Hypertrophy of granule cells of cerebellum with enhanced NF content	58
Peripheral neuropathies	NF accumulations in peripheral axons that can be induced by various toxic agents, such as IDPN, hexanedione, and acrylamide	59

Note: ALS, amyotrophic lateral sclerosis; *IDPN,* β,β′-immodipropinitrile; *NF,* neurofilament.

ed in Alzheimer's disease.[16] Whether the changes in neurofilament mRNA levels in these diseases are due to alterations in transcription and/or mRNA stability remains to be determined.

Aging, a predisposition factor for many neurodegenerative diseases, is accompanied by changes in neurofilament metabolism. Parhad et al.[17] detected dramatic declines of 60–70% in neurofilament mRNA levels during aging in the rat, whereas Gou et al.[18] reported a progressive hyperphosphorylation of NF-H protein with aging. It is conceivable that these changes during aging could affect assembly and transport of neurofilament proteins resulting in their abnormal accumulations in neurofilament-rich neurons, a phenomenon that may be more pronounced in neurodegenerative situations.

MOUSE MODELS WITH NEUROFILAMENT PATHOLOGIES

Even though the neurofilament accumulations have been considered an early pathological hallmark in ALS,[19] until recently there was no clear indication that neurofilaments could play a pathogenic role in human disease. Such neurofilament depositions were widely viewed as a consequence of neuronal dysfunction, perhaps reflecting defects in axonal transport. The first evidence that neurofilaments can play a causative role in neurodegenerative disease came from studies with transgenic mice that developed abnormal neurofilament accumulations as a result of overexpression of human NF-H or mouse NF-L proteins (Table 6.2). A modest 2–3-fold overexpression of human NF-H protein in mice provoked a late-onset motor neuron disease characterized by the presence of aberrant neurofilament accumulations in spinal motor neurons.[20] The motor dysfunction in these NF-H transgenic mice progresses during aging, and it is accompanied by the atrophy and slow degeneration of axons distal to the neurofilamentous swellings. Secondary muscle degeneration occurs in these transgenics as a result of axonal degeneration.

TABLE 6.2   Mouse Models with Neurofilament Pathology

Transgene	Neuronal populations affected	Motor neuron disease	References
Neurofilament transgenics			
Human NF-H	Spinal motor neurons and DRG neurons	Yes	20, 31
Mouse NF-H	Spinal motor neurons and DRG neurons	No	28
Mouse NF-H/lacZ	Perikarya of central nervous system and peripheral nervous system neurons	No	3
MSV/mouse NF-L	Spinal motor neurons and DRG neurons	Yes	21
MSV/mutant NF-L	Spinal motor neurons and DRG neurons	Yes	22
Human NF-L	Thalamic neurons and cortical neurons	No	26
Human NF-M	Cortical neurons and forebrain neurons	No	27
MSV/mouse NF-M	Spinal motor neurons and DRG neurons	No	24, 25
Others			
BAPG1 knock-out or (*dt/dt*) mice	Degeneration of DRG neurons Neurofilament aggregates in sensory axons	No	34
Wobbler mouse	Motor neurons Increased levels of NF-M mRNA expression by 3–4 fold	Yes	35
Transgenic expressing SOD1 mutants	Prominent vacuoles in motor neurons and neurofilamentous spheroids in proximal motor axons	Yes	43, 24, 25, 44, 49

Note: DRG, dorsal root ganglia; *MSV,* murine sarcoma virus; *NF,* neurofilament.

An elevated expression of the wild-type mouse NF-L gene in mice achieved through the use of the strong viral promoter from murine sarcoma virus (MSV) provoked an early-onset motor neuron disease accompanied by massive accumulation of neurofilaments in spinal motor neurons and by muscle atrophy.[21] A more severe phenotype was produced by expressing an assembly-disrupting NF-L mutant having a leucine to a proline substitution residue near the end of the conserved rod domain.[22] Although no such NF-L mutations have been reported in human ALS, similar mutations in keratins are the cause of the severe forms of genetic skin disease.[23] Expression of this mutant NF-L protein at only 50% the endogenous NF-L levels was sufficient to induce within 4 weeks massive neurofilament accumulations in spinal motor neurons accompanied by a selective death of motor neurons, neuronophagia, and severe denervation atrophy of the skeletal muscle.

The cellular selectivity of disease has remained a puzzling issue of ALS. In this regard, it is remarkable that in these transgenic mice neurodegeneration was found to occur specifically in spinal motor neurons even though the NF-L or NF-H transgenes were expressed throughout the nervous system. One simple explanation for the specific vulnerability of spinal motor neurons is their high content of neurofilaments. It is also noteworthy that these transgenic mouse models are not absolute replicas of human ALS in that motor neurons from the brain cortex did not develop neurofilamentous pathology. This is likely due to the smaller size and thereby lower neurofilament content of these neurons in mice as compared to the corresponding neurons in humans. Clearly, additional unknown cellular factors may also contribute to the poisoning effects of neurofilament deposits, as large neurofilament accumulations were well tolerated within dorsal root ganglia (DRG) neurons.[20, 21, 22] Some kinds of neurofilament accumulations appear to be less toxic than others, as many lines of transgenic mice with neurofilament abnormalities did not exhibit overt phenotypes (see Table 6.2). Massive neurofilament aggregates sequestered in perikarya of spinal motor neurons were observed in transgenic mice expressing a mouse NF-H/lacZ fusion gene but with little neurodegeneration.[3] The perikaryal neurofilamentous swellings in DRG and motor neurons of transgenic mice overexpressing NF-M transgenes under the control of the MSV promoter did not result in overt phenotypes or axonal degeneration.[24, 25] Also, overt phenotypes occurred in transgenic mice with moderate overexpression of the human NF-L[26] or NF-M genes,[27] although a detailed analysis of the mice revealed the presence of abnormal perikaryal immunoreactivity in some brain neurons. It has not yet been documented whether these accumulations of neurofilaments result in neurodegeneration.

Overexpression of the wild-type mouse NF-H proteins resulted in a retarded axonal transport of neurofilaments, accumulation of neurofilaments in motor neurons and atrophy of myelinated axons.[28] However, unlike transgenic mice overexpressing by 2 fold the human NF-H,[20] the transgenic mice overexpressing by 4–5 fold the wild-type mouse NF-H proteins did not develop a motor neuron disease.[28] There are considerable sequence differences between the phosphorylation tail domain of the mouse NF-H and human NF-H proteins. Perhaps, this may account for the different effects of these two proteins. In the mouse NF-H, the number of KSPXK, a consensus site for the cdk5 kinase, occurs only 9 times, whereas in human NF-H it occurs 34 times.[29] It is also striking that the human NF-H exhibits a high degree of homology with the NF-H from the rabbit, an animal species highly susceptible to neurofilamentous pathology induced by aluminum.[30] Thus, the enhanced poisoining effect of the human NF-H in mice is likely due to the properties of its tail phosphorylation domain, a region that can affect filament packing density and axonal transport.

INTERFERENCE OF AXONAL TRANSPORT BY DISORGANIZED NEUROFILAMENTS

Axonal transport studies with transgenic mice overexpressing human NF-H provided clues on how disorganized neurofilaments can contribute to neurodegeneration. Pulse-labeling with [^{35}S]-methionine of spinal motor neurons in NF-H transgenics demonstrated slower rates of axonal transport not only for neurofilament proteins but also for other axonal components.[31] One possible explanation is that the abnormal accumulation of neurofilaments can physically block the delivery of components required for axonal maintenance, including cytoskeletal elements and mitochondria. An alternative interpretation is that disorganized neurofilament proteins could provoke dysfunction of other cellular components required for axonal transport. For instance, it is conceivable that neurofilament abnormalities might alter microtubule dynamics and function, resulting in a retardation of axonal transport.

A block of axonal transport by disorganized neurofilaments is a pathological mechanism consistent with several aspects of ALS. It can explain in part the cellular selectivity of the disease. Large motor neurons represent a class of neurons more susceptible to develop neurofilament abnormalities because of their high synthesis of neurofilament proteins. This mechanism of disease is in agreement with the recent report of a marked increase of neurofilaments, mitochondria, or lysosomes in the axon hillock of motor neurons in ALS patients, whereas a decreased content of cytoplasmic organelles was detected in the initial axon segment from chromatolytic neurons.[32] Moreover, there is a retardation in the slow axonal transport of cytoskeletal elements during aging,[33] a factor that can predispose to the disease. While ALS is a disease of many etiological factors, the neurofilament depositions may be key intermediates contributing to neurodegeneration. Many factors can potentially lead to the accumulation of neurofilaments including deregulation of neurofilament protein synthesis, defective transport, abnormal phosphorylation and proteolysis, and other protein modifications. A list of potential factors is provided in Table 6.3. A disorganization of neurofilaments can also result from defects in proteins interacting with neurofilaments as revealed recently by the analysis of mice lacking the neuronal isoform of BPAG1, a protein that cross-links actin and neurofilaments[34] (see Table 6.2). The BPAG1 belongs to a group of large coiled-coil proteins that includes desmoplakin and plectin. A neuronal splice form, BPAG1n, is expressed specifically in sensory neurons. The targeted disruption of the BPAG1 gene in mice produced a selective degeneration and death of sensory neurons identical to *dystonia musculorum (dt/dt),* an autosomal recessive mouse mutant. The BPAG1 knock-out or *dt/dt* mice are characterized by the presence of abnormal neurofilament aggregates in sensory axons.[34] Nonetheless, it remains unknown to what extent the disorganization of neurofilaments contributes to the degeneration process in these mouse mutants.

There is a mouse mutant of progressive motor neuron disease, the wobbler mouse, that develops neurofilament accumulations in spinal motor neurons (see Table 6.2). Of particular interest is the recent finding of increased levels of NF-M mRNA by 3–4 fold in affected neurons of the wobbler mouse.[35] The wobbler gene has not yet been identified but it does not map to any of the neurofilament gene loci.[36] This mouse mutant will provide another useful model to investigate neurofilaments as possible intermediates of disease.

TABLE 6.3   Factors That Can Potentially Lead to Accumulation of Neurofilaments

Factor	Evidence	References
Neurofilament gene mutations	Codon deletions in the NF-H gene of ALS patients	37
	Severe motor neuron disease in transgenic mice expressing a NF-L assembly-disrupting mutant	22
Deregulation of neurofilament gene expression	Transgenic mice overexpressing NF-L, NF-M, or NF-H	20, 21, 27
	Decreased ratio of NF-L to NF-H mRNAs in a dog model HCSMA	52
SOD1 mutations	Neurofilament inclusions in FALS and in transgenic mice expressing SOD1 mutants	48, 49
Alterations in neurofilament-associated proteins	Degeneration of sensory neurons in mice lacking a BPAG1 isoform	34
Toxic agents such as IDPN, hexanedione and acrylamide	Neurofilament accumulations in animal models and cultured cells	60
Aluminum intoxication	Neurofibrillary pathology in rabbits	30
Altered neurofilament phosphorylation	Aggregation of neurofilaments modulated by PKC inhibitor	61
	Hyperphosphorylation of NF-H protein in neuronal perikarya by stress-activated protein kinase	40

Note: ALS, amyotrophic lateral sclerosis; *FALS,* familial amyotrophic lateral sclerosis; *HCSMA,* hereditary canine spinal muscular atrophy; *IDPN,* β,β′-immodipropinitrile; *NF,* neurofilaments; *PKC,* protein kinase C; *SOD1,* superoxide dismutase.

NF-H MUTATIONS IN AMYOTROPHIC
LATERAL SCLEROSIS

Further compelling evidence for neurofilament involvement in ALS came from the discovery of codon deletions in the Lys-Ser-Pro (KSP) phosphorylation domain of NF-H from cases of ALS. The combined results from the screening of 562 sporadic ALS cases by two groups have revealed codon deletions in the NF-H gene of 7 unrelated ALS patients diagnosed as sporadic cases. In 3 ALS patients, the deletions consisted of 102 bp, 24 bp, and 18 bp, resulting in the loss of 1 or more KSP phosphorylation motifs.[37] Four other ALS cases shared the same 3-bp deletion for the codon AAG (K) in the phosphorylation sequence SPVKEE. No mutations were found in ~500 control samples. So far, the search for mutations in the KSP repeat domain of NF-H[38] or in all coding regions of NF-L, NF-M, or NF-H[39] has failed to reveal mutations linked to disease in >100 familial amyotrophic lateral sclerosis (FALS) cases. Thus, mutations in neurofilament genes as primary causes of disease appear to be responsible for only a small fraction of sporadic or familial ALS cases.

Although the mechanism underlying disease induced by NF-H mutations remains to be elucidated, our working hypothesis is that codon deletions in the NF-H tail domain could alter the properties of neurofilaments. In any case, the mutations in the KSP-repeat domain emphasize again the importance of the NF-H phosphorylation domain. It is noteworthy that in four ALS cases the loss of a lysine residue in the sequence SPVKEE converted this consensus phosphorylation sequence for the cdk5 kinase[29] as a phosphorylation site pertinent to a stress-activated kinase (SAPK).[40] Agents that activate SAPKγ are capable to induce abnormal hyperphosphorylation of NF-H protein in the perikarya of cultured DRG neurons.[40] This finding raises the possibility that phosphorylation by stress may lead to neurofilament-induced pathology. However, it remains to be established that activation of SAPK can provoke in vivo the abnormal phosphorylation and accumulation of neurofilaments.

A LINK BETWEEN NEUROFILAMENTS AND
OXIDATIVE STRESS?

Mutations in the copper–zinc superoxide dismutase (SOD1) gene are responsible for ~20% of familial ALS cases.[41, 42] The SOD1 is a metalloprotein that forms homodimers. Copper is the essential cofactor for the conversion of the superoxide anion ($\cdot O_2^-$) to hydrogen peroxide (H_2O_2), which is then detoxified to water and oxygen by catalase and glutathione peroxidase. So far, more than 40 different SOD1 mutations have been identified in familial ALS with the majority of these mutations inherited in an autosomal dominant fashion.[42] Most of the SOD1 mutations are missense point mutations, and the current view is that SOD1 mutations cause ALS through a mechanism involving not a loss of superoxide

dismutase activity but rather by a gain of detrimental function. Compelling evidence for this view came from transgenic mouse studies. Transgenic mice overexpressing the G93A,[43] G37R,[24, 25] and G85R[44] SOD1 mutants developed motor neuron disease, whereas the SOD activities in mice were not reduced. In addition, mice homozygous for the targeted disruption of the SOD1 gene did not develop motor neuron disease.[45] To date, two abnormal SOD1 activities have been proposed for the toxicity of mutations. Beckman et al.[46] suggested a gain-of-function mechanism where the mutations would render the copper in the active site of SOD1 more accessible to peroxynitrite to form a nitronium-like intermediate that can nitrate proteins on tyrosine residues. In another mechanism, the SOD1 mutations increase the peroxidase activity, producing more hydroxyl radicals from hydrogen peroxide as substrate than wild-type SOD1.[47] The raises in protein nitration or in peroxidase activity resulting from SOD1 mutations are two nonmutually exclusive mechanisms that can contribute to damage to various cellular components.

There is growing evidence of a link between SOD1 mutations and the formation of neurofilament depositions. The presence of neurofilament accumulations in motor neurons have recently been described in ALS families with different SOD1 mutations,[48] and neurofilament inclusions resembling those found in ALS have been observed in transgenic mice expressing SOD1 mutants[49] (see Table 6.2). Moreover, it is logical that abundant proteins with long half-lives in motor neurons such as neurofilament proteins would be favoured targets of damage by SOD1 mutants. This view is supported by the immunohistochemical detection of nitrotyrosine residues in NF-L associated with neurofilamentous spheroids within motor neurons of familial ALS cases.[50] It has been hypothesized that tyrosine nitration of NF-L may impede its assembly and lead to formation of disorganized filaments.[50] Reactive nitrogen species such as peroxynitrite might also create cross-links by the formation of dityrosine and thereby induce neurofilament aggregation. Other types of oxidative damage by altered SOD1 activity are also possible. For instance, protein cross-links may occur through a copper-mediated oxidation of sulfhydryl groups or a production of carbonyls on lysine residues. Indeed, carbonyl-related modification of NF-H protein has been reported in the neurofibrillary pathology of Alzheimer's disease,[51] but it remains to be demonstrated in ALS.

FUTURE DIRECTIONS

It is expected that the recent lines of evidence for a key role of neurofilaments in motor neuron disease will stimulate research on the potential factors and signaling pathways that could lead to their abnormal accumulations in neurons. Factors that can potentially induce neurofilament disorganization are listed in Table 6.3. These include factors that can affect the regulation of expression, assembly, transport, proteolysis, phosphorylation, and other posttranslational

modifications of neurofilament proteins. Although the current evidence suggests that the toxicity of neurofilament depositions involves a block of axonal transport, it remains to be explored why certain types of neurofilament accumulations are more toxic than others in specific cell types. The species differences in the NF-H properties and the location of mutations found in the NF-H of ALS cases highlighted the importance of the KSP phosphorylation domain. In this regard, the transgenic mouse approach should be useful to address the effects of mutations in the NF-H phosphorylation domain on the organization and metabolism of neurofilaments.

Another aspect that merits further attention is the apparent paradox that the neuronal levels of neurofilament mRNAs are reduced dramatically in neurodegenerative diseases associated with neurofilament accumulations (see Table 6.1). A decreased ratio of NF-L to NF-H mRNAs has also been reported in a dog model of motor neuron disease.[52] These findings raise the possibility that alterations in the stoichiometry of the NF-L, NF-M, and NF-H subunits might lead to disorganization of neurofilaments. It is also possible that the declines of neurofilament mRNAs in diseases and to a lower extent during aging could increase neuronal vulnerability to oxidative stress. Of particular importance will be investigations aiming to clarify the specific oxidative modifications and the in vivo effects of neurofilament proteins in disease caused by SOD1 mutations. In this regard, the use of knock-out mice for neurofilament genes should provide a powerful approach to study how reduced levels of subunits can affect neurofilament metabolism and neuronal function. Moreover, the breeding of knock-out mice that lack neurofilament proteins with mouse models of neurodegeneration, such as transgenic mice expressing SOD1 mutants, BPAG1 knock-out mice, and wobbler mice that are listed in Table 6.2, should provide a means to assess the contribution of neurofilament proteins in diseases of different etiologies.

ACKNOWLEDGMENTS

This work was supported by the Medical Research Council of Canada (MRC) and the ALS association (USA). J.-P. J. has an MRC senior scholarship.

REFERENCES

1. Hoffman, P. N., & Lasek, R. J. (1975). The slow component of axonal transport: Identification of the major structural polypeptides of the axon and their generality among mammalian neurons. *J. Cell Biol., 66,* 351–366.
2. Yamasaki, H., Itakura, C., & Mizutant, M. (1991). Hereditary hypotrophic axonopathy with neurofilament deficiency in a mutant strain of the Japanese quail. *Acta. NeuropathOL., 82,* 427–434.
3. Eyer, J., & Peterson, A. (1994). Neurofilament-deficient axons and perikaryal aggregates in viable transgenic mice expressing a neurofilament-β-galactosidase fusion protein. *Neuron, 12,* 389–405.

4. Zhu, Q., Couillard-Després, S., & Julien, J.-P. (1997). Delayed maturation of regenerating myelinated axons in mice lacking neurofilaments. *Exp. Neurol.*,

5. Steinert, P. M. & Roop, D. R. (1988). Molecular and cellular biology of intermediate filaments. *Annu. Rev. Biochem., 57,* 593–625.

6. Ching, G., & Liem, R. (1993). Assembly of type IV neuronal intermediate filaments in nonneuronal cells in the absence of preexisting cytoplasmic intermediate filaments. *J. Cell Biol., 122,* 1323–1335.

7. Lee, M. K., Xu, Z., Wong, P. C., & Cleveland, D. W. (1993). Neurofilaments are obligate heteropolymers in vivo. *J. Cell Biol., 122,* 1337–1350.

8. Hisanaga, S., & Hirokawa, N. (1990). Molecular architecture of neurofilament: In vitro reassembly process of neurofilament-L protein. *J. Mol. Biol., 211,* 871–882.

9. Julien, J. P., & Mushynski, W. E. (1982). Multiple phosphorylation sites in mammalian neurofilament polypeptides. *J. Biol. Chem., 257,* 10467–10470.

10. Julien, J. P., Cote, F., Beaudet, L., Sidky, M., Flavell, D., Grosveld, F., & Mushynski, W. (1988). Sequence and structure of the mouse gene coding for the largest neurofilament subunit. *Gene, 68,* 307–314.

11. Lees, J. F., Shneidman, P. S., Skuntz, S. F., Carden, M. J., & Lazzarini, R. A. (1988). The structure and organization of the human heavy neurofilament subunit (NF-H) and the gene encoding it. *EMBO J., 7,* 1947–1955.

12. de Waegh, S. M., Lee, V. M. Y., & Brady, S. T. (1992). Local modulation of neurofilament phosphorylation, axonal caliber, and slow axonal transport by myelinating Schwann cells. *Cell, 68,* 451–463.

13. Cole, J. S., Messing, A., Trojanowski, J. Q., & Lee, V. M. (1994). Modulation of axon diameter and neurofilaments by hypomyelinating Schwann cells in transgenic mice. *J. Neurosci., 14,* 6956–6966.

14. Bergeron, C., Beric-Maskarel, K., Muntasser, S., Weyer, L., Somerville, M. J., & Percy, M. (1994). Neurofilament light and polyadenylated mRNA levels are decreased in amyotrophic lateral sclerosis motor neurons. *J. Neuropathol. Exp. Neurol., 53,* 221–230.

15. Hill, W. D., Arai, M., Cohen, J. A., & Trojanowski, J. Q. (1993). Neurofilament mRNA is reduced in Parkinson's disease substantia nigra pars compacta neurons. *J. Comp. Neurol., 329,* 328–336.

16. Crapper McLachlan, D. R., Lukiw, W. J., Wong, L., Bergeron, C., & Bech-Hansen, N. T. (1988). Selective messenger RNA reduction in Alzheimer's disease. *Mol. Brain Res. 3,* 255–262.

17. Parhad, I. M., Scott, J. N., Cellars, L. A., Bains, J. S., Krekoski, C. A., & Clark, A. W. (1995). Axonal atrophy in aging is associated with a decline in neurofilament gene expression. *J. Neurosci. Res., 41,* 355–366.

18. Gou, J. P., Eyer, J., & Leterrier, J. F. (1995). Progressive hyperphosphorylation of neurofilament heavy subunits with aging. *Biochem. Biophys. Res. Commun., 215,* 368–377.

19. Hirano, A., Donnenfeld, H., Sasaki, S., & Nakano, I. (1984). Fine structural observations of neurofilamentous changes in amyotrophic lateral sclerosis. *J. Neuropathol. Exp. Neurol., 43,* 461–470.

20. Côté, F., Collard, J. F., & Julien, J. P. (1993). Progressive neuronopathy in transgenic mice expressing the human neurofilament heavy gene: A mouse model of amyotrophic lateral sclerosis. *Cell, 73,* 35–47.

21. Xu, Z., Cork, L .C., Griffin, J. W., & Cleveland, D. W. (1993). Increased expression of neurofilament subunit NF-L produces morphological alterations that resemble the pathology of human motor neuron disease. *Cell, 73,* 23–33.

22. Lee, M. K., Marszalek, J. R., & Cleveland, D. W. (1994). A mutant neurofilament subunit causes massive, selective motor neuron death: Implications for the pathogenesis of human motor neuron disease. *Neuron, 13,* 975–988.

23. Fuchs, E. (1994). Intermediate filaments and disease: Mutations that cripple strength. *J. Cell Biol., 125,* 511–516.

24. Wong, P. C., Marszalek, J., Crawford, T. O., Xu, Z., Hsieh, S. T., Griffin, J. W., & Cleveland, D. W. (1995). Increasing neurofilament subunit NF-M expression reduces axonal NF-H, inhibits radial growth, and results in neurofilamentous accumulation in motor neurons. *J. Cell Biol., 130,* 1413–1422.

25. Wong, P. C., Pardo, P. C., Borchelt, D. R., Lee, M. K., Copeland, N. G., Jenkins, N. A., Sisodia, S. S., Cleveland, D. W., & Price, D. L. (1995). An adverse property of a familial ALS-linked SOD1 mutation causes motor neuron disease characterized by vacuolar degeneration of mito-chondria. *Neuron, 14,* 1105–1116.

26. Ma, D., Descarries, L., Julien, J. P., & Doucet, G. (1995). Abnormal perikaryal accumulation of neurofilament light protein in the brain of mice transgenic for human NF-L: Sequence of post-natal development. *Neuroscience, 68,* 135–149.

27. Vickers, J. C., Morrison, J. H., Friedrich, V. L., Elder, G. A., Perl, D. P., Katz, R. N., & Lazzarini, R. A. (1994). Age-associated and cell-type-specific neurofibrillary pathology in transgenic mice expressing the human midsized neurofilament subunit. *J. Neuroscience, 14,* 5603–5612.

28. Marszalek, J. R., Williamson, T. L., Lee, M. K., Xu, Z., Hoffman, P. N., Crawford, T. O., & Cleveland, D. W. (1996). Neurofilament subunit NF-H modulates axonal diameter by selective-ly slowing neurofilament transport. *J. Cell Biol., 135,* 711–724.

29. Pant, H., & Veeranna. (1995). Neurofilament phosphorylation. *Biochem. Cell Biol., 73,* 575–592.

30. Strong, M. J. (1994). Aluminium neurotoxicity: An experimental approach to the induction of neurofilamentous inclusions. *J. Neurol. Sci., 124,* 20–26.

31. Collard, J. P., Côté, F., & Julien, J. P. (1995). Deficient axonal transport in a transgenic mouse model of ALS. *Nature, 375,* 61–64.

32. Sasaki, S., & Iwata, M. (1996). Impairment of fast axonal transport in the proximal axons of anterior horn neurons in amyotrophic lateral sclerosis. *Neurology, 47,* 535–540.

33. McQuarrie, I. G., Brady, S. T., & Lasek, R. J. (1989). Retardation in the slow axonal transport of cytoskeletal elements during maturation and aging. *Neurobiol. Aging, 10,* 359–365.

34. Yang, Y., Dowling, J., Yu, Q.-C., Kouklis, P., Cleveland, D. W., & Fuchs, E. (1996). An essential cytoskeletal linker protein connecting actin filaments to intermediate filaments. *Cell, 86,* 655–665.

35. Pernasalonso, R., et al. (1996). Early upregulation of medium neurofilament gene expression in developing spinal cord of the wobbler mouse mutant. *Mol. Brain Res., 38,* 267–275.

36. Des Portes, V., Coulpier, M., Melki, J., & Dreyfus, P. A. (1994). Early detection of mouse wob-bler mutation: A model of pathological motoneurone death. *Neuroreport, 5,* 1861–1864.

37. Julien, J. P. (1997). Neurofilaments and motor neuron disease. *Trends Cell Biol., 7,* 243–249.

38. Rooke, K., Figlewicz, D. A., Han, F., & Rouleau, G. A. (1996). Analysis of the KSP repeat of the neurofilament heavy subunit in familial amyotrophic lateral sclerosis. *Ann. Neurol., 46,* 789–790.

39. Vechio, J. D., Bruijn, L. I., Xu, Z., Brown, R. H., & Cleveland, D. W. (1996). Sequence vari-ants in human neurofilament proteins: Absence of linkage to familial amyotrophic lateral scle-rosis. *Ann. Neurol., 40,* 603–610.

40. Giasson, B. I., & Mushynski, W. E. (1996). Aberrant stress-induced phosphorylation of perikaryal neurofilaments. *J. Biol. Chem., 271,* 30404–30409.

41. Rosen, D. R., et al. (1993). Mutations in Cu/Zn superoxide dismutase gene are associated with familial amyotrophic lateral sclerosis. *Nature, 362,* 59–62.

42. Brown, R. H. (1995). Amyotrophic lateral sclerosis: Recent insights from genetics and trans-genic mice. *Cell, 80,* 687–692.

43. Gurney, M. E., Pu, H., Chiu, A. Y., Canto, M. C. D., Polchow, C. Y., Alexander, D. D., Caliendo, J., Hentati, A., Kwon, Y. W., Deng, H.-X., Chen, W., Zhai, P., Sufit, R. L., & Siddique, T. (1994). Motor neuron degeneration in mice expressing a human Cu Zn superoxide dismutase mutation. *Science, 264,* 1772–1775.

44. Ripps, M. E., Huntley, G. W., Hof, P. R., Morrison, J. H., & Gordon, J. W. (1995). Transgenic

mice expressing an altered murine superoxide dismutase gene provide an animal model of amyotrophic lateral sclerosis. *Proc. Natl. Acad. Sci. USA, 92,* 689–693.

45. Reaume, A. G., Elliott, J. L., Hoffman, E. K., Kowall, N. W., Ferrante, R. J., Siwek, D. F., Wilcox, H. M., Flood, D. G., Beal, M. F., Brown, R. H., Scott, R. W., & Snider, W. D. (1996). Motor neurons in Cu/Zn superoxide dismutase-deficient mice deveop normally but exhibit enhanced cell death after axonal injury. *Nature Genet., 13,* 43–47

46. Beckman, J. S., Carson, M., Smith, C. D., & Koppenol, W. H. (1993). ALS SOD and peroxynitrite. *Nature, 364,* 584.

47. Wiedau-Pazos, M., Goto, J. J., Rabizadeh, S., Gralla, E. B., Roe, J. A., Lee, M. K.,Valentine, J. S., & Bredesen, D. E. (1996). Altered reactivity of superoxide dismutase in familial amyotrophic lateral sclerosis. *Science, 271,* 515–518.

48. Rouleau, G. A., Clark, A. W., Rooke, K., Pramatarova, A., Krizus, A., Uchowersky, O., Julien, J. P., & Figlewicz, D. (1996). SOD1 mutation is associated with accumulation of neurofilaments in amyotrophic lateral sclerosis. *Ann. Neurol., 39,* 113–117.

49. Tu, P. H., Raju, O., Robinson, K. A., Gurney, M. E., Trojanowski, J. Q., & Lee, V. M. Y. (1995). Transgenic mice carrying a human superoxide dismutase transgene develop neuronal cytoskeletal pathology resembling human amyotrophic lateral sclerosis lesions. *Proc. Natl. Acad. Sci. USA, 93,* 3155–3160.

50. Chou, S. M., Wang, H. S., & Komai, K. (1996). Colocalization of NOS and SOD1 in neurofilament accumulation within motor neurons of amyotrophic lateral sclerosis: An immunohistochemical study. *J. Chem. Neuroanat., 10,* 249–258.

51. Smith, M. A., Rudnicka-Nawrot, M., Richey, P. L., Praprotnik, D., Mulvihill, P., Miller, C. A., Sayre, L., & Perry, G. (1995). Carbonyl-related posttranslational modification of neurofilament protein in the neurofibrillary pathology of Alzheimer's disease. *J. Neurochem., 64,* 2660–2666.

52. Muma, N. A., & Cork, L. C. (1993). Alternations in neurofilament mRNA in hereditary canine spinal muscular atrophy. *Lab. Invest., 69,* 436–442.

53. Carpenter, S. (1968). Proximal enlargement in motor neuron diseases. *Neurology, 18,* 841–851.

54. Schmidt, M. L., Murray, J. M., Lee, V. M.-Y., Hill, W. D., Wertkin, A., & Trojanowski, J. Q. (1991). Epitope map of neurofilament protein domains on cortical and peripheral nervous system Lewy bodies. *Am. J. Pathol., 139,* 53–65.

55. Perry, G., Kawai, M., Tabaton, M., Onorato, M., Mulvihill, P., Richey, P., Morandi, A., Connolly, H. A., & Gambetti, P. (1991). Neuropil threads of Alzheimer's disease show a marked alteration of the neuronal cytoskeleton. *J. Neurosci., 11,* 1748–1755.

56. Rodgers-Johnson, P., Garruto, R. M., Yanagihara, R., Chen, K.-M., Gajdusek, D. C., & Gibbs, C. J., Jr. (1986). Amyotrophic lateral sclerosis and parkinsonism-dementia on Guam: A 30-year evaluation of clinical and neuropathological trends. *Neurology, 36,* 7–13.

57. Klymkowsky, M. W., & Plummer, D. J. (1985). Giant axonal neuropathy: A conditional mutation affecting cytoskeletal organization. *J. Cell Biol., 100,* 245–250.

58. Yachnis, A. T., Trojanowski, J. Q., Memmo, M., & Schlaepfer, W. W. (1988). Expression of neurofilament proteins in the hypertrophic granule cells of Lhermitte-Duclos disease: An explanation for the mass effect and the myelation of parallel fibers in the disease state. *J. Neuropathol. Exp. Neurol., 47,* 206–216.

59. Brimijioin, S. (1984). The role of axonal transport in nerve disease. In P. J. Dyk, P. K. Thomas, E. H. Lambert, & R. Bung, (Eds.), *Perpheral neuropathy: Vol. 1* (pp. 477–493). Philadelphia: Saunders.

60. Griffin, J. W., Fahnestock, K. E., Price, D. L., & Cork, L. C. (1983). Cytoskeletal disorganization induced by local application of β,β'-iminodipropionitrile and 2,5-hexanedione. *Ann. Neurol., 14,* 55–61.

61. Carter, J. E., Gallo, J.-M., Anderson, V. E. R., Anderton, B. H., & Robertson, J. (1996). Aggregation of neurofilaments in NF-L transfected neuronal cells. Regeneration of the filamentous network by a protein kinase C inhibitor. *J. Neurochem., 67,* 1997–2004.

7

TRANSGENIC MODELS OF AMYOTROPHIC LATERAL SCLEROSIS AND ALZHEIMER'S DISEASE

PHILIP C. WONG, DAVID R. BORCHELT,
MICHAEL K. LEE, GOPAL THINAKARAN,
SANGRAM S. SISODIA,* AND
DONALD L. PRICE†

Departments of Pathology (all authors), Neurology,† Neuroscience,†
and the Division of Neuropathology (all authors),
The Johns Hopkins University School of Medicine
Baltimore, Maryland 21205*

Among the most puzzling and devastating illnesses in medicine are the neurodegenerative disorders, a group of age-related, chronic, progressive diseases. Clinical signs reflect the vulnerabilities of specific populations of neurons. For example, in amyotrophic lateral sclerosis (ALS), damage to upper and lower motor neurons results in spasticity and weakness/muscle atrophy, respectively; in Alzheimer's disease (AD), the involvement of a variety of brain regions–neuronal populations is reflected in loss of memory, cognitive–behavioral impairments, and, eventually, profound dementia. These diseases may be inherited in an autosomal dominant fashion: Some cases of familial amyotrophic lateral sclerosis (FALS) are linked to mutations in superoxide dismutase 1 (SOD1); some individuals with familial Alzheimer's disease (FAD) have mutations in the genes encoding the amyloid precursor protein (APP) or one of the presenilins (PS1 and PS2). The products of these mutant genes impact upon specific subsets of neural cells, causing characteristic clinical manifestations. For the most part, these illnesses are fatal, and treatments are symptomatic.

For neurological disorders, it is particularly important to establish model systems that faithfully reproduce disease. In the past, surrogate animal models (i.e., spontaneously occurring or experimentally induced) were used to study mechanisms that lead to degeneration of specific populations of neurons. With the recent identification of gene mutations linked to FALS and FAD, transgenic (Tg) strategies have allowed investigators to reproduce features of genetic human neurodegenerative disorders in mice.

In this review, we briefly describe Tg technologies and discuss the clinical, genetic, and neuropathological features of ALS and AD, with particular emphasis on the autosomal dominant forms of these diseases. For each disease, we discuss the results of our analyses of several selected Tg models and the implications of these findings for understanding the pathogenetic mechanisms of disease. We anticipate that the approaches illustrated in these examples will be increasingly valuable for modeling other genetically determined human diseases.

TRANSGENIC APPROACHES

Direct pronuclear injections of exogenous DNA into fertilized zygotes[1] and injections of genetically modified embryonic stem cells into mouse blastocysts[2, 3] have become standard procedures for the generation of Tg mice for in vivo studies of the properties associated with mutant gene products. Mutations of genes involved in dominantly inherited diseases can be introduced into mice by overexpressing transgenes that bear mutations or by using embryonic stem cells and gene-targeting approaches to introduce mutations at a gene locus. The pronuclear injection method has been particularly valuable in studies of Tg mice designed to examine the consequences of dominantly inherited mutations linked to familial neurodegenerative diseases. Several Tg mice that overexpress mutant transgenes develop features of the human disorders, and investigations of these models have begun to disclose some of the pathogenic mechanisms associated with mutations in specific genes.

AMYOTROPHIC LATERAL SCLEROSIS AND TRANSGENIC MICE WITH SOD1 MUTATIONS

CLINICAL FEATURES

ALS and FALS are characterized by paralysis, muscular atrophy, spasticity, and a variety of other motor signs, with relative preservation of eye movements, potency, and sphincters.[4] Electrodiagnostic studies disclose fibrillations, fasciculations, and giant polyphasic potentials; muscle biopsies demonstrate denervation atrophy with many small angulated fibers and fiber-type grouping.

GENETICS

It has been estimated that 10% of adult-onset cases of ALS are familial with autosomal dominant inheritance and age-dependent penetrance[5] Approximately 20% of cases of FALS are linked to mutations in SOD1,[6] a member of a family of metalloenzymes that acts as a free radical scavenger by catalyzing the formation of H_2O_2 through the dismutation of O_2.[7] Mutations are located in exons 1, 3, 5, and 6 and, with the exception of the H46R and H48Q mutations, largely spare the metal-binding portion of the protein. Moreover, the widespread mutations are not concentrated at the dimer interface or the active site. Initially, the finding of SOD1 mutations in some cases of FALS was interpreted to indicate that these mutations reduced enzyme activity, leading to O_2-mediated damage of motor neurons.[8] However, recent studies suggest that cell injury is associated with the gain of a toxic property by mutant SOD1.

NEUROPATHOLOGY

Lesions of upper motor neurons are associated with spasticity, hyperreflexia, and extensor plantar signs, whereas abnormalities of large β-motor neurons of the brain stem and spinal cord cause weakness and atrophy. Lower motor neurons show a variety of abnormalities including ubiquitin and phosphorylated neurofilament (NF) immunoreactivity in cell bodies, swollen axons (spheroids) with maloriented arrays of NF, and fragmentation of the Golgi apparatus.[9–13] In some cases of SOD1-linked FALS, motor neurons may also contain SOD1-immunoreactive intracytoplasmic inclusions.[14, 15] Eventually, in both ALS and FALS, motor axons undergo Wallerian degeneration, and denervated muscles show grouped atrophy. In end-stage disease, numbers of motor neurons are reduced in the spinal cord, brain stem nuclei, and motor cortex.[13]

TRANSGENIC MODELS OF SOD1-LINKED FALS

Over the past decade, studies of a variety of animal models of ALS have provided important information concerning the biological function and dysfunction/death of motor neurons.[16–28] Recently, several groups of investigators produced Tg mice with FALS-linked SOD1 mutations.[29–31]

Lines of Tg mice that express the G37R HuSOD1 mutation at 3–12x levels of endogenous SOD1 in the spinal cord invariably develop progressive motor neuron disease.[31] At 4 months of age, these Tg mice begin to show reduced spontaneous movements, difficulty moving their hindlimbs, and muscle wasting. Eventually, the forelimbs become weak, and hindlimbs are completely paralyzed. Electromyographic patterns and muscle biopsy results are consistent with ALS.

G37R HuSOD1 mice have significantly elevated levels of SOD1, and activity gels show increases in SOD1 activity, confirming that the G37R mutation synthesized in vivo retains full specific activity.[31] In mice that express mutant SOD1, motor axons accumulate SOD1 immunoreactivity.[32] In motor neurons, SOD1 is transported anterograde in axons as part of the slow component.[32] In the early preclinical period, these axons, as well as some dendrites, exhibit very small vacuoles, usually associated with enlarged degenerating mitochondria and swollen endoplasmic reticulum.[31] Subsequently, cell bodies exhibit abnormal patterns of ubiquitin and phosphorylated NF immunoreactivities. Motor axons, some of which show abnormalities of cytoskeletal elements, undergo Wallerian degeneration; muscle fibers exhibit evidence of denervation.

A very similar motor neuron disease has been documented in a line of G93A HuSOD1 Tg mice.[29] These mice, which also express the transgene at relatively high levels, show a neuropathology similar to that reported in G37R SOD1 Tg mice.[33–35] Moreover, the administration of vitamin E and selenium to G93A HuSOD1 Tg mice was associated with a very modest delay in both the onset and progression of the disease without affecting survival; in contrast, the two antiexcitotoxic agents did not influence the onset or progression of the disease but did increase survival slightly.[36]

PATHOGENESIS

The observation that Tg mice expressing mutant SOD1 show no loss of enzymatic activity yet develop disease[29–31, 37] strongly suggests that, at least for these mutations, SOD1-linked FALS is caused by the gain of a toxic property by the mutant enzyme. This concept is supported by several other lines of evidence: Some mutant SOD1 possess near-normal levels of enzyme activity/stability in vitro.[38] and/or restore SOD1 null yeast to the wild-type (wt) phenotype but accelerate the death of nerve cells in vitro[39]; SOD1 mutations do not have a dominant negative effect on wt SOD1[40]; and wt HuSOD1 Tg mice do not develop overt motor neuron disease.[31] SOD1 null mice develop normally; although these mice do not develop a FALS-like disease, motor neurons show enhanced susceptibility to retrograde degeneration after nerve transection,[41] as well as some evidence of a distal motor neuropathy.[42]

Major questions in the field include the nature of the toxic properties acquired by mutant SOD1 and the mechanisms of cell injury. In addition to its enzymatic activities, this abundant protein may have other functions (i.e., Cu^{2+} buffering/shielding).[43] In FALS, mutations are widespread throughout the enzyme but for the most part do not appear to have significant influences on the metal-binding domains of SOD1 (i.e., Cu^{2+} remains bound as a cofactor, and the enzyme retains significant activity).[38] The widespread mutations may open the structure of the enzyme to allow a variety of substrates to access the Cu^{2+}-containing portion of the enzyme. One hypothesis is that a peroxidase activity of mutant SOD1 catalyzes conversion to H_2O_2 to OH, which is capable of oxidizing a variety of targets.[44] In vitro, A4V and G93A HuSOD1, but not free Cu^{2+} or the SOD1 apoenzyme, have enhanced the abilities to catalyze the oxidation of a spin-trap compound and to induce cell death in vitro[44]; these effects are attenuated by treatment with Cu^{2+} chelators.[44] A second hypothesis[45] is that mutant SOD1 has the enhanced ability to utilize peroxynitrite to form nitronium ions that can nitrate tyrosine residues. To test this possibility, levels of nitrotyrosine were determined in one FALS-linked SOD1 line of mice. In spinal cords of G37R mice, 3-nitrotyrosine levels were elevated 2- to 3-fold as compared to mice that express high levels of wt HuSOD1 or non-Tg mice.[46] These results are consistent with the view that tyrosine nitration is one in vivo aberrant property of this FALS-linked mutant SOD1. The nitration of proteins important in the biology of motor neurons (glutamate transporters, neurofilaments, trkB, etc.) could have a significant impact on the structure/functions of these cells. Both of these hypotheses are consistent with the premise that FALS mutations alter protein structure in such a way that mutant SOD1 has a toxic effect on substrates critical for the survival of motor neurons. Because SOD1 is abundant in spinal motor neurons.[47] and is transported anterograde in axons,[32] the toxic mutant protein could damage a variety of molecular targets with significant consequences for motor neurons.

ALZHEIMER'S DISEASE AND TRANSGENIC MICE WITH APP OR PS1 MUTATIONS

CLINICAL FEATURES

AD is the most common cause of dementia that occurs in mid-to-late life.[48, 49] Clinical signs include memory loss, impairment of higher cortical functions, behavioral/psychiatric abnormalities, and, eventually, profound dementia.[50] Because of increases in life expectancy, the number of people with AD will triple over the next 25 years as will the cost of treating these individuals.[51]

GENETICS OF AUTOSOMAL DOMINANT FAD

Approximately 10% of cases of AD are familial and show autosomal dominant inheritance; subsets of these cases are linked to mutations of specific genes,[52, 53] including the APP, PS1, and PS2 genes.

Amyloid Precursor Protein

In a small fraction of cases of FAD, mutations have been documented in the APP gene (chromosome 21), which encodes APP-695, -751, and -770.[54, 55] APP, a type I integral membrane glycoprotein,[54, 55] contains the 4-kilodalton (kDa) Aβ peptide comprised of 28 amino acids of the ectodomain and 11–14 amino acids of the adjacent transmembrane domain.[54, 56] Some APP are cleaved endo-proteolytically by APP β-secretase, a membrane-bound endoprotease,[57] between positions 16 and 17 of Aβ.[58, 59] Aβ is formed by cleavages at the N-terminus (by β-secretase) and C-terminus (by γ-secretase).[60, 61] APP mutations, in prox-imity to the sequence that encodes the Aβ domain,[62–65] influence the amount, length, or fibrillogenic properties of Aβ42, the most fibrillogenic and toxic peptide species.

In a small fraction of autosomal dominant cases, mutations have been docu-mented in the APP gene.[63–65] In some families, the valine residue at position 717 (four amino acids downstream of the C-terminus of Aβ42) is substituted with either isoleucine, glycine, or phenylalanine.[62, 63] Cells that express APP harbor-ing 717 mutations secrete a higher fraction of longer Aβ peptides (i.e., extending to Aβ residue 42) relative to cells that express wt APP.[66] In two large, related, early-onset AD families from Sweden, Lys-Met codons immediately N-terminal to Asp 1 of Aβ are substituted within Asn-Leu.[65] Cells that express APP harbor-ing the "Swedish" substitutions (APPswe) secrete higher levels of Aβ40 as com-pared to cells that express wt constructs.[60, 67]

APP is expressed in many cell types including neurons; in nerve cells, APP is transported by rapid anterograde and retrograde systems to axons and nerve ter-minals.[68–70] It is likely that APP, transported to nerve terminals, is one source of the Aβ deposited in the brain.

Presenilins

A significant number of cases of early-onset FAD are linked to mutations in the PS1 gene located on chromosome 14,[71–73] whereas cases of autosomal dom-inant AD in Volga German kindreds and in an Italian pedigree show mutations in the PS2 gene, localized on chromosome 1.[74–77] PS1 and PS2 are highly homolo-gous proteins[73] predicted to contain eight transmembrane helices.[78, 79] A cyto-plasmic loop domain appears to interact with cytosolic constituents. PS1 is endo-proteolytically processed to an N-terminal 28-kDa fragment and a C-terminal 18-kDa fragment. Because of the paucity of accumulated full-length PS and the generality of PS processing across tissues, we have suggested that PS fragments are functional units in vivo.[79, 80]

PS1 mRNA protein is expressed in the brain, particularly in neurons.[79, 81] Although the biological functions of PS are not yet well understood, a PS homolog in *Caenorhabditis elegans,* termed sel-12, has been described that facil-itates the lin2/notch signaling pathway in the determination of cell fates during development in *C. elegans*.[82] An egg-laying defect in *C. elegans* lacking func-tional sel-12 can be rescued by PS1 and PS2, indicating that the homologous pro-teins are functionally interchangeable.[83] These studies suggest that PS may be important in developmental processes in mammalian tissues. The results of gene-targeting studies of PS1 are consistent with the view that PS1 facilitates notch signaling during mouse embryonic development.[84, 85]

Although the mechanisms by which mutations in PS1 and PS2 predispose individuals to autosomal dominant FAD are not clear, the absence of deletions or truncation mutations in PS1 and PS2 favor the idea that the disease is the result of a toxic effect of the mutant PS protein. Significantly, PS1 mutations influence the production of a highly fibrillogenic and pathogenic form of Aβ, termed Aβ42.

NEUROPATHOLOGY

AD selectively affects neurons in certain brain regions and populations of neurons including the hippocampus, neocortex, and basal forebrain cholinergic system.[86–88] Many vulnerable nerve cells accumulate abnormal straight and paired helical filaments, comprised principally of hyperphosphorylated tau, in cell bodies and proximal dendrites (neurofibrillary tangles), neurites (axons/terminals), and neuropil threads (distal dendrites).[89] In vitro, the exposure of nonphosphorylated three- and four-repeat recombinant tau to sulphated glycosaminoglycans leads to the formation of straight and paired helical filaments, respectively.[90] These findings suggest that interactions of tau and glycosaminoglycans may play an important role in the formation of neurofibrillary tangles. In addition to the presence of tau epitopes, tangle-bearing cells show ubiquitin, neurofilaments, and tyrosine immunoreactivity.[89] Eventually, tangle-bearing cells die, possibly by mechanisms that involve apoptotic pathways.[91, 92]

TRANSGENIC MODELS OF FAD

Many groups have attempted to create Tg mice with the features of AD.[93–99] Several lines of Tg mice that express mutant transgenes have shown Aβ deposits.[100–102]

In one line of mice,[100] the PDGF β-promoter was used to drive the expression of a HuAPP minigene encoding the FAD-linked APP (717V F) mutation; the construct contained portions of APP introns 6–8 to allow alternative splicing of exons 7 and 8. The resulting transcripts encode the three major APP isoforms, particularly those containing the Kunitz-type serine protease inhibitor domain. Levels of HuAPP mRNA and protein were significantly greater (4–5x) than endogenous levels. Aβ was present in diffuse deposits and in plaques within hippocampus and cerebral cortex. Extracellular Aβ fibrils were abundant, and dystrophic neurites were often present in proximity to plaques.

Using a vector containing the prion promoter, Hsiao et al.[101] overexpressed APP-695 containing APPswe in mice. At 3 months of age, these Tg mice showed normal learning and memory in spatial reference memory and alternation tasks; however, at 9–10 months of age, animals were impaired on these tasks. Levels of Aβ40 and Aβ42 were increased 5 fold and 14 fold, respectively, and Aβ plaques were present in hippocampus and cortex.

Recently, Tg mice that express wt or mutant PS1 transgenes have been produced.[103–105] Tg mice with high levels of wt or mutant A246E HuPS1 transgene products accumulate two PS1 fragments in brain.[105] Mice that express the mutant PS1 transgene (12 months of age) have not yet developed a FAD phenotype. To determine whether PS1 mutations influence APP/Aβ processing, particularly levels of the toxic Aβ42, A246E PS1 Tg mice were mated with APPswe Tg animals. The brains of young Tg animals that coexpress mutant PS1 and mutant APP showed elevated ratios of Aβ42:Aβ40 as compared to a line of APPswe Tg mice or to mice that coexpress wt PS1 and APPswe.[103] Recently, we demonstrated that mice coexpressing A246E HuPS1 and APPswe develop amyloid deposits at much earlier ages than mice expressing APPswe alone. These investigations, in concert with studies of sera from patients with PS mutations, media from fibroblasts from these individuals, and culture media from mutant PS-transfected cells, which show elevated Aβ42 levels in these settings, indicate that one mechanism by which mutant PS1 can cause AD is the acquisition (or enhancement) of properties that lead to an increase in extracellular concentrations of Aβ42. We conclude that a principal pathway by which mutant PS1 predisposes individuals to FAD is to accelerate the deposition of Aβ by altering APP metabolism to generate higher levels of highly fibrillogenic Aβ42 peptides.

Similar Tg approaches can be used to examine the effects of apoE isoforms; apoE3 or E4 alleles can be introduced into Tg mice that express wt or mutant APP transgenes. These Tg models will undoubtedly prove to be of extraordinary value in studies designed to clarify the cellular and biochemical mechanisms of disease and to test novel therapies.

PATHOGENESIS

Recent studies have begun to clarify the processes by which mutant genes promote the cellular abnormalities that occur in AD. Missense mutations at position 717 (of APP-770) influence APP processing in a manner that results in the secretion of higher levels of Aβ42.[66] On the other hand, cells that harbor the APPswe secrete higher levels of Aβ40 and Aβ42.[60, 67, 80] Other mutations within the Aβ peptide sequence increase the microheterogeneity and hydrophobicity of secreted Aβ species[64, 106–108] or increase the propensity of Aβ to aggregate into fibrils.[109]

Aβ42, a highly fibrillogenic peptide,[110, 111] is deposited early in the brains of individuals with AD and Down syndrome.[112–115] Several lines of evidence indicate that PS1 mutations also impact on levels of Aβ42: Plasma and conditioned media from fibroblasts obtained from carriers of mutations have elevated levels of Aβ42 species as compared to samples from unaffected family members[116]; in vitro studies of transfected cell lines that express mutations show increased ratios of Aβ42/Aβ40 as compared to ratios in media of cells that express wt PS1.[103] Moreover, as outlined above, the brains of Tg mice that overexpress mutant APP or that coexpress a chimeric APP and FAD-linked PS1 variant show elevated Aβ42/Aβ40 ratios and Aβ deposits in brain.[101, 103]

In concert, studies of APP and PS mutations are consistent with the idea that the mutant protein promotes the production of Aβ42. Aggregated Aβ42 peptides are neurotoxic,[117] but the mechanisms whereby toxic peptide species damage cells are not yet fully established.[118–122]

CONCLUSIONS

ALS and AD are prototypes of human neurodegenerative diseases. In ALS, weakness and atrophy reflect dysfunction and death of motor neurons; in AD, memory loss and dementia are the result of neurofibrillary tangles, Aβ42 amyloid deposits, and death of neurons in cortex and hippocampus. In the past, investigators have relied on experimentally induced and spontaneously occurring models, but, more recently, genetically engineered Tg mice have provided model systems that are proving to be very useful for examining mechanisms of disease in vivo. Subsets of cases of FALS and FAD often show dominant inheritance: some cases of FALS are linked to mutations in the SOD1 gene, and some pedigrees with FAD exhibit mutations in genes encoding either APP or PS1 and PS2. Much has been learned about the biology of these mutant transgene products by recent in vitro and in vivo studies. In vitro, some mutant SOD1 show normal enzyme activity but acquire neurotoxic properties; similarly, in vitro studies have provided new information concerning APP and PS processing and the effects of mutations on the generation of Aβ. Recently, exciting advances have been made using Tg strategies that allow investigators to reproduce features of these human disorders in mice. Tg mice with SOD1 mutations develop weakness and muscle atrophy associated with degenerative changes in motor neurons that result from the acquisition of toxic properties by mutant SOD1. Similarly, Tg mice that express mutant HuFAD-linked genes show behavioral impairments Aβ42 deposits associated with dystrophic neurites. The formation of Aβ42, derived by the aberrant processing of APP, is influenced by APP mutations that increase the amount, length, and fibrillogenic properties of Aβ. Mutations of PS1 increase levels of Aβ42. These Tg models are of great value for investigations of mechanisms of disease and will be critical for testing novel therapies that, if efficacious in model systems, can be introduced rapidly into clinical trials.

ACKNOWLEDGMENTS

The authors thank Drs. Don Cleveland, Edward Koo, Neal Copeland, Nancy Jenkins, Bruce Lamb, Carlos Pardo, Mark Becher, and John Gearhart for their contributions to some of the work mentioned in this text.

This work was supported by grants from the U.S. Public Health Service (NIH NS 20471, AG

05146), as well as the Metropolitan Life Foundation, the Adler Foundation, the Alzheimer's Association, the Develbiss Fund, the American Health Assistance Foundation, the Amyotrophic Lateral Sclerosis Association, and Merck, Sharp & Dohme. Dr. Price is the recipient of a Javits Neuroscience Investigator Award (NIH NS 10580) and a Leadership & Excellence in Alzheimer's Disease (LEAD) Award (NIH AG 07914). Dr. Sisodia is the recipient of a Zenith Award from the Alzheimer's Association.

REFERENCES

1. Palmiter, R. D., & Brinster, R. L. (1986). Germ-line transformation of mice. *Annu. Rev. Genet., 20,* 465–499.
2. Capecchi, M. R. (1989). Altering the genome by homologous recombination. *Science, 244,* 1288–1299.
3. Takahashi, J. S., Pinto, L. H., & Hotz Vitaterna, M. (1994). Forward and reverse genetic approaches to behavior in the mouse. *Science, 264,* 1724–1733.
4. Kuncl, R. W., Crawford, T. O., Rothstein, J. D., & Drachman, D. B. (1992). Motor neuron diseases. In A. K. Asbury, G. M. McKhann, W. I. McDonald (Eds.), *Diseases of the nervous system* (pp. 1179–1208). Philadelphia: Saunders.
5. Siddique, T., Figlewicz, D. A., Pericak-Vance, M. A., Haines, J. L., Rouleau, G., Jeffers, A. J., Sapp, P., Hung, W.-Y., Bebout, J., McKenna-Yasek, D., Deng, G., Horvitz, H. B., Gusella, J. F., Brown, R. H., Jr., Roses, A. D., Roos, R. P., Williams, D. B., Mulder, D. W., Watkins, P. C., Noore, R., Nicholson, G., Reed, R., Brooks, B. R., Festoff, B., Antel, J. P., Tandan, R., Munsat, T. L., Laing, N. G., Halperin, J. J., Norris, F. H., Van den Bergh, R., Swerts, L., Tanzi, R. E., Jubelt, B., Mathews, K. D., & Bosch, E. P. (1991). Linkage of a gene causing familial amyotrophic lateral sclerosis to chromosome 21 and evidence of genetic-locus heterogeneity. *N. Engl. J. Med., 324,* 1381–1384.
6. Rosen, D. R., Siddique, T., Patterson, D., Figlewicz, D. A., Sapp, P., Hentati, A., Donaldson, D., Goto, J., O'Regan, J. P., Deng, H.-X., Rahmani, Z., Krizus, A., McKenna-Yasek, D., Cayabyab, A., Gaston, S. M., Berger, R., Tanzi, R. E., Halperin, J. J., Herzfeldt, B., Van den Bergh, R., Hung, W.-Y., Bird, T., Deng, G., Mulder, D. W., Smyth, C., Laing, N. G., Soriano, E., Pericak-Vance, M. A., Haines, J., Rouleau, G. A., Gusella, J. S., Horvitz, H. R., & Brown, R. H., Jr. (1993). Mutations in Cu/Zn superoxide dismutase gene are associated with familial amyotrophic lateral sclerosis. *Nature, 362,* 59–62.
7. Fridovich, I. (1986). Superoxide dismutases. *Adv. Enzymol. Relat. Areas Mol. Biol., 58,* 61–97.
8. Deng, H.-X., Hentati, A., Tainer, J. A., Iqbal, Z., Cayabyab, A., Hung, W.-Y., Getzoff, E. D., Hu, P., Herzfeldt, B., Roos, R. P., Warner, C., Deng, G., Soriano, E., Smyth, C., Parge, H. E., Ahmed, A., Roses, A. D., Hallewell, R. A., Pericak-Vance, M. A., & Siddique, T. (1993). Amyotrophic lateral sclerosis and structural defects in Cu,Zn superoxide dismutase. *Science, 261,* 1047–1051.
9. Carpenter, S. (1968). Proximal axonal enlargement in motor neuron disease. *Neurology, 18,* 841–851.
10. Schmidt, M. L., Carden, M. J., Lee, V. M.-Y., & Trojanowski, J. Q. (1987). Phosphate dependent and independent neurofilament epitopes in the axonal swellings of patients with motor neuron disease and controls. *Lab. Invest., 56,* 282–294.
11. Chou, S. M. (1992). Pathology-light microscopy of amyotrophic lateral sclerosis. In R. A. Smith (Ed.), *Handbook of amyotrophic lateral sclerosis* (pp. 133–181). New York: Marcel Dekker.
12. Gonatas, N. K., Stieber, A., Mourelatos, Z., Chen, Y., Gonatas, J. O., Appel, S. H., Hays, A. P., Hickey, W. F., & Hauw, J.-J. (1992). Fragmentation of the Golgi apparatus of motor neurons in amyotrophic lateral sclerosis. *Am. J. Pathol., 140,* 731–737.
13. Oppenheimer, D. R., & Esiri, M. M. (1992). Diseases of the basal ganglia, cerebellum and motor neurons. In J. H. Adams & L. W. Duchen (Eds.), *Greenfield's neuropathology* (pp. 988–1045). New York: Oxford University Press.
14. Rouleau, G. A., Clark, A. W., Rooke, K., Pramatarova, A., Krizus, A., Suchowersky, O., Julien,

J.-P., & Figlewicz, D. (1996). SOD1 mutation is associated with accumulation of neurofilaments in amyotrophic lateral sclerosis. *Ann. Neurol., 39*, 128–131.

15. Shibata, N., Hirano, A., Kobayashi, M., Siddique, T., Deng, H.-X., Hung, W.-Y., Kato, T., & Asayama, K. (1996). Intense superoxide dismutase-1 immunoreactivity in intracytoplasmic hyaline inclusions of familial amyotrophic lateral sclerosis with posterior column involvement. *J. Neuropathol. Exp. Neurol., 55*, 481–490.

16. Griffin, J. W., Hoffman, P. N., Clark, A. W., Carroll, P. T., & Price, D. L. (1978). Slow axonal transport of neurofilament proteins: Impairment by β,β'-iminodipropionitrile administration. *Science, 202*, 633–635.

17. Cork, L. C., Griffin, J. W., Munnell, J. F., Lorenz, M. D., Adams, R. J., & Price, D. L. (1979). Hereditary canine spinal muscular atrophy. *J. Neuropathol. Exp. Neurol., 38*, 209–221.

18. Cork, L. C., Griffin, J. W., Choy, C., Padula, C. A., & Price, D. L. (1982). Pathology of motor neurons in accelerated hereditary canine spinal muscular atrophy. *Lab. Invest., 46*, 89–99.

19. Griffin, J. W., Fahnestock, K. E., Price, D. L., & Cork, L. C. (1983). Cytoskeletal disorganization induced by local application of β,β'-iminodipropionitrile and 2,5-hexanedione. *Ann. Neurol., 14*, 55–61.

20. Troncoso, J. C., Hoffman, P. N., Griffin, J. W., Hess-Kozlow, K. M., & Price, D. L. (1985). Aluminum intoxication: A disorder of neurofilament transport in motor neurons. *Brain Res., 342*, 172–175.

21. Troncoso, J. C., Sternberger, N. H., Sternberger, L. A., Hoffman, P. N., & Price, D. L. (1986). Immunocytochemical studies of neurofilament antigens in the neurofibrillary pathology induced by aluminum. *Brain Res., 364*, 295–300.

22. Cote, F., Collard, J.-F., & Julien, J.-P. (1993). Progressive neuronopathy in transgenic mice expressing the human neurofilament heavy gene: A mouse model of amyotrophic lateral sclerosis. *Cell, 73*, 35–46.

23. Xu, Z., Cork, L. C., Griffin, J. W., & Cleveland, D. W. (1993). Increased expression of neurofilament subunit NF-L produces morphological alterations that resemble the pathology of human motor neuron disease. *Cell, 73*, 23–33.

24. Lee, M. K., Marszalek, J. R., & Cleveland, D. W. (1994). A mutant neurofilament subunit causes massive, selective motor neuron death: Implications for the pathogenesis of human motor neuron disease. *Neuron, 13*, 975–988.

25. Price, D. L., Cleveland, D. W., & Koliatsos, V. E. (1994). Motor neurone disease and animal models. *Neurobiol. Dis., 1*, 3–11.

26. Strong, M. J., & Garruto, R. M. (1994). Experimental paradigms of motor neuron degeneration. In M. L. Woodruff, A. J. Nonneman (Eds.), *Toxin-induced models of neurological disorders* (pp. 39–88). New York: Plenum Press.

27. Collard, J.-F., Côté, F., & Julien, J.-P. (1995). Defective axonal transport in a transgenic mouse model of amyotrophic lateral sclerosis. *Nature, 375*, 61–64.

28. Marszalek, J. R., Williamson, T. L., Lee, M. K., Xu, Z., Hoffman, P. N., Becher, M. W., Crawford, T. O., & Cleveland, D. W. (1996). Neurofilament subunit NF-H modulates axonal diameter by selectively slowing neurofilament transport. *J. Cell Biol., 135*, 711–724.

29. Gurney, M. E., Pu, H., Chiu, A. Y., Dal Canto, M. C., Polchow, C. Y., Alexander, D. D., Caliendo, J., Hentati, A., Kwon, Y. W., Deng, H.-X., Chen, W., Zhai, P., Sufit, R. L., & Siddique, T. (1994). Motor neuron degeneration in mice that express a human Cu,Zn superoxide dismutase mutation. *Science, 264*, 1772–1775.

30. Ripps, M. E., Huntley, G. W., Hof, P. R., Morrison, J. H., & Gordon, J. W. (1995). Transgenic mice expressing an altered murine superoxide dismutase gene provide an animal model of amyotrophic lateral sclerosis. *Proc. Natl. Acad. Sci. USA, 92*, 689–693.

31. Wong, P. C., Pardo, C. A., Borchelt, D. R., Lee, M. K., Copeland, N. G., Jenkins, N. A., Sisodia, S. S., Cleveland, D. W., & Price, D. L. (1995). An adverse property of a familial ALS-linked SOD1 mutation causes motor neuron disease characterized by vacuolar degeneration of mitochondria. *Neuron, 14*, 1105–1116.

32. Borchelt, D. R., Wong, P. C., Becher, M. W., Pardo, C. A., Lee, M. K., Zu, Z.-S., Thinkaran, G.,

Jenkins, N. A., Copeland, N. G., Sisodia, S. S., Cleveland, D. W., Price, D. L., & Hoffman, P. N. (1996). Early axonal abnormalities and axonal transport of mutant superoxide dismutase 1 in a transgenic model of familial amyotrophic lateral sclerosis. *J. Neurosci.,*   .

33. Dal Canto, M. C., & Gurney, M. E. (1994). The development of central nervous system pathology in a murine transgenic model of human amyotrophic lateral sclerosis. *Am. J. Pathol., 145,* 1–9.

34. Dal Canto, M. C., & Gurney, M. E. (1994). Development of central nervous system pathology in a murine transgenic model of human amyotrophic lateral sclerosis. *Am. J. Pathol., 145,* 1271–1280.

35. Tu, P.-H., Raju, P., Robinson, K. A., Gurney, M. E., Trojanowski, J. Q., & Lee, V. M.-Y. (1996). Transgenic mice carrying a human mutant superoxide dismutase transgene develop neuronal cytoskeletal pathology resembling human amyotrophic lateral sclerosis lesions. *Proc. Natl. Acad. Sci. USA, 93,* 3155–3160.

36. Gurney, M. E., Cuttings, F. B., Zhai, P., Doble, A., Taylor, C. P., Andrus, P. K., & Hall, E. D. (1996). Benefit of vitamin E, riluzole, and gabapentin in a transgenic model of familial amyotrophic lateral sclerosis. *Ann. Neurol., 39,* 147–157.

37. Bruijn, L. I., Becher, M. W., Lee, M. K., Anderson, K. L., Jenkins, N. A., Copeland, N. G., Sisodia, S. S., Rothstein, J. D., Borchelt, D. R., Price, D. L., & Cleveland, D. W. (1997). ALS-linked SOD1 mutant G85R mediates damage to astrocytes and promotes rapidly progressive disease with SOD1-containing inclusions. *Neuron, 18,* 327–338.

38. Borchelt, D. R., Lee, M. K., Slunt, H. H., Guarnieri, M., Xu, Z.-S., Wong, P. C., Brown, R. H., Jr., Price, D. L., Sisodia, S. S., & Cleveland, D. W. (1994). Superoxide dismutase 1 with mutations linked to familial amyotrophic lateral sclerosis possesses significant activity. *Proc. Natl. Acad. Sci. USA, 91,* 8292–8296.

39. Rabizadeh, S., Butler Gralla, E., Borchelt, D. R., Gwinn, R., Selverstone Valentine, J., Sisodia, S., Wong, P., Lee, M., Hahn, H., & Bredesen, D. E. (1995). Mutations associated with amyotrophic lateral sclerosis convert superoxide dismutase from an antiapoptotic gene to a proapoptotic gene: Studies in yeast and neural cells. *Proc. Natl. Acad. Sci. USA, 92,* 3024–3028.

40. Borchelt, D. R., Guarnieri, M., Wong, P. C., Lee, M. K., Slunt, H. S., Xu, Z., Sisodia, S. S., Price, D. L., & Cleveland, D. W. (1995). Superoxide dismutase 1 subunits with mutations linked to familial amyotrophic lateral sclerosis do not affect wild-type subunit function. *J. Biol. Chem., 270,* 3234–3238.

41. Reaume, A. G., Elliott, J. L., Hoffman, E. K., Kowall, N. W., Ferrante, R. J., Siwek, D. F., Wilcox, H. M., Flood, D. G., Beal, M. F., Brown, R. H., Jr., Scott, R. W., & Snider, W. D. (1996). Motor neurons in Cu/Zn superoxide dismutase-deficient mice develop normally but exhibit enhanced cell death after axonal injury. *Nature Genet., 13,* 43–47.

42. Scott, R. (personal observation). Cephalon. West Chester, PA.

43. Liu, X. F., & Cizewski Culotta, V. (1994). The requirement for yeast superoxide dismutase is bypassed through mutations in BSD2, a novel metal homeostasis gene. *Mol. Cell Biol., 14,* 7037–7045.

44. Wiedau-Pazos, M., Goto, J. J., Rabizadeh, S., Gralla, E. B., Roe, J. A., Lee, M. K., Valentine, J. S., & Bredesen, D. E. (1996). Altered reactivity of superoxide dismutase in familial amyotrophic lateral sclerosis. *Science, 271,* 515–518.

45. Beckman, J. S., Carson, M., Smith, C. D., & Koppenol, W. H. (1993). ALS, SOD and peroxynitrite. *Nature, 364,* 584.

46. Bruijn, L. I., Beal, M. E., Becher, M. W., Schulz, J. B., Wong, P. C., Price, D. L., & Cleveland, D. W. (1997). Elevated free nitrotyrosine levels, but not protein-bound nitrotyrosine or hydroxyl radicals, throughout amyotrophic lateral sclerosis (ALS)-like disease implicate tyrosine nitration as an aberrant in vivo property of one familial ALS-linked superoxide dismutase 1 mutant. *Proc. Natl. Acad. Sci. USA, 94,* 7606–7611.

47. Pardo, C. A., Xu, Z., Borchelt, D. R., Price, D. L., Sisodia, S. S., & Cleveland, D. W. (1995). Superoxide dismutase is an abundant component in cell bodies, dendrites, and axons of motor neurons and in a subset of other neurons. *Proc. Natl. Acad. Sci. USA, 92,* 954–958.

48. McKhann, G., Drachman, D., Folstein, M., Katzman, R., Price, D., & Stadlan, E. M. (1984). Clinical diagnosis of Alzheimer's disease: Report of the NINCDS-ADRDA Work Group under the auspices of the Department of Health and Human Services Task Force on Alzheimer's Disease. *Neurology, 34,* 939–944.

49. Evans, D. A., Funkenstein, H. H., Albert, M. S., Scherr, P. A., Cook, N. R., Chown, M. J., Hebert, L. E., Hennekens, C. H., & Taylor, J. O. (1989). Prevalence of Alzheimer's disease in a community population of older persons. Higher than previously reported. *JAMA, 262,* 2551–2556.

50. Price, D. L. (1995). Aging and age-associated neurodegenerative disorders. *Curr. Opin. Neurol., 8,* 253–255.

51. Ernst, R. L., & Hay, J. W. (1994). The US economic and social costs of Alzheimer's disease revisited. *Am. J. Public Health, 84,* 1261–1264.

52. Schellenberg, G. D. (1995). Progress in Alzheimer's disease genetics. *Curr. Opin. Neurol., 8,* 262–267.

53. Schellenberg, G. D. (1995). Genetic dissection of Alzheimer's disease, a heterogeneous disorder. *Proc. Natl. Acad. Sci. USA, 92,* 8552–8559.

54. Glenner, G. G., & Wong, C. W. (1984). Alzheimer's disease: Initial report of the purification and characterization of a novel cerebrovascular amyloid protein. *Biochem. Biophys. Res. Commun.,* 885–890.

55. Kang, J., Lemaire, H.-G., Unterbeck, A., Salbaum, J. M., Masters, C. L., Grzeschik, K.-H., Multhaup, G., Beyreuther, K., & Müller-Hill, B. (1987). The precursor of Alzheimer's disease amyloid A4 protein resembles a cell-surface receptor. *Nature, 325,* 733–736.

56. Masters, C. L., Simms, G., Weinman, N. A., Multhaup, G., McDonald, B. L., & Beyreuther, K. (1985). Amyloid plaque core protein in Alzheimer disease and Down syndrome. *Proc. Natl. Acad. Sci. USA, 82,* 4245–4249.

57. Sisodia, S. S. (1992). β-Amyloid precursor protein cleavage by a membrane-bound protease. *Proc. Natl. Acad. Sci. USA, 89,* 6075–6079.

58. Esch, F. S., Keim, P. S., Beattie, E. C., Blacher, R. W., Culwell, A. R., Oltersdorf, T., McClure, D., & Ward, P. J. (1990). Cleavage of amyloid β peptide during constitutive processing of its precursor. *Science, 248,* 1122–1124.

59. Sisodia, S. S., Koo, E. H., Beyreuther, K., Unterbeck, A., & Price, D. L. (1990). Evidence that β-amyloid protein in Alzheimer's disease is not derived by normal processing. *Science, 248,* 492–495.

60. Citron, M., Oltersdorf, T., Haass, C., McConlogue, L., Hung, A. Y., Seubert, P., Vigo-Pelfrey, C., Lieberburg, L., & Selkoe, D. J. (1992). Mutation of the β-amyloid precursor protein in familial Alzheimer's disease increases β-protein production. *Nature, 360,* 672–674.

61. Haass, C., Schlossmacher, M. G., Hung, A. Y., Vigo-Pelfrey, C., Mellon, A., Ostaszewski, B. L., Lieberburg, I., Koo, E. H., Schenk, D., Teplow, D. B., & Selkoe, D. J. (1992). Amyloid β-peptide is produced by cultured cells during normal metabolism. *Nature, 359,* 322–325.

62. Chartier-Harlin, M.-C., Crawford, F., Houlden, H., Warren, A., Hughes, D., Fidani, L., Goate, A., Rossor, M., Roques, P., Hardy, J., & Mullan, M. (1991). Early-onset Alzheimer's disease caused by mutations at codon 717 of the β-amyloid precursor protein gene. *Nature, 353,* 844–846.

63. Goate, A., Chartier-Harlin, M.-C., Mullan, M., Brown, J., Crawford, F., Fidani, L., Giuffra, L., Haynes, A., Irving, N., James, L., Mant, R., Newton, P., Rooke, K., Roques, P., Talbot, C., Pericak-Vance, M., Roses, A., Williamson, R., Rossor, M., Owen, M., & Hardy, J. (1991). Segregation of a missense mutation in the amyloid precursor protein gene with familial Alzheimer's disease. *Nature, 349,* 704–706.

64. Hendricks, L., van Duijn, C. M., Cras, P., Cruts, M., Van Hul, W., van Harskamp, F., Warren, A., McInnis, M. G., Antonarakis, S. E., Martin, J.-J., Hofman, A., & Van Broeckhoven, C. (1992). Presenile dementia and cerebal haemorrhage linked to a mutation at codon 692 of the β-amyloid precursor protein gene. *Nature Genet., 1,* 218–221.

65. Mullan, M., Crawford, F., Axelman, K., Houlden, H., Lillius, L., Winblad, B., & Lannfelt, L.

(1992). A pathogenic mutation for probable Alzheimer's disease in the APP gene at the N-terminus of β-amyloid. *Nature Genet., 1,* 345–347.

66. Suzuki, N., Cheung, T. T., Cai, X.-D., Odaka, A., Otvos, L., Jr., Eckman, C., Golde, T. E., & Younkin, S. G. (1994). An increased percentage of long amyloid β protein secreted by familial amyloid β protein precursor (βAPP717) mutants. *Science, 264,* 1336–1340.

67. Alzheimer's Disease Collaborative Group. (1995). The structure of the presenilin 1 (S182) gene and identification of six novel mutations in early onset AD families. *Nature Genet., 11,* 219–222.

68. Koo, E. H., Sisodia, S. S., Archer, D. R., Martin, L. J., Weidemann, A., Beyreuther, K., Fischer, P., Masters, C. L., & Price, D. L. (1990). Precursor of amyloid protein in Alzheimer disease undergoes fast anterograde axonal transport. *Proc. Natl. Acad. Sci. USA, 87,* 1561–1565.

69. Sisodia, S. S., Koo, E. H., Hoffman, P. N., Perry, G., & Price, D. L. (1993). Identification and transport of full-length amyloid precursor proteins in rat peripheral nervous system. *J. Neurosci., 13,* 3136–3142.

70. Yamazaki, T., Selkoe, D. J., & Koo, E. H. (1995). Trafficking of cell surface β-amyloid precursor protein: Retrograde and transcytotic transport in cultured neurons. *J. Cell Biol., 129,* 431–442.

71. St. George-Hyslop, P. H., Haines, P., Rogaev, E., Mortilla, M., Vaula, G., Pericak-Vance, M., Foncin, J.-F., Montesi, M., Bruni, A., Sorbi, S., Rainero, I., Pinessi, L., Pollen, D., Polinsky, R., Nee, L., Kennedy, J., Macciardi, F., Rogaeva, E., Liang, Y., Alexandrova, N., Lukiw, W., Schlumpf, K., Tanzi, R., Tsuda, T., Farrer, L., Cantu, J.-M., Duara, R., Amaducci, L., Bergamini, L., Gusella, J., Roses, A., & Crapper McLachlan, D. (1992). Genetic evidence for a novel familial Alzheimer's disease locus on chromosome 14. *Nature Genet., 2,* 330–334.

72. Cai, X.-D., Golde, T. E., & Younkin, S. G. (1993). Release of excess amyloid β protein from a mutant amyloid β protein precursor. *Science, 259,* 514–516.

73. Sherrington, R., Rogaev, E. I., Liang, Y., Rogaeva, E. A., Levesque, G., Ikeda, M., Chi, H., Lin, C., Li, G., Holman, K., Tsuda, T., Mar, L., Foncin, J.-F., Bruni, A. C., Montesi, M. P., Sorbi, S., Rainero, I., Pinessi, L., Nee, L., Chumakov, I., Pollen, D., Brookes, A., Sanseau, P., Polinsky, R. J., Wasco, W., Da Silva, H. A. R., Haines, J. L., Pericak-Vance, M. A., Tanzi, R. E., Roses, A. D., Fraser, P. E., Rommens, J. M., & St. George-Hyslop, P. H. (1995). Cloning of a gene bearing missense mutations in early-onset familial Alzheimer's disease. *Nature, 375,* 754–760.

74. Levy-Lahad, E., Wasco, W., Poorkaj, P., Romano, D. M., Oshima, J., Pettingell, W. H., Yu, C.-E., Jondro, P. D., Schmidt, S. D., Wang, K., Crowley, A. C., Fu, Y.-H., Guenette, S. Y., Galas, D., Nemens, E., Wijsman, E. M., Bird, T. D., Schellenberg, G. D., & Tanzi, R. E. (1995). Candidate gene for the chromosome 1 familial Alzheimer's disease locus. *Science, 269,* 973–977.

75. Levy-Lahad, E., Wijsman, E. M., Nemens, E., Anderson, L., Goddard, K. A. B., Weber, J. L., Bird, T. D., & Schellenberg, G. D. (1995). A familial Alzheimer's disease locus on chromosome 1. *Science, 269,* 970–973.

76. Li, J., Ma, J., & Potter, H. (1995). Identification and expression analysis of a potential familial Alzheimer disease gene on chromosome 1 related to AD3. *Proc. Natl. Acad. Sci. USA, 92,* 12180–12184.

77. Rogaev, E. I., Sherrington, R., Rogaeva, E. A., Levesque, G., Ikeda, M., Liang, Y., Chi, H., Lin, C., Holman, K., Tsuda, T., Mar, L., Sorbi, S., Nacmias, B., Piacentini, S., Amaducci, L., Chumakov, I., Cohen, D., Lannfelt, L., Fraser, P. E., Rommens, J. M., & St. George-Hyslop, P. H. (1995). Familial Alzheimer's disease in kindreds with missense mutations in a gene on chromosome 1 related to the Alzheimer's disease type 3 gene. *Nature, 376,* 775–778.

78. Doan, A., Thinakaran, G., Borchelt, D. R., Slunt, H. H., Ratovitsky, T., Podlisny, M., Selkoe, D. J., Seeger, M., Gandy, S. E., Price, D. L., & Sisodia, S. S. (1996). Protein topology of presenilin 1. *Neuron, 17,* 1023–1030.

79. Lee, M. K., Slunt, H. H., Martin, L. J., Thinakaran, G., Kim, G., Gandy, S. E., Seeger, M., Koo, E., Price, D. L., & Sisodia, S. S. (1996). Expression of presenilin 1 and 2 (PS1 and PS2) in human and murine tissues. *J. Neurosci., 16,* 7513–7525.

80. Thinakaran, G., Teplow, D. B., Siman, R., Greenberg, B., & Sisodia, S. S. (1996). Metabolism of the "Swedish" APP variant in Neuro2A (N2a) cells: Evidence that cleavage at the "β-secretase" site occurs in the Golgi apparatus. *J. Biol. Chem., 271,* 9390–9397.

81. Kovacks, D. M., Fausett, H. J., Page, K. J., Kim, T.-W., Moir, R. D., Merriam, D. E., Hollister, R. D., Hallmark, O. G., Mancini, R., Felsenstein, K. M., Hyman, B. T., Tanzi, R. E., & Wasco, W. (1996). Alzheimer-associated presenilins 1 and 2: Neuronal expression in brain and localization to intracellular membranes in mammalian cells. *Nature Med., 2,* 224–229.

82. Levitan, D., & Greenwald, I. (1995). Facilitation of lin-12-mediated signalling by sel-12, a *Caenorhabditis elegans* S182 Alzheimer's disease gene. *Nature, 377,* 351–354.

83. Levitan, D., Doyle, T. G., Brousseau, D., Lee, M. K., Thinakaran, G., Slunt, H. H., Sisodia, S. S., & Greenwald, I. (1996). Assessment of normal and mutant human presenilin function in *Caenorhabditis elegans. Proc. Natl. Acad. Sci. USA, 93,* 14940–14944.

84. Wong, P. C., Zheng, H., Chen, H., Becher, M. W., Sirinathsinghji, D. J. S., Trumbauer, M. E., Chen, H. Y., Price, D. L., Van der Ploeg, L. H. T., & Sisodia, S. S. (1997). Presenilin 1 is required for Notch1 and Dll1 expression in the paraxial mesoderm. *Nature, 387,* 288–292.

85. Shen, J., Bronson, R. T., Chen, D. F., Xia, W., Selkoe, D. J., & Tonegawa, S. (1997). Skeletal and CNS defects in presenilin-1-deficient mice. *Cell, 89,* 629–639.

86. Arnold, S. E., Hyman, B. T., Flory, J., Damasio, A. R., & Van Hoesen, G. W. (1991). The topographical and neuroanatomical distribution of neurofibrillary tangles and neuritic plaques in the cerebral cortex of patients with Alzheimer's disease. *Cereb. Cortex., 1,* 103–116.

87. Braak, H., & Braak, E. (1994). Pathology of Alzheimer's disease. In D. B. Calne (Ed.), *Neurodegenerative diseases* (pp. 585–613). Philadelphia: Saunders.

88. Price, D. L., Kawas, C. H., & Sisodia, S. S. (1996). Aging of the brain and dementia of the Alzheimer's type. In E. R. Kandel, J. H. Schwartz, &T. M. Jessell (Eds.), *Principles of neural science.* New York: Elsevier.

89. Lee, V. M.-Y. (1995). Disruption of the cytoskeleton in Alzheimer's disease. *Curr. Opin. Neurobiol., 5,* 663–668.

90. Goedert, M., Jakes, R., Spillantini, M. G., Hasegawa, M., Smith, M. J., & Crowther, R. A. (1996). Assembly of microtubule-associated protein tau into Alzheimer-like filaments induced by sulphated glycosaminoglycans. *Nature, 383,* 550–553.

91. Cotman, C. W., & Su, J. H. (1996). Mechanisms of neuronal death in Alzheimer's disease. *Brain Pathol., 6,* 493–506.

92. Troncoso, J. C., Sukhov, R. R., Kawas, C. H., & Koliatsos, V. E. (1996). In situ labeling of dying cortical neurons in normal aging and in Alzheimer's disease: Correlations with senile plaques and disease progression. *J. Neuropathol. Exp. Neurol., 55,* 1134–1142.

93. Higgins, L. S., Holtzman, D. M., Rabin, J., Mobley, W. C., & Cordell, B. (1994). Transgenic mouse brain histopathology resembles early Alzheimer's disease. *Ann. Neurol., 35,* 598–607.

94. Price, D. L., & Sisodia, S. S. (1994). Cellular and molecular biology of Alzheimer's disease and animal models. *Annu. Rev. Med., 45,* 435–446.

95. Howland, D. S., Savage, M. J., Huntress, F. A., Wallace, R. E., Schwartz, D. A., Loh, T., Melloni, R. H., Jr., DeGennaro, L. J., Greenberg, B. D., Siman, R., Swanson, M. E., & Scott, R. W. (1995). Mutant and native human β-amyloid precursor proteins in transgenic mouse brain. *Neurobiol. Aging, 16,* 685–699.

96. Hsiao, K. K., Borchelt, D. R., Olson, K., Johannsdottir, R., Kitt, C., Yunis, W., Xu, S., Eckman, C., Younkin, S., Price, D., Iadecola, C., Clark, H. B., & Carlson, G. (1995). Age-related CNS disorder and early death in transgenic FVB/N mice overexpressing Alzheimer amyloid precursor proteins. *Neuron, 15,* 1203–1218.

97. LaFerla, F. M., Tinkle, B. T., Bieberich, C. J., Haudenschild, C. C., & Jay, G. (1995). The Alzheimer's Aβ peptide induces neurodegeneration and apoptotic cell death in transgenic mice. *Nature Genet., 9,* 21–30.

98. Moran, P. M., Higgins, L. S., Cordell, B., & Moser, P. C. (1995). Age-related learning deficits in transgenic mice expressing the 751-amino acid isoform of human β-amyloid precursor protein. *Proc. Natl. Acad. Sci. USA, 92,* 5341–5345.

99. Oster-Granite, M. L., McPhie, D. L., Greenan, J., & Neve, R. L. (1996). Age-dependent neuronal and synaptic degeneration in mice transgenic for the C terminus of the amyloid precursor protein. *J. Neurosci., 16,* 6732–6741.

100. Games, D., Adams, D., Alessandrini, R., Barbour, R., Berthelette, P., Blackwell, C., Carr, T., Clemens, J., Donaldson, T., Gillespie, F., Guido, T., Hagopian, S., Johnson-Wood, K., Khan, K., Lee, M., Leibowitz, P., Lieberburg, I., Little, S., Masliah, E., McConologue, L., Montoya-Zavala, M., Mucke, L., Paganini, L., Penniman, E., Power, M., Schenk, D., Seubert, P., Snyder, B., Soriano, F., Tan, H., Vitale, J., Wadsworth, S., Wolozin, B., & Zhao, J. (1995). Alzheimer-type neuropathology in transgenic mice overexpressing V717F β-amyloid precursor protein. *Nature, 373,* 523–527.

101. Hsiao, K., Chapman, P., Nilsen, S., Eckman, C., Harigaya, Y., Younkin, S., Yang, F., & Cole, G. (1996). Correlative memory deficts, Aβ elevation and amyloid plaques in transgenic mice. *Science, 274,* 99–102.

102. Borchelt, D. R., Ratovitski, T., Van Lare, J., Lee, M. K., Gonzales, V. B., Jenkins, N. A., Copeland, N. G., Price, D. L., & Sisodia, S. S. (1997). Accelerated amyloid deposition in the brains of transgenic mice co-expressing mutant presenilin 1 and amyloid precursor proteins. *Neuron, 19,* 939–945.

103. Borchelt, D. R., Thinakaran, G., Exkman, C. B., Lee, M. K., Davenport, E., Ratovitsky, T., Prada, C.-M., Kim, G., Seekins, S., Yager, D., Slunt, H. H., Wang, R., Seeger, M., Levey, A. I., Gandy, S. E., Copeland, N. G., Jenkins, N. A., Price, D. L., Younkin, S. G., & Sisodia, S. S. (1996). Familial Alzheimer's disease-linked presenilin 1 variants elevate Aβ1-42/1-40 ratio in vitro and in vivo. *Neuron, 17,* 1005–1013.

104. Duff, K., Eckman, C., Zehr, C., Yu, X., Prada, C.-M., Perez-Tur, J., Hutton, M., Buee, L., Harigaya, Y., Yager, D., Morgan, D., Gordon, M. N., Holcomb, L., Refolo, L., Zenk, B., Hardy, J., & Younkin, S. (1996). Increased amyloid-β42(43) in brains of mice expressing mutant presenilin 1. *Nature, 383,* 710–713.

105. Thinakaran, G., Borchelt, D. R., Lee, M. K., Slunt, H. H., Spitzer, L., Kim, G., Ratovitski, T., Davenport, F., Nordstedt, C., Seeger, M., Hardy, J., Levey, A. I., Gandy, S. E., Jenkins, N., Copeland, N., Price, D. L., & Sisodia, S. S. (1996). Endoproteolysis of presenilin 1 and accumulation of processed derivatives in vivo. *Neuron, 17,* 181–190.

106. Levy, E., Carman, M. D., Fernandez-Madrid, I. J., Power, M. D., Lieberburg, I., van Duinen, S. G., Bots, G. T. A. M., Luyendijk, W., & Frangione, B. (1990). Mutation of the Alzheimer's disease amyloid gene in hereditary cerebral hemorrage, Dutch type. *Science, 248,* 1124–1126.

107. Van Broeckhoven, C., Haan, J., Bakker, E., Hardy, J. A., Van Hul, W., Wehnert, A., Vegter-Van der Vlis, M., & Roos, R. A. C. (1990). Amyloid β protein precursor gene and hereditary cerebral hemorrhage with amyloidosis (Dutch). *Science, 248,* 1120–1122.

108. Haass, C., Hung, A. Y., Selkoe, D. J., & Teplow, D. B. (1994). Mutations associated with a locus for familial Alzheimer's disease result in alternative processing of amyloid β-protein precursor. *J. Biol. Chem. 269,* 17741–17748.

109. Wisniewski, T., Ghiso, J., & Frangione, B. (1991). Peptides homologous to the amyloid protein of Alzheimer's disease containing a glutamine for glutamic acid substitution have accelerated amyloid fibril formation. *Biochem. Biophys. Res. Commun.,. 179,* 1247–1254.

110. Burdick, D., Soreghan, B., Kwon, M., Kosmoski, J., Knauer, M., Henschen, A., Yates, J., Cotman, C., & Glabe, C. (1992). Assembly and aggregation properties of synthetic Alzheimer's A4/β amyloid peptide analogs. *J. Biol. Chem., 267,* 546–554.

111. Jarrett, J. T., & Lansbury, P. T., Jr. (1993). Seeding "one-dimensional crystallization" of amyloid: A pathogenic mechanism in Alzheimer's disease and scrapie? *Cell, 73,* 1055–1058.

112. Iwatsubo, T., Odaka, A., Suzuki, N., Mizusawa, H., Nukina, N., & Ihara, Y. (1994). Visualization of Aβ42(43)-positive and Aβ40-positive senile plaques with end-specific Aβ-monoclonal antibodies: Evidence that an initially deposited Aβ species is Aβ1-42(43). *Neuron, 13,* 45–53.

113. Iwatsubo, T., Mann, D. M. A., Odaka, A., Suzuki, N., & Ihara, Y. (1995). Amyloid b protein (Aβ) deposition: Aβ42(43) preceeds Aβ40 in Down syndrome. *Ann. Neurol., 37,* 294–299.

114. Shinkai, Y., Yoshimura, M., Ito, Y., Odaka, A., Suzuki, N., Yanagisawa, K., & Ihara, Y. (1995). Amyloid β-proteins 1–40 and 1–42(43) in the soluble fraction of extra- and intracranial blood vessels. *Ann. Neurol., 38,* 421–428.

115. Lemere, C. A., Blusztajn, J. K., Yamaguchi, H., Wisniewski, T., Saido, T. C., & Selkoe, D.J. (1996). Sequence of deposition of heterogeneous amyloid β-peptides and APO E in Down syndrome: Implications for initial events in amyloid plaque formation. *Neurobiol. Dis., 3,* 16–32.

116. Scheuner, D., Eckman, C., Jensen, M., Song, X., Citron, M., Suzuki, N., Bird, T. D., Hardy, J., Hutton, M., Kukull, W., Larson, E., Levy-Lahad, E., Viitanen, M., Peskind, E., Poorkaj, P., Schellenberg, G., Tanzi, R., Wasco, W., Lannfelt, L., Selkoe, D., & Younkin, S. (1996). Secreted amyloid β-protein similar to that in the senile plaques of Alzheimer's disease is increased in vivo by the presenilin 1 and 2 and APP mutations linked to familial Alzheimer's disease. *Nature Med., 2,* 864–852.

117. Pike, C. J., Burdick, D., Walencewicz, A. J., Glabe, C. G., & Cotman, C. W. (1993). Neurodegeneration induced by β-amyloid peptides in vitro: The role of peptide assembly state. *J. Neurosci., 13,* 1676–1687.

118. Koh, J. Y., Yang, L. L., & Cotman, C. W. (1990). β-Amyloid protein increases the vulnerability of cultured cortical neurons to excitotoxic damage. *Brain Res., 533,* 315–320.

119. Loo, D. T., Copani, A., Pike, C. J., Whittemore, E. R., Walencewicz, A. J., & Cotman, C. W. (1993). Apoptosis is induced by β-amyloid in cultured central nervous system neurons. *Proc. Natl. Acad. Sci. USA, 90,* 7951–7955.

120. Mattson, M. P., Barger, S. W., Cheng, B., Lieberburg, I., Smith-Swintosky, V. L., & Rydel, R. E. (1993). β-Amyloid precursor protein metabolites and loss of neuronal Ca^{2+} homeostasis in Alzheimer's disease. *Trends Neurosci., 16,* 409–414.

121. Good, P. F., Werner, P., Hsu, A., Olanow, C. W., & Perl, D. P. (1996). Evidence for neuronal oxidative damage in Alzheimer's disease. *Am. J. Pathol., 149,* 21–28.

122. Yankner, B. A. (1996). Mechanisms of neuronal degeneration in Alzheimer's disease. *Neuron, 16,* 921–932.

8

TOWARD A GENETIC ANALYSIS OF UNUSUALLY SUCCESSFUL NEURAL AGING

GEORGE M. MARTIN

Departments of Pathology and Genetics and
The Alzheimer's Disease Research Center
University of Washington, Seattle, Washington 98195

The genetic analysis of neural function during aging has been dominated by an interest in susceptibility to late life-disease, particularly to dementias of the Alzheimer type. Might it be possible to identify genetic factors responsible for unusually robust retention of various neural structures and functions during aging? Recent progress in both molecular and statistical genetics suggests that neurobiologists might now profitably address this question. Inferences from the evolutionary biological theory of aging would argue that measurements of function beginning at around 45 years of age would be relevant, thus providing potential immediate access to the DNA of three generations and relevant aging phenotypes of two generations. There are formidable difficulties inherent in such research, however, notably the likelihood that the phenotypes of greatest interest, such as aspects of declarative memory, are subject to highly polygenic controls.

Genetic approaches to problems in biology have the potential to address primary mechanisms. My own neuroscience-related research (and, in fact, much of my more general gerontological research), as well as the research activities of the great majority of my colleagues in medical genetics, has been focused upon gene action that leads to premature functional declines. These efforts have been exemplified by the surprisingly rapid progress on the genetics of dominant forms of early-onset dementias of the Alzheimer type (DAT). By contrast, there has been a relative paucity of research on the genetic basis of unusually well-preserved neural function. The purpose of this brief overview is to consider several approaches to this important question and to encourage wider and deeper research efforts in this neglected but exceedingly basic area of neurobiology.

A BRIEF PRIMER OF EVOLUTIONARY BIOLOGICAL ASPECTS OF RELEVANT GENE ACTION

Although increasingly embraced by biogerontologists, the principles of evolutionary biology and of population genetics have been largely ignored by the community of researchers interested in genetic contributions to late-life cognitive functions. It will be useful to review these principles as a basis for a fuller understanding of why neurologists, psychiatrists, neuropsychologists, and neuropathologists are presented with such striking variations in the patterns of neural aging.

Evolutionary biologists believe that they have a perfectly plausible explanation as to *why* we age, an explanation that is largely independent of the specific mechanisms of *how* we age. Simply put, we age because the collections of senescent phenotypes that unfold late in the life course have escaped the force of natural selection. As background for this conclusion, let us consider the evolution of the life histories of two different mammalian populations: One that has evolved a comparatively short life span and a closely related species that has evolved a significantly longer life span. Like most mammals, we shall assume that both species are iteroparous (i.e., they experience repeated bouts of reproduction). This leads to age-structured populations such that, at any given time, there are individuals of various ages with the potential to contribute to the gene pool of subsequent generations. The differences between the life history traits of these two species can be attributed to the ecological circumstances of their evolutionary history. Short-lived creatures will have evolved under conditions of high environmental hazards, such as predation, starvation, droughts, accidents, and infection. These conditions will select for genetic variation that leads to such life history traits as rapid development and sexual maturation and the generation of large numbers of progeny over short periods of time. In the wild, only a very few individuals would be likely to survive for several years. As such, any gene variant (good or bad) that does not reach phenotypic expression for several years would be essentially ignored by natural selection. Any potential contribution to the gene pool of such alleles by these rare individuals would thus be swamped by the very large numbers of alleles contributed by the numerous younger members of the population. Given the opportunity, nature would indeed act to select for alleles that contribute to the enhanced maintenance of function late in the life course and against those alleles that lead to reduced function. Such an opportunity arises when the environmental hazards are decreased, allowing greater numbers of individuals to live for longer periods of time. Field observations by Austad[1] on two sibling species of Virginia opossums (*Didelphis virginiana*), one the primordial mainland species and the other a derivative that had evolved under a relatively protected insular environment over a period of approximately 4000 years, gave results entirely consistent with this formulation. The insular species was found to have an actuarial rate of aging approximately half that of the mainland species, together with other life history differences.

There are two main types of gene action that are thought to be the basis for both the interspecific and intraspecific variations that one observes in late-life phenotypes such as altered neural functions. The first type is often referred to as "mutation accumulation" and was elucidated by Medawar[2] after suggestions by Haldane. Medawar was referring to constitutional mutations, not somatic mutations. Moreover, these represent a special class of constitutional mutations— those whose effects do not reach a phenotypic threshold of expression until late in life, at ages when the force of natural selection has become negligible. Mutations are predominately random and rare. Their prevalence would therefore be expected to be a function of genetic drift and their effects quite different. Medawar imagined that certain of these mutations might have initially interfered with reproductive fitness and thus have been subject to natural selection. He suggested, however, that there would be selective pressure for the emergence of mutations and/or polymorphisms at other loci leading to the postponement of deleterious effects such that these phenotypes would eventually have escaped the force of natural selection. (Geneticists refer to such alleles as "suppressors.") These constitutional mutations and their putative suppressors, being rare and random, are idiosyncratic. Their frequencies can be expected to vary substantially among different ethnic-geographic groups. I therefore like to refer to them as leading to "*private*" modulations of senescence; the initial rare mutations would enhance a particular aspect of senescence, while the secondary rare suppressors would be protective.

The second major type of gene action is an example of an evolutionary "trade-off." It has been given the tongue-twisting names of "antagonistic pleiotropy" or "negative pleiotropy." It involves gene actions early in the life span that lead to enhanced reproductive fitness (as is the case for most alleles), but which have deleterious "downstream" late effects (actions that presumably would be the case for only a minor subset of these alleles). Such alleles would be likely to become prevalent within a number of different populations, given their contributions to reproductive fitness. It would be difficult for a population to lose such alleles once they have appeared. One could therefore consider this class of gene action as leading to "*public*" modulations of senescence. This type of gene action may be more likely to involve polymorphic forms of a gene, rather than mutations of a gene. Polymorphisms can be defined as the regular and simultaneous occurrence in the same population of two or more alleles, each with a frequency greater than that which one would expect to result from recurrent mutation. By convention, the less frequent allele (for the case of biallelic polymorphisms) should have a frequency of about 1% or 2%. This is in contrast to the frequencies of de novo constitutional mutations, which may range from around 1 per 1000 to around 1 per 10 million. A typical frequency would be circa 1 per million gametes.

SOME IMPLICATIONS OF THE EVOLUTIONARY BIOLOGICAL THEORY OF AGING FOR HUMAN NEUROBIOLOGY

The first point to make is a most optimistic observation. Evolutionary theory predicts that life span potential is highly plastic. Laboratory experiments involving both direct and indirect selections for enhanced life spans have provided dramatic evidence in support of that theory. These increments of both mean and maximum life spans must surely have been accompanied by various improvements in the efficiency and durability of the maintenance of macromolecular integrity and of physiological homeostatic mechanisms, including those of importance to central nervous system (CNS) function. If we can learn more about the mechanisms whereby natural selection achieves such enhanced function, then there is the potential for positive interventions in our own species.

A second point, however, does not lead to such optimism. Despite the views of some developmental biologists, there is no evidence that aging is simply a logical continuation of development, by which there are a few "master" sequential switches in gene expression that can be defined as determinative and "programmed." There are no "aging genes" or "master switches" leading to aging (or, more properly, to senescence). To a considerable extent, we must be dealing with stochastic processes. Indeed, experimental gerontologists regularly observe enormous variations in function and in longevity amongst genetically defined organisms maintained under apparently identical environments. Moreover, for any complex phenotype, like those associated with the differential maintenance of various neural functions, a very large number of genes are likely to be involved, both those of the constitutional mutation class and those of the "trade-off" class described above. About 20 years ago, I[3] tabulated the known or provisional genetic loci involved in the emergence of dementias and/or relevant types of neuropathology in man. There were 55 such loci among the total of 2336 listed in the 1975 edition of Victor McKusick's *Mendelian Inheritance of Man.*[4] The proportion of such genes among all known or provisional genes was therefore 2.35%. Assuming we have around 100,000 genes, this would give an estimate of some 2350 genes with the potential to modulate our propensity to develop various types of dementias and associated neuropathologies. For each deleterious mutation or polymorphism at such loci, there may exist exceptionally "good" alleles that provide exceptional protection against precipitous declines in particular domains of cognitive function.

The fact that allelic variation at many loci can modulate a particular phenotype does not necessarily lead to the conclusion that numerous independent mechanisms are involved in the genesis of those phenotypes. Take the free radical theory of aging, for example (now generally referred to as the oxidative damage or oxidative stress theory of aging, because not all reactive oxygen species are free radicals). One can make a list of hundreds of genes that potentially modulate the relative success an organism may have in dealing with this by-product of our aerobic style of lifestyle. These genes can be divided into seven broad categories: (1) those modulating the rates of genesis of free radicals; (2) those that code for enzymes that scavenge free radicals; (3) those that regulate the metabolism of free radical scavengers that are not enzymes; (4) those that determine the degree of redundancy of macromolecular targets of free radical damage; (5) those that determine the molecular structure and, hence, the relative vulnerability of the targets of these free radicals; (6) those that are involved in the reversal, repair, and tolerance of macromolecular damage; and (7) those that modulate the cellular proliferative responses to free radical damage.

It is possible that the genetic complexity underlying such basic aspects of gene action (actions that many believe to be highly relevant to the maintenance of human cognitive function) can be unraveled by experiments with model organisms such as yeast, nematodes and fruit flies. Indeed, current evidence does indicate that mutants selected for enhanced life spans share certain phenotypes, notably an enhanced resistance to oxidative stress, thermal stress, and ultraviolet light. There are grounds for caution, however. Let us return to the notion of antagonistic pleiotropy. We know, for example, that there are striking differences, even among mammalian species, in mating behaviors and, hence, in the particular subset of genetic loci, variation at which is of major importance in determining the success of mating. Thus, the negative late-life effects of such different genes are likely also to be species specific. By the same token, alleles whose effects are less deleterious also are likely to be species specific. In the last analysis, then, we may be faced with the proposition that "the proper study of man is man," at least with respect to the types of gene action of greatest interest to the subject of our workshop, namely the higher cognitive functions of man.

The final implication of the evolutionary theory of aging to consider is the prediction that there are likely to be a variety of *intraspecific* patterns of aging. We have already seen evidence of this in the above analysis of the numerous loci associated with dementia and degenerative neuropathologies. Which of these many loci deserve our special attention? One could argue that we should focus upon gene variations that lead to "public" modulations of the phenotypes of interest, as these would obviously impact upon a much larger proportion of the population. If one accepts the logic of the arguments that were developed earlier in this essay, that would mean concentrating upon polymorphisms rather than upon rare constitutive mutations. In fact, with the outstanding exception of the *APOE* polymorphism, most of us (including myself) are doing just the opposite. In so doing, we may be leading ourselves down the proverbial primrose path.

The reason we are so enamored of the *APP, PS1,* and *PS2* mutations, of course, is that we believe that these experiments of nature will elucidate the mechanisms underlying the vastly more common late-onset sporadic forms of DAT. This idea has been reinforced by the evidence implicating alternative processing of the β-amyloid precursor protein as the integrative pathogenic mechanism for all three mutations. This may well yet prove to be the case. It remains entirely possible, however, that in most forms of DAT, β-amyloid is a *result* of cell injury rather than a *cause* of cell injury, or if it is indeed a cause of cell injury, it may only be a secondary cause superimposed upon a more fundamental mechanism.

NATURE–NURTURE INTERACTIONS

Genes do not act in a vacuum, of course. Perhaps the most dramatic example, for gerontology, comes from the results of the now classic experiments in which fruit flies with substantially enhanced longevities were obtained after serial selection for females who exhibited unusually good fecundity late in the life course. Such selection most surely had to result in greater retention of function in a variety of organ systems, including the CNS. A surprising result, however, was that this altered life history was only observed under conditions of larval crowding. An environment in which the larval densities were carefully controlled did not lead to such altered life histories.

The most recent (and very exciting) research on the role of genes in the maintenance of various domains of cognitive function late in the life span, by McClearen et al.,[5] utilized Swedish same-sex twin pairs aged 80 years or more. Comparisons of identical with fraternal twins provided evidence for surprisingly large influences of heredity. Estimates for heritability (the proportion of phenotypic variance attributable to genetic variance) were 62% for general cognitive ability, 55% for verbal ability, 32% for spatial ability, 62% for speed of processing, and 52% for memory (using the forward and backward Digit Span Test and the Picture Memory Test). The test procedures included a modified Wechsler Adult Intelligence Scale, Verbal Meaning and Information, Figure Logic and Block Design, Symbol Digits, Digit Span, and Picture Memory. On the assumption that errors of measurements accounted for only about 10% of the nongenetic variance (the tests were chosen, in part, because of evidence for relatively high reliability), the authors estimated that the amounts of variance resulting from nonshared environments ranged from 15% for the case of verbal ability to 38% for the case of memory.

APPROACHES TO THE GENETIC ANALYSIS OF UNUSUALLY WELL-PRESERVED COGNITIVE FUNCTION

The phenotypes of interest are, in principle, amenable to quantitative trait analysis (QTL). This assumes, however, that the genetic basis for the quantitative variation of the specific phenotype of interest (the more specific the better) is not attributable to small effects of allelic variation at many dozens or hundreds of loci. There is no way to predict, a priori, the likelihood that a few (e.g., 5 or 6) major gene effects will be discovered for any given phenotype. But it is an effort worth making, especially given the spectacular recent improvements in the methodologies of molecular genetic analysis. The old sib-pair methodologies of Penrose[6] have been rediscovered and greatly embellished with statistical innovations such as extremely discordant or extremely concordant sib-pairs. There have also been major advances in genetic association methodologies. Ingenious hybrids of both parametric and nonparametric methods of genetic analysis have been made widely available via user-friendly computer programs. All of these have been made possible by the new abundance of informative polymorphic genetic markers, large numbers of candidate genes, attractive animal models (including recombinant inbred strains of mice), and improved statistical methods. The stage now seems set for an intelligent reexamination of the genetic basis of unusually robust structure and function of late life history traits.

WHICH PHENOTYPES DESERVE OUR INITIAL ATTENTION?

In making this decision, we should consider at least the following set of criteria:

1. The phenotype of interest should be a highly discrete aspect of neural function. The more discrete the function, the more likely one might be able to discover a few major genes controlling differential rates of decline in aging populations.
2. The assays should be relatively cheap, fast, noninvasive, objective, quantitative, and reliable.
3. They should have been standardized with a large population of subjects, should exhibit high-order sensitivity, and should reflect a wide range of function, such that differential performance at both the high end and low end of the functional potential in nondiseased populations may be assayed.
4. The assays should be amenable to longitudinal studies. For human subjects, differential rates of decline should be detectable within a period of 5–10 years.

5. For the case of human subjects, these variable declines should be detectable during middle age (circa 45–60 years). There are three rationales behind this particular criterion. There is first of all the evidence from population geneticists that, for human populations, the force of natural selection is essentially nil for alleles that do not reach phenotypic expression by age 45 years (Fig. 8.1). Secondly, the choice of such an age range will make possible, in at least a proportion of cases, the availability of DNA from three generations; such DNA should expedite many types of genetic studies. Thirdly, middle-aged subjects are less likely to have developed diseases that might influence the phenotypes of interest.

6. Ideally, one should be able to assess a comparable phenotype in mice, to permit experimental genetic approaches, such as a formal QTL analysis and a search for suppressors and enhancers of major gene effects.

For the peripheral nervous system, a candidate phenotype that might fulfill many of these criteria would be the velocity and amplitude of peripheral nerve conduction. There are good data for various co-variates that would have to be

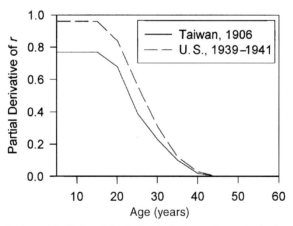

FIGURE 8.1   A graphic display of the decline in the force of natural selection as a function of age. The equation on which the graph is based (given in Martin et al.[9]) addresses the question of the extent to which an individual of age X contributes to the ancestry of future generations. On the Y-axis is plotted the partial derivative of the Malthusian parameter r, a measure of that contribution. Note that this parameter approaches zero by about the middle of the fifth decade. The shapes of the curves are comparable for populations with relatively high mortality, such as occurred in Taiwan in the early twentieth century, and in a more recent U. S. population characterized by relatively low mortality rates. One could conclude that alleles that do not reach some significant level of phenotypic expression by about 45 or 50 years essentially escape the force of natural selection. This could constitute one definition of the age at which senescent processes begin to manifest themselves. The figure was modified after Brian Charlesworth (referenced in Martin, Austad, and Johnson, 1996) and is reproduced with permission of the publishers of *Nature Genetics*.

taken into consideration, such as height and skin temperature, thanks to a cross-sectional study of some 4462 Vietnam-era veterans by the U.S. Centers for Disease Control.[7] Rates of decline (in cross-sectional studies) appear to be particularly steep for the case of the sural nerve and can be readily ascertained during middle age.

The possibility of discovering a few major genes responsible for unusually robust retention of higher cognitive functions in aging human subjects would seem remote. These phenotypes are likely to be under highly polygenic controls, but there could be some surprises here. *Drosophila* geneticists have shown that quantitative traits may be modulated by a large number of genes, but that the major contributions to the genetic variance can be attributable to only a few genes; these can be mapped and cloned.

A battery of psychometric tests has been administered as part of a cross-sectional and longitudinal study of adult intellectual development of members of a Seattle Health Maintenance Organization. The study already has 35 years of data, much of it summarized in a recent monograph by Schaie.[8] For the longitudinal component, the steepest rates of decline over the widest range of adult ages (25–88 years) were for tests of "perceptual speed." These assessed the speed and accuracy with which one could find figures, make comparisons, and perform other simple tasks that required visual perception. These and many other psychometric tests suffer from the potential problem of a "rehearsal" bias, however. This would be a particular problem for the case of the otherwise robust story recall tests of simple immediate memory span. Such tests would have to be suitably modified for longitudinal studies designed to assess differential rates of change.

CONCLUSIONS

This brief essay on approaches to the discovery of the genetic basis for unusually well-preserved neural function during aging merely serves as a point of departure for future discussion. Although the goals of such proposed research are laudable, its implementation, at least in the short term, seems unlikely for a variety of reasons, including economic factors not discussed in the text. A dialog between geneticists and neuropsychologists would be highly desirable, with the goal of critiquing extant testing procedures for their suitability for such research. The set of criteria for the selection of such tests, suggested in the preceding, should also be critiqued and expanded.

FURTHER READINGS

Gu, C., & Rao, D. C. (1997). A linkage strategy for detection of human quantitative-trait loci. I. Generalized relative risk ratios and power of SIB pairs with extreme values. *Am. J. Human Genet., 61,* 200–210.

Gu, C. & Rao, D. C. (1997). A linkage strategy for detection of human quantitative-tract loci. II

Optimization of study designs based on extreme SIB pairs and generalized relative risk ratios. (1997). *Am. J. Human Genet., 61,* 211–222.

Johnson, T.E., DeFries, J. C., & Markel, P. D. (1992). Mapping quantitative trait loci for behavioral traits in the mouse. *Behav. Genet., 22,* 635–653.

Kruglyak, L., Daly, M. J., Reeve-Daly, M. P., & Lander, E. S. (1996). Parametric and nonparametric linkage analysis: A unified multipoint approach. *Am. J. Human Genet., 58,* 1347–1363.

Lezak, M. D. (1995). Neuropsychological assessment (3rd ed.). Oxford: Oxford University Press.

Zhang, H., & Risch, N. (1996). Mapping quantitative-trait loci in humans by use of extreme concordant SIB pairs: Selected sampling by parental phenotypes. *Am. J. Human Genet., 59,* 951–157. [Published erratum appears in (1997). *Am. J. Human Genet., 60,* 748–749.]

REFERENCES

1. Austad, S. N. (1993). Retarded senescence in an insular population of Virginia *(Didelphis virginiana)* oppossums. *J. Zool. London, 229,* 695–708.
2. Medawar, P. B. (1952). *An unsolved problem in biology.* London: H. K. Lewis.
3. Martin, G. M. (1978). Genetic syndromes in man with potential relevance to the pathobiology of aging. *Birth Defects 14,* 5–39.
4. McKusick, V. (1975). *Mendelian inheritance of man.* Baltimore: Johns Hopkins University Press.
5. McClearn, G. E., Johannson, B., Berg, S., Pedersen, N. L., Ahern, F., Petrill, S. A., & Plomin, R. (1997). Substantial genetic influence on cognitive abilities in twins 80 or more years old. *Science, 276,* 1560–1563.
6. Penrose, L. S. (1935). The detection of autosomal linkage in data which consist of pairs of brothers and sisters of unspecified parentage. *Ann. Eugen., 6,* 133–138.
7. Letz, R. & Gerr F. (1994). Covariates of human peripheral nerve function. I. Nerve conduction velocity and amplitude. *Neurotoxicol. Teratol., 16,* 95–104.
8. Schaie, K. W. (1996). *Intellectual development in adulthood. The Seattle Longitudinal Study.* Cambridge: Cambridge University Press.
9. Martin, G. M., Austad, S. N., & Johnson, T. E. (1996). Genetic analysis of ageing: Role of oxidative damage and environmental stresses. *Nature Genetics, 13,* 25–34.

9

THE ROLE OF THE PRESENILINS IN ALZHEIMER'S DISEASE

RUDOLPH E. TANZI

Genetics and Aging Unit
Massachusetts General Hospital and Harvard Medical School
Charlestown, Massachusetts 02129

Alzheimer's disease (AD) is the predominant cause of dementia in the elderly. This progressive neurodegenerative disorder is characterized pathologically by the presence of intracellular neurofibrillary tangles and extracellular neuritic plaques. The major component of the neuritic (senile) plaques and other deposits of β-amyloid is the amyloid β-peptide (Aβ), which is 39–43 amino acids in length and proteolytically derived from the much larger amyloid β-protein precursor (APP). The etiology of AD includes a large genetic component involving either "causative"–"determinsitic" gene defects with virtually 100% penetrance (<10% of cases) or "genetic risk factors," which simply confer increased risk for the disease. Early-onset familial Alzheimer's disease (FAD) occurs in persons 60 years of age and younger and is inherited as an autosomal dominant disorder due to mutations in one of at least three different genes. Rare mutations for early-onset FAD were first identified in the gene encoding APP, which is located on chromosome 21 (for review see Tanzi et al.[1] Meanwhile up to 50% of early-onset FAD is caused by defects in the presenilin 1 (PS1) gene on chromosome 14.[2] A homolog of PS1, presenilin 2 (PS2), on chromosome 1[3, 4] accounts for a much smaller proportion of FAD. Forty-five different PS1 and two FAD mutations in PS2 have been described to date.[1, 5, 6] Some of the greatest clues to how the mutations in the presenilins cause AD have been derived from studies of the cellular processing of these proteins.

PRESENILIN METABOLISM AND FUNCTION

Much has been learned about presenilin biology through studies of their processing and degradation pathways. In mammalian cells transfected with presenilin cDNAs, PS1 is observed as a 43~45 kDa polypeptide,[7–9] whereas PS2 appears as a roughly 53~55 kDa protein.[10] However, in native cell lines and human brain, very few full-length PS1s and PS2s are detected, and two endoproteolytic cleavage fragments are primarily observed for each presenilin. These fragments are saturable and thus generated via a highly regulated pathway.[7, 11] They have also been shown to form oligomeric complexes in vitro as well as in vivo.[12] PS1 C-terminal fragments have also been shown to be phosphorylated by protein kinase C (PKC).[12, 13] We have recently shown that the proteasome pathway is utilized to degrade excess PS2 holoproteins in the endoplasmic reticulum (ER) and this may be used to regulate the amount of presenilin that is cleaved by the endoproteolytic processing pathway.[10] Since the presenilins exist in vivo mainly as cleavage fragments, it is possible that the accumulation of excessive amounts of full-length presenilin cause potentially harmful effects to cells. Evidence for this possibility is discussed in detail in the following based on studies in which overexpression of the presenilins led to apoptosis.

Clues regarding the biological function of the presenilins have come from the presenilin homolog, SEL-12 (50% identity), in *Caenorhabditis elegans,*[14] which facilitates the nematode Notch receptor, LIN-12. It has recently been shown that both human presenilins can rescue a *sel-12* mutant phenotype in the nematode, suggesting that these homologs are functionally interchangeable with SEL-12.[15, 16] Null mice for PS1 gene die either in utero or shortly after birth and exhibit axial skeleton defects, faulty somite segmentation, cerebral hemorrhage,[17, 18] and neuronal loss.[19] Similar defects have been demonstrated for Notch1 and Dll1 null mice. Information regarding the function of the presenilins has also been sought by attempts to identify molecules that interact with presenilins. Recently PS1 was shown to interact with members of the catenin family, mammalian homologs of the *Drosophila* armadillo family that participate in the Wingless pathway. PS1 was found to interact in brain with a novel catenin, named δ-catenin and in peripheral cell lines with β-catenin.[20] The Notch signaling pathway and Wingless signaling pathways have previously been shown to be functionally associated in a mutually inhibitory relationship that centers around the protein, "disheveled." Interestingly, the gene for the human homolog of "disheveled" resides on a region of chromosome 3 to which we have mapped a novel putative late-onset AD gene. Thus, this gene is an appealing candidate AD locus.

PRESENILIN MUTATIONS AND APP METABOLISM

Plasma and fibroblasts from patients and at-risk carriers carrying presenilin mutations have been shown to contain elevated levels of Aβ42, the longer and more amyloidogenic form of Aβ.[21] Increased Aβ42:Aβ40 ratios have also been reported for transfected cell lines and transgenic mice expressing FAD mutant forms of both PS1 and PS2.[11, 22–25] Finally, neuropathological studies have shown that it is Aβ42 that is predominantly deposited in β-amyloid found in the brains of FAD patients carrying PS1 or PS2 mutations.[26–29] These data indicate that presenilin FAD mutations somehow influence APP processing by the enzyme(s) known as γ-secretase, which cleaves at the carboxyl-terminus of Aβ at position 40 or 42.

It has recently been suggested that misprocessing of APP in the presence of mutant presenilins involves a direct physical association between immature APP and the presenilins in the ER.[30, 31] Since recent studies have shown that the N-terminus, the large hydrophilic loop between transmembrane domains 6 and 7, and the C-terminus of PS1 are oriented toward the cytoplasm[32, 33] and that the cytoplasmic domain of APP is not required for the putative APP–PS interaction,[31] it is most likely the luminal domains of the presenilins that mediate complex formation with the ectodomain of APP. Other recent studies have demonstrated that Aβ42 is generated primarily in the ER,[34] (V. M.-Y. Lee, personal communication), whereas Aβ40 is generated through endocytic pathways. Since PS1 and PS2 have previously been localized to the ER,[35–38] a potential APP–PS interaction within the ER may directly influence the rate of Aβ42 generation.

A ROLE FOR THE PRESENILINS IN APOPTOSIS

Apoptotic cell death has been implicated to play a role in the neuropathogenesis of AD.[39–42] In recent studies, a cDNA for the C-terminus of mouse PS2 was reported to prevent apoptosis in T cells, whereas overexpression of PS2 in differentiated PC12 cells was shown to enhance apoptosis. Meanwhile, transfection with antisense PS2 rescued cells from apoptosis.[43, 44] In other studies, the N141I FAD mutation in PS2 led to enhanced apoptosis,[45] and the L286V FAD mutation in PS1 was reported to sensitize neuronal cells to apoptosis.[46] These data have suggested mutant presenilins carry the constitutive ability to make neurons more vulnerable to neuronal injury by fostering the onset of apoptosis.

More recently, we have reported that the presenilins can undergo "alternative" endoproteolysis at a site distal to the normal regulated cleavage site.[47] Alternative endoproteolysis of presenilins was shown to be mediated by a member of the apoptotic cysteine protease family known as caspases, which are activated during apoptosis and have been shown to play a critical role in apoptotic cell death. Since the alternative cleavage of the presenilins can be inhibited either by the treatment with a broad-spectrum caspase inhibitor (e.g., zVAD) or by a more selective inhibitor of CPP32-like caspases, the enzyme responsible for the alternative clip of the presenilins is most likely a member of the caspase 3 family of proteases. This is further supported by data showing that the substitution of Asp residue(s) at consensus caspase 3 cleavage sites (Asp-X) located distal to the normal cleavage sites. Together, these findings suggest that the presenilins serve as cell death substrates and further raise the possibility that the caspase 3–mediated presenilin cleavage fragments are proapoptotic effectors.[47]

A potential role for caspase-mediated endoproteolysis of the presenilins in AD neuropathogenesis has been supported by the observation that the ratio of the alternative-to-normal C-terminal cleavage fragments of PS2 was significantly elevated (approximately 3-fold) in stably transfected inducible cell lines expressing a PS2 containing the N141I FAD mutation relative to those expressing wild-type PS2.[47] An interesting analogy can be found with huntingtin, a protein that is defective in the neurodegenerative disorder, Huntington's disease (HD). This protein has also been shown to be cleaved by a caspase 3 type protease (apopain). In this case, an N-terminal polyglutamine tract that is expanded in HD patients has been reported to modulate the rate of caspase 3–mediated cleavage.[48] Along similar lines, a conformational change in PS2 caused by the N141I FAD mutation could be envisaged to contribute to the enhanced susceptibility of mutant PS2 to caspase cleavage. The observation that PS2 carrying a FAD mutation leads to enhanced alternative endoproteolysis by a caspase 3–type protease associated with apoptosis suggests that altered endoproteolysis of both PS1 and PS2 carrying any of more than 40 different FAD mutations described to date, may contribute to the etiology of FAD. Another report has recently shown that alternative cleavage of the presenilins appears to occur in the normal brain sporadically at around the fifth decade of life. Thus, it is conceivable that caspase-mediated cleavage of the presenilins also underlies sporadic forms of AD,[49] which exponentially increases in prevalence with advanced age. Caspase may also be activated due to insults and injuries that occur in older brains (e.g., due to problems with energy metabolism, cerebral blood flow, or transient ischemic events). It is not exactly clear how alternative cleavage of the presenilins, induction of apoptosis, and increased generation of Aβ42 interplay to bring about neuronal cell death in the brains of AD patients. However, attempts to treat AD by blocking the caspase 3 family member that alternatively cleaves the presenilins represent a potentially promising strategy for future drug design.

SUMMARY

Over 40 different mutations in the genes encoding the presenilins cause up to half of all early-onset cases of AD. The identification of the presenilin genes has provided an unprecedented opportunity to shed new light on the molecular mechanism of the etiological and neuropathological events underlying AD. Here, I review recent progress regarding our and other laboratories' attempts to increase our understanding of the normal and pathological functions of the presenilins. Particular emphasis is placed on the effects of presenilin FAD mutations on the processing and metabolism of APP and on the presenilins, themselves, under normal and apoptotic conditions.

REFERENCES

1. Tanzi, R. E., Kovacs, D. M., Kim, T.-W., Moir, R. D., Guenette, S. Y., & Wasco, W. (1996). The gene defects responsible for familial Alzheimer's disease. *Neurobiol. Dis., 3,* 159–168.
2. Sherrington, R., Rogaev, E. I., Liang, Y., Rogaeva, E. A., Levesque, G., Ikeda, M., Chi, H., Lin, C., Li, G., Holman, K., Tsuda, T., Mar, L., Foncin, J.-F., Bruni, A. C., Montesi, M. P., Sorbi, S., Rainero, I., Pinessi, L., Nee, L., Chumakov, Y., Pollen, D., Wasco, W., Hainus, J. L., Da Silva, R., Pericak-Vance, M., Tanzi, R. E., Roses, A. D., Fraser, P. E., Rommens, J. M., & St. George-Hyslop, P. H. (1995). Cloning a novel gene bearing missense mutations in early onset familial Alzheimer disease. *Nature, 375,* 754–760.
3. Levy-Lahad, E., Wasco, W., Poorkaj, P., Romano, D. M., Oshima, J. M., Pettingell, W. H., Yu, C., Jondro, P. D., Schmidt, S. D., Wang, K., Crowley, A. C., Fu, Y.-H., Guenette, S. Y., Galas, D., Nemens, E., Wijsman, E. M., Bird, T. D., Schellenberg, G. D., & Tanzi, R. E. (1995). Candidate gene for the chromosome 1 familial Alzheimer's disease locus. *Science, 269,* 973–977.
4. Rogaev, E. I., Sherrington, R., Rogaeva, E. A., Levesque, G., Ikeda, M., Liang, Y., Chi, H., Lin, C., Hallman, K., Tsuda, T., Mar, L., Sorbi, S., Nacimias, B., Piacentini, S., Amaducci, L., Chumakov, I., Cohen, D., Lanfelt, L., Fraser, P. E., Rommens, J. M., & St. George-Hyslop, P. H. (1995). Familial Alzheimer's disease in kindreds with missense mutations in a gene on chromosome 1 related to the Alzheimer's disease type 3 gene. *Nature, 376,* 775–778.
5. Hardy, J. (1997). Amyloid, the presenilins and Alzheimer's disease. *Trends Neurosci., 20,* 154–159.
6. Haass, C. (1997). Presenilins: Genes for life and death. *Neuron, 18,* 687–690.
7. Thinakaran, G., Borchelt, D. R., Lee, M. K., Slunt, H. H., Spitzer, L., Kim, G., Ratovitsky, T., Davenport, F., Nordstedt, C., Seger, M., Hardy, J., Levey, A. I., Gandy, S. E., Jenkins, N. A., Copeland, N. G., Price, D. L., & Sisodia, S. S. (1996). Endoproteolysis of presenilin 1 and accumulation of processed derivatives in vivo. *Neuron, 17,* 181–190.
8. Mercken, M., Takahashi, H., Honda, T., Sato, K., Murayama, M., Nakazato, Y., Noguchi, K., Imahori, K., & Takashima, A. (1996). Characterization of human presenilin 1 using N-terminal specific monoclonal antibodies: Evidence that Alzheimer mutations affect proteolytic processing. *FEBS Lett., 389,* 297–303.
9. Podlisny, M. B., Citron, M., Amarante, P., Sherrington, R., Xia, W., Zhang, J., Diehl, T., Levesque, G., Fraser, P., Haass, C., Koo, E. H. M., Seubert, P., St. George-Hyslop, P., Teplow, D. B., & Selkoe, D. J. (1997). Presenilin proteins undergo heterogenous endoproteolysis between Thr291 and Ala299 and occur as stable N- and C-terminal fragments in normal and Alzheimer brain tissue. *Neurobiol. Dis., 3,* 325–337.
10. Kim, T.-W., Pettingell, W. H., Hallmark, O. G., Moir, R. D., Wasco, W., & Tanzi, R. E. (1997). Endoproteolytic processing and proteasomal degradation of presenilin 2 in transfected cells. *J. Biol. Chem., 272,* 11006–11010.
11. Borchelt, D. R., Thinakaran, G., Eckman, C. B., Lee, M. K., Davenport, F., Ratovitsky, T., Prada, C.-M., Kim, G., Seekins, S., Yager, D., Slunt, H. H., Wang, R., Seeger, M., Levey, A. I., Gandy, S. E., Copeland, N. G., Jenkins, N. A., Price, D. L., Younkin, S. G., & Sisodia, S. S. (1996). Familial Alzheimer's disease-linked presenilin 1 variants elevate Aβ1-42/1-40 ratio in vitro and in vivo. *Neuron, 17,* 1005–1013.
12. Seeger, M., Nordstedt, C., Petanceska, S., Kovacs, D. M., Gouras, G. K., Hahne, S., Fraser, P., Levesque, L., Czernik, A. J., St. George-Hyslop, P., Sisodia, S. S., Thinakaran, G., Tanzi, R. E., Greengard, P., & Gandy, S. (1997). Evidence for phosphorylation and oligomeric assembly of presenilin 1. *Proc. Natl. Acad. Sci. USA, 94,* 5090–5094.

13. Walter, J., Grünberg, J., Capell, A., Pesold, B., Schindzielorz, A., Citron, M., Mendla, K., St. George-Hyslop, P., Multhaup, G., Selkoe, D. J., & Haass, C. (1997). Proteolytic processing of the Alzheimer disease–associated presenilin-1 gene generates an in vivo substrate for protein kinase C. *Proc. Natl. Acad. Sci. USA, 94,* 5349–5354.

14. Levitan, D., & Greenwald, I. (1995). Facilitation of lin-12 mediated signalling by sel-12, a *Caenorhabditis elegans* S182 Alzheimer's disease gene. *Nature, 377,* 351–354.

15. Levitan, D., Doyle, T. G., Brousseau, D., Lee, M. K., Thinakaran, G., Slunt, H. H., Sisodia, S. S., & Greenwald, I. (1996). Assessment of normal and mutant human presenilin function in *Caenorhabditis elegans. Proc. Natl. Acad. Sci. USA, 93,* 14940–14944.

16. Baumeister, R., Leimer, U., Zweckbronner, J., Jakubek, C., Gruenberg, J., & Haas, C. (1997). The Sel-12 mutant phenotype of *C. elegans* is rescued independent of proteolytic processing by Wt but not mutant presenilin. *Genes Funct., 1,* 149–159.

17. Wong, P., Zheng, H., Chen, H., Becher, M. W., Sirinathsinghji, D. J. S., Trumbauer, M. E., Proce, D. L., Van der Ploeg, L. H. T., & Sisodia, S. S. (1997). Presenilin 1 is required for Notch1 and Dll1 expression in the paraxial mesoderm. *Nature, 387,* 288–292.

18. Shen, J., Bronson, R. T., Chen, D. F., Xia, W., Selkoe, D. J., & Tonegawa, S. (1997). Skeletal and CNS defects in presenilin-1-deficient mice. *Cell, 89,* 629–639.

19. Berezovska, O., Xia, M., Page, K., Wasco, W., Tanzi, R., & Hyman, B. (1997). Developmental regulation of presenilin mRNA expression parallels notch expression. *J. Neuropathol. Exp. Neurol., 56,* 40–44.

20. Zhou, J., Lyanage, U., Medina, M., Ho, C., Simmons, A. D., Lovett, M., & Kosik, K. S. (1997). Presenilin 1 interacts in brain with a novel member of the Armadillo family. *NeuroReport, 8,* 1489–1494.

21. Scheuner, D., Eckman, C., Jensen, M., Song, X., Citron, M., Suzuki, N., Bird, T. D., Hardy, J., Hutton, M., Kukull, W., Larson, E., Levy-Lahad, E., Viitanen, M., Peskind, E., Poorkaj, P., Schellenberg, G., Tanzi, R. E., Wasco, W., Lannfelt, L., Selkoe, D., & Younkin, S. (1996). Aβ42(43) is increased in vivo by the PS1/2 and APP mutations linked to familial Alzheimer's disease. *Nature Med., 2,* 864–870.

22. Citron, M., Westaway, D., Xia, W., Carlson, G., Diehl, T., Levesque, G., Johnson-Wood, K., Lee, M., Seubert, P., Davis, A., Kholodenko, D., Motter, R., Sherrington, R., Perry, B., Yao, H., Strome, R., Lieberburg, I., Rommens, J., Kim, S., Schenk, D., Fraser, P., St. George-Hyslop, P., & Selkoe, D. J. (1996). Mutant presenilins of Alzheimer's disease increase production of 42-residue amyloid beta-protein in both transfected cells and transgenic mice. *Nature Med., 3,* 67–72.

23. Duff, K., Eckman, C., Zehr, C., Yu, X., Prada, C.-M., Perez-Tur, J., Hutton, M., Buee, L., Harigaya, Y., Yager, D., Morgan, D., Gordon, M. N., Holcomb, L., Refolo, L., Zenk, B., Hardy, J., & Younkin, S. (1996). Increased amyloid-β42(43) in brains of mice expressing mutant presenilin 1. *Nature, 383,* 710–713.

24. Tomita, T., Maruyama, K., Saido, T. C., Kume, H., Shinozaki, K., Tokuhiro, S., Capell, A., Walter, J., Grünberg, J., Haass, C., Iwatsubo, T., & Obata, K. (1997). The presenilin 2 mutation (N141I) linked to familial Alzheimer disease (Volga German families) increases the secretion of amyloid β protein ending at the 42nd (or 43rd) residue. *Proc. Natl. Acad. Sci. USA, 94,* 2025–2030.

25. Xia, W., Zhang, J., Kholodenko, D., Citron, M., Podlisney, M. B., Teplow, D. B., Haass, C., Seubert, P., Koo, E. H., & Selkoe, D. J. (1997). Enhanced production and oligomerization of the 42-residue amyloid beta protein by Chinese hamster ovary cells stably expressing mutant presenilins. *J. Biol. Chem., 272,* 7977–7982.

26. Lemere, C. A., Lopera, F., Kosik, K. S., Lendon, C. L., Ossa, J., Saido, T. C., Yamaguchi, H., Ruiz, A., Martinez, A., Madrigal, L., Hincabie, L., Arango, L. J. C., Anthony, D. C., Koo, E. H., Goate, A. M., Selkoe, D. J., & Arango, V. J. C. (1996). The E280A presenilin 1 Alzheimer mutation produces increased Aβ42 deposition and severe cerebellar pathology. *Nature Med., 2,* 1146–1150.

27. Gomez-Isla, T., Wasco, W., Pettingell, W. P., Garubhagavatula, S., Schmidt, D. D., Jondro, P. D.,

McNamara, M., Rodes, L. A., DiBlasi, T., Growdon, W. B., Seubert, P., Schenk, D., Growdon, J. H., Hyman, B. T., & Tanzi, R. E. (1997). Novel presenilin 1 gene mutation: Increased β-amyloid and neurofibrillary changes. *Ann. Neurol., 41*, 809–813.

28. Mann, D. M. A., Iwatsubo, T., Cairns, N. J., Lantos, P. L., Nochlin, D., Sumi, S. M., Bird, T. D., Poorkaj, P., Hardy, J., Hutton, M., Prihar, G., Cook, R., Rossor, M. N., & Haltia, M. (1996). Amyloid β protein (Aβ) deposition in chromosome 14-linked Alzheimer's disease: Predominance of AβB42(43). *Ann. Neurol., 49*, 149–156.

29. Mann, D. M. A., Iwatsubo, T., Nochlin, D., Sumi, S. M., Levy-Lahad, E., & Bird, T. D. (1997). Amyloid (Aβ) deposition in chromosome 1-linked Alzheimer's disease: The Volga German families. *Ann. Neurol., 41*, 52–57.

30. Weidemann, A., Paliga, K., Dürrwang, U., Czech, D., Evin, G., Masters, C. L., & Beyreuther, K. (1997). Formation of stable complexes between two Alzheimer's disease gene products: Presenilin-2 and β-amyloid precursor protein. *Nature Med., 3*, 328–332.

31. Xia, W., Zhang, J., Koo, E. H., & Selkoe, D. J. (1997). In vivo interaction between amyloid precursor protein and presenilins in mammalian cells: Implication for the pathogenesis of Alzheimer's disease. *Proc. Natl. Acad. Sci. USA, 94*, 8208–8213.

32. Doan, A., Thinakaran, G., Borchelt, D. R., Slunt, H. H., Ratovitsky, T., Podlisny, M., Selkoe, D. J., Seeger, M., Gandy, S. E., Price, D. L., & Sisodia, S. S. (1996). Protein topology of presenilin 1. *Neuron, 17*, 1023–1030.

33. Lehmann, S., Chiesa, R., & Harris, D. A. (1997). Evidence for a six-transmembrane domain structure of presenilin 1. *J. Biol. Chem., 272*, 12047–12051.

34. Wild-Bode, C., Yamazaki, T., Capell, A., Leimer, U., Steiner, H., Ihara, Y., & Haass, C. (1997). Intracellular generation and accumulation of amyloid β-peptide terminating at amino acid 42. *J. Biol. Chem., 272*, 16085–16088

35. Kovacs, D. M., Fausett, H. J., Page, K. J., Kim, T.-W., Moir, R. D., Merriam, D. E., Hollister, R. D., Hallmark, O. G., Mancini, R., Felsenstein, K. M., Hyman, B. T., Tanzi, R. E., & Wasco, W. (1996). Alzheimer associated presenilins 1 and 2: Neuronal expression in brain and localization to intracellular membranes in mammalian cells. *Nature Med., 2*, 224–229.

36. Walter, J., Capell, A., Grünberg, J., Pesold, B., Schindzielorz, A., Prior, R., Podlisny, M. B., Fraser, P., St. George Hyslop, P., Selkoe, D. J., & Haass, C. (1997). The Alzheimer's disease–associated presenilins are differentially phosphorylated proteins located predominantly within the endoplasmic reticulum. *Mol. Med., 2*, 673–691.

37. Cook, D. G., Sung, J. C., Golde, T. E., Felsenstein, K. M., Wojczk, B. S., Tanzi, R. E., Trojanowski, J. Q., Lee, V. M.-Y., & Doms, R. W. (1996). Expression and analysis of presenilin 1 in a human neuronal system: Localization in cell bodies and dendrites. *Proc. Natl. Acad. Sci. USA, 93*, 9223–9228.

38. Strooper, B. D., Beullens, M., Contreras, B., Levesque, L., Craessaerts, K., Cordell, B., Moechars, D., Bollen, M., Fraser, P., St. George-Hyslop, P., & Van Leuven, F. (1997). Phosphorylation, subcellular localization, and membrane orientation of the Alzheimer's disease–associated presenilins. *J. Biol. Chem., 272*, 3590–3598.

39. Su, J., Anderson, A., Cummings, B., & Cotman, C. (1994). Immunohistochemical evidence for apoptosis in Alzheimer's disease. *NeuroReport, 5*, 2529–2533.

40. Johnson, E. M. (1994). Possible role of neuronal apoptosis in Alzheimer's disease. *Neurobiol. Aging, 15*, 5187–5189.

41. Cotman, C. W., & Anderson, A. J. (1995). A potential role for apoptosis in neurodegeneration and Alzheimer's disease. *Mol. Neurobiol., 10*, 19–45.

42. LeBlanc, A. (1996). Apoptosis and Alzheimer's disease. In W. Wasco, & R. E. Tanzi (Eds). *Molecular mechanism of dementia* (pp. 57–71). Totowa, NJ: Humana Press.

43. Vito, P., Lancana, E., & D'Adamio, L. (1996). Interfering with apoptosis: Ca^{2+}-binding protein ALG-2 and Alzheimer's disease gene ALG-3. *Science, 271*, 521–524.

44. Wolozin, B., Iwasaki, K., Vito, P., Ganjei, K., Lacana, E., Sunderland, T., Zhao, B., Kusiak, J. W., Wasco, W., & D'Adamio, L. (1996). PS2 participates in cellular apoptosis: Constitutive activity conferred by Alzheimer mutation. *Science, 274*, 1710–1713.

45. Deng, G., Pike, C. J., & Cotman, C. W. (1996). Alzheimer-associated presenilin-2 confers increased sensitivity to apoptosis in PC12 cells. *FEBS Lett., 397,* 50–54.
46. Guo, Q., Furukawa, K., Sopher, B. L., Pham, D. G., Xie, J., Robinson, N., Martin, G. M., & Mattson, M. P. (1996). Alzheimer's PS-1 mutation perturbs calcium homeostasis and sensitizes PC12 cells to death induced by amyloid β-peptide. *NeuroReport, 8,* 379–383.
47. Kim, T.-W., Pettingell, W. H., Jung, Y. K., Kovacs, D. M., & Tanzi, R. E. (1997). Alternative cleavage of Alzheimer-associated presenilins during apoptosis by a caspase-3 family protease. *Science, 277,* 373–376.
48. Goldberg, Y. P., Nicholson, D. W., Rasper, D. M., Kalchman, M. A., Koide, H. B., Graham, R. K., Bromm, M., Kazemi-Esfarjani, P., Thornberry, N. A., Vailancourt, J. P., & Hyden, M. R. (1996). Cleavage of huntingtin by apopain, proapoptotic cysteine protease, is modulated by the polyglutamine tract. *Nature Genet., 13,* 441–449.
49. Hartmann, J., Busciglio, J., Baumann, K.-H., Staufenbiel, M., & Yankner, B. A. (1997). Developmental regulation of presenilin-1 processing in the brain suggests a role in neuronal differentiation. *J. Biol. Chem., 2172,* 14505–14508.

10

MECHANISMS OF NEURON DEATH IN NEURODEGENERATIVE DISEASES OF THE ELDERLY

ROLE OF THE LEWY BODY

JOHN Q. TROJANOWSKI,* JAMES E. GALVIN,*†
M. LUISE SCHMIDT,* PANG-HSIEN TU,*
TAKESHI IWATSUBO,‡ AND VIRGINIA M.-Y. LEE*

*Center for Neurodegenerative Disease Research, Department of Pathology and
Laboratory Medicine, University of Pennsylvania School of Medicine, and
†Department of Neurology, Allegheny University of the Health Sciences
Philadelphia, Pennsylvania 19104
‡Department of Neuropathology and Neuroscience
University of Tokyo, Tokyo, Japan 113-0052

Lewy bodies (LBs) are hallmark lesions of Parkinson's disease (PD), and they typically occur in dopaminergic neurons of the substantia nigra pars compacta in association with a marked loss of these neurons and gliosis.[1-5] However, LBs also occur in other populations of neurons in the PD brain that undergo selective degeneration.[1-5] LBs are formed from masses of 7–25 nm in diameter filaments, but the composition of these filaments remains to be clarified, and it is uncertain how LBs contribute to the degeneration of affected neurons in PD. However, studies of LBs by immunohistochemistry revealed that neurofilament (NF) proteins[6-11] and ubiquitin (Ub)[12-14] are prominent components of LBs, but other cytoskeletal and noncytoskeletal proteins also have been detected in LBs [15-18] (for a review, see Pollanen et al.[5]).

In addition to PD, LBs also occur in cortical neurons of elderly demented patients, including a subgroup of patients with Alzheimer's disease (AD) who exhibit parkinsonian features, and this is known as the LB variant of AD (LBVAD)[19–24] Further, when the brains of patients with an AD-like dementia are shown to contain numerous cortical and subcortical intraneuronal LBs, but only rare neurofibrillary tangles (NFTs) or amyloid-rich senile plaques (SPs), this disorder is termed diffuse Lewy body disease (DLBD) or dementia with LBs (DLB).[25–32] Notably, DLB may be the second most common dementing neurodegenerative disorder in the elderly after AD,[14] and the number of cortical LBs correlates with the severity of the dementia in DLB patients.[23] LBs are rare in normal elderly individuals, but LB-like inclusions also occur in amyotrophic lateral sclerosis (ALS) and a few other neurodegenerative disorders.[2, 3, 5, 30]

To gain insight into the pathobiology of LBs, methods for the purification of LBs from the brains of patients with PD and DLB have been developed thereby allowing the generation of monoclonal antibodies (MAbs) to LB proteins.[33, 34] These reagents should facilitate elucidation of the biochemical composition of LBs and the role LBs play in the degeneration of affected neurons in PD, LBVAD, and DLB. Further, anti-LB MAbs could be exploited for the development of early antemortem diagnostic tests for LB disorders. Although information on the biological consequences of LB formation is incomplete, some of the more commonly detected neuron-specific proteins in LBs are the triplet of NF subunits[35, 36] known as the high (NF-H), middle (NF-M), and low (NF-L) molecular weight NF polypeptides. The notion that NF-rich inclusions in human LB disorders may lead to the death of affected neurons is based on circumstantial evidence from studies of human LB disorders, as well as on data from studies of transgenic mice that develop LB-like intraneuronal inclusions followed by the age-dependent degeneration of some of these of neurons (see Tu et al.[37]) Below, we review efforts to dissect the molecular composition of LBs and elucidate the role these lesions play in neurodegenerative diseases of the elderly.

MORPHOLOGICAL AND
IMMUNOHISTOCHEMICAL STUDIES OF LBs

Typically, LBs in substantia nigra neurons of PD patients are composed of a central meshwork of filaments surrounded by a radial array of filaments together with membranous profiles and amorphous material.[5, 10, 38, 39] However, LBs also may be irregularly shaped and extend into neuronal processes. The filaments in LBs resemble NFs, but LB filaments lack the projecting "side-arms" typical of NFs in normal axons.[10, 35, 38–42] Cortical LBs closely resemble the LBs in PD substantia nigra neurons, but they lack a well-defined central core and a peripheral corona.[5, 11, 39, 41–45]

Immunohistochemical studies have contributed the most information on the composition of LBs, and the first group of known neuronal proteins detected in

LBs using immunohistochemistry were NF proteins[5, 7–9] (i.e., the subunits that form the heteropolymeric intermediate filaments that are the major cytoskeletal elements of axons).[35, 36, 39, 41, 42] NF-L forms the shaft of the NF into which NF-M and NF-H are incorporated, but each subunit includes a head, a tail, and a central rod domain. The rod domain is required for filament assembly, and the tail domains of NF-M and NF-H contain multiple tandem repeats of KSP motifs that are extensively phosphorylated[46] by incompletely defined kinases[35] that may include calcium–calmodulin–dependent kinase[47] and cyclin-dependent kinase 5.[48] NF-M and NF-H are phosphorylated most extensively in axons[49] that may enable NFs to regulate the caliber of large axons.[35, 36, 41, 50]

Immunohistochemical studies of cortical and subcortical LBs using a large library of domain-specific MAbs have documented epitopes spanning nearly the entire extent of each NF subunit, and one MAb (RMO32) specific for a highly phosphorylated epitope in the tail domain of NF-M[11, 46, 51] has been shown to label numerous LBs, but no other neurodegenerative disease lesions.[11] However, this MAb is of limited use for studies of archival tissue samples because it detects LBs most effectively in ethanol-fixed tissues. In contrast to antibodies to NF proteins, which primarily stain a small subset of LBs, anti-Ub antibodies detect numerous LBs in archival formalin-fixed tissues, including cortical LBs that are difficult to visualize in hematoxylin and eosin stained preparations.[11–14] Ubiquitin (Ub) is a 76–amino acid protein that is conjugated to other polypeptides through an isopeptide bond between its carboxy terminal glycine and the ε-amino group of lysine in the target protein.[13] However, Ub binds to many different proteins targeted for adenosine triphosphate (ATP)–dependent, nonlysosomal degradation including components of NFTs, SPs, Rosenthal fibers, Mallory bodies, Pick bodies, etc.[12] Thus, the detection of Ub immunoreactivity in a lesion is not specific for LBs. Although considerable attention has focused on the role of NF proteins and Ub in the pathogenesis of LBs, a large number of other neuronal proteins have been detected in LBs and LB-like inclusions by immunohistochemistry[5, 7, 15–18, 52–67]; but it is unclear how they contribute to the formation of LBs, since many of these proteins may be trapped in LBs nonspecifically.

GENERATION AND CHARACTERIZATION OF MABS TO PURIFIED LBS

Efforts to isolate and characterize LB proteins from the postmortem brain began several years ago,[44, 45, 60] and some of these studies confirmed earlier in situ immunohistochemical data showing the NF proteins are major constituents of LBs.[44, 45] More recently, an alternative method for the isolation of LBs from DLB brains was developed to generate MAbs to purified LBs and to dissect the composition of LB filaments.[34] This method enabled the purification of about 2×10^6 LBs (approximately 25 μg protein) from 200 g of DLB cortex, and these LBs were used as immunogens to generate >181 hybridomas from three separate fusions. Notably, an MAb (LB112)[34] generated by one of these hybridomas strongly labeled LBs in unfixed LB smears and in ethanol-fixed frozen sections of DLB cortex. Although LB112 weakly stained NFTs and neuropil threads, it did not stain normal structures. Further investigation demonstrated that LB112 did not recognize free Ub or Ub conjugating enzymes. Instead, LB112 reacted with high molecular weight polyubiquitin chains in Western blots of purified LB proteins, which is in contrast to the monoubiquitinated proteins found in NFTs.[13, 34] Three other anti-LB MAbs (LB48, LB202, and LB204) also were characterized, and all three of these MAbs were specific for LBs as evidenced by the fact that they did not stain other neurodegenerative disease lesions (e.g., tau-rich NFTs, amyloid-rich SPs) or normal structures.[33] Further, in Western blots, all three MAbs strongly labeled major immunobands around 160 kDa, which is consistent with NF-M, and a minor immunoband around 50 kDa, suggesting that these MAbs cross-react weakly with tubulin.[33] Since none of these new MAbs stained LBs in formalin-fixed tissue, these antibodies may be of limited use in the diagnostic evaluation of routinely prepared archival brain samples. Nonetheless, these studies are significant because they confirm immunohistochemical data suggesting that the major components of LBs are Ub and NF proteins and because they demonstrate the feasibility and potential utility of methods to purify and analyze LBs.

INSIGHT INTO THE PATHOBIOLOGY OF LBs FROM TRANSGENIC MICE

The accumulation of NF-rich intraneuronal LBs in a specific subset of late-life neurodegenerative disorders suggests that LBs play a mechanistic role in neuronal death in these diseases, but this remains to be proven.[10, 36, 37, 39, 68] Thus, it was significant in this regard that transgenic mice overexpressing NF-L or NF-H developed NF-rich inclusions in spinal cord motor neurons and a clinical phenotype similar to ALS.[69–73] Further, Tu et al.[74] demonstrated that mice overexpressing a mutant form of the human Cu/Zn superoxide dismutase (SOD1) gene (G93A transgenic mice) like that found in familial ALS patients also developed LB-like inclusions in spinal cord motor neurons together with other phenotypic features of human ALS. In addition, a point mutation in the NF-L gene was shown to cause massive selective degeneration of anterior horn cells accompanied by accumulations of NFs when this mutant gene was expressed in other transgenic mice.[75] More recently, Tu et al.[37] investigated the biological significance of LB-like NF-rich inclusions in central nervous system (CNS) neurons using transgenic mice developed by Eyer and Peterson[76] that overexpress an NF-H–β-galactosidase fusion protein (NF-H/lacZ) and develop LB-like inclusions. Significantly, cerebellar Purkinje cells developed inclusions that contained numerous entrapped cellular organelles (termed type II inclusions), whereas hippocampal CA1 neurons developed inclusions that were devoid of organelles (termed type I inclusions). The type II inclusions had a deleterious long-term consequence on the Purkinje cells,[37] since these neurons degenerated in an age-related fashion while the neurons with the type I inclusions did not. This suggests that the sequestration of organelles within NF inclusions may functionally isolate these organelles, compromising the viability of the affected neurons.[37] This contrasts with other transgenic mouse models where neuronal degeneration was ascribed to a blockage of axonal transport by NF inclusions in the proximal axon.[68] Presumably, this blockage could result from the accumulation of NF due to the overexpression of NF genes as well as to the overexpression of mutant human SOD1 (reviewed in Tu et al.[77]). However, in the NF-H/lacZ mice, the selective degeneration of the Purkinje cells probably results from the entrapment of vital cellular organelles.[37] Significantly, this degeneration of selectively vulnerable neurons recapitulates that seen in human neurodegenerative diseases such as PD and DLB in that the degenerative process is age dependent and protracted.

SUMMARY

The biological significance LBs in patients with LB disorders is unclear, but this situation is beginning to be rectified by investigators pursuing the lines of research summarized in the preceding. For example, the transgenic mouse models that were analyzed in the studies reviewed here suggest that NF inclusions may induce neuronal death by at least two alternative mechanisms: (1) an impairment of axonal transport by NF aggregates in proximal axons; (2) the entrapment and sequestration of cellular organelles within perikaryal type II NF inclusions. Further, the recent report of a mutation in the α-synuclein gene leading to an Ala to Thr substitution at position 53 (Ala53Thr) in the α-synuclein protein in several kindreds with familial PD[78] is likely to provide additional opportunities for elucidating mechanisms of LB formation and the role of LB in neurodegenerative diseases characterized by LBs, including AD where fragments (termed non-Aβ component of AD amyloid plaques or NAC) of α-synuclein (termed NACP) have been shown to accumulate in AD amyloid plaques.[79]

Although the abnormal accumulation of normal NFs into pathological LBs represents a marker of the disease state, this process also could compromise the function and viability of neurons thereby contributing to neuron dysfunction and death. For example, axons of LB-containing nigral and cortical neurons could undergo a "dying back" process due to blocked axonal transport, thereby disconnecting the substantia nigra neurons from the striatum and one cortical region from another. Thus, efforts to elucidate pathobiology of LBs are likely to lead to improved strategies for the antemortem diagnosis of LB disorders, as well as to the development of novel therapeutic interventions for the treatment of PD, DLB, and LBVAD.

ACKNOWLEDGMENTS

This work is supported in part from grants from the National Institute on Aging of the National Institutes of Health. Dr. Galvin is supported through a training grant in Neurorehabilitation from NIH/NINDS.

REFERENCES

1. Cronford, M. E., Chang, L., & Miller, B. L. (1995). The neuropathology of parkinsonism: An overview. *Brain Cognition, 28,* 321–341.
2. Forno, L. S. (1996). Neuropathology of Parkinson's disease. *J. Neuropathol. Exp. Neurol., 55,* 259–272.
3. Hughes, A. J., Daniel, S. E., Blankson, S., & Lees, A. J. (1993). A clinicopathologic study of 100 cases of Parkinson's disease. *Arch. Neurol., 50,* 140–148.
4. Gibb, W. R. G., & Poewe, V. V. H. (1986). The centenrary of Friederich H. Lewy. *Neuropathol. Appl. Neurobiol., 12,* 217–221.
5. Pollanen, M. S., Dickson, D. W., & Bergeron, C. (1993). Pathology and biology of the Lewy body. *J. Neuropathol. Exp. Neurol., 52,* 183–191.
6. Forno, L. S., Sternberger, L. A., Sternberger, N. H., Strefling, A. M., Swanson, K., & Eng, L. F. (1986). Reaction of Lewy bodies with antibodies to phosphorylated and nonphosphorylated neurofilaments. *Neurosci. Lett., 64,* 253–258.
7. Galloway, P. G., Mulvihill, P., & Perry, G. (1992). Filaments of Lewy bodies contain insoluble cytoskeletal elements. *Am. J. Pathol., 140,* 809–822.
8. Goldman, J. E., & Yen, S. H. (1986). Cytoskeletal protein abnormalities in neurodegenerative diseases. *Ann. Neurol., 19,* 209–223.
9. Goldman, J. E., Yen, S. H., Chiu, F. C., & Peress, N. S. (1993). Lewy bodies of Parkinson's disease contain neurofilament antigens. *Science, 221,* 1082–1084.
10. Hill, W. D., Lee, V. M. Y., Hurtig, H. I., Murray, J. M., & Trojanowski, J. Q. (1991). Epitopes located in spatially separate domains of each neurofilament subunit are present in the Lewy bodies of Parkinson's disease. *J. Comp. Neurol., 109,* 150–160.
11. Schmidt, M. L., Murray, J., Lee, V. M. Y., Hill, W. D., Wertkin, A., & Trojanowski, J. Q. (1991). Epitope map of neurofilament protein domains in cortical and peripheral nervous system Lewy bodies. *Am. J. Pathol., 139,* 53–65.
12. Lowe, J., Blanchard, A., Morrell, K., Lennox, G., Reynolds, L., Billet, M., Landon, M., & Mayer, R. J. (1988). Ubiquitin is a common factor in intermediate filament inclusion bodies of diverse type in man, including those of Parkinson's disease, Pick's disease and Alzheimer's disease as well as Rosenthal fibers in cerebellar astrocytomas, cytoplasmic bodies in muscle, and Mallory bodies in alcoholic liver disease. *J. Pathol., 155,* 9–15.
13. Mayer, R. J., Lowe, J., Lennox, G., Landon, M., MacLennan, K., & Doherty, F. J. (1989). Intermediate filament-ubiquitin disease: Implication for cell sanitization. *Biochem. Soc. Symp., 55,* 193–201.
14. McKeith, I. G., Galasko, D., Kosaka, K., Perry, E. K., Dickson, D. W., Hansen, L. A., Salmon, D. P., Lowe, J., Mirra, S. S., Byrne, E. J., Lennox, G., Quinn, N. P., Edwardson, J. A., Ince, P. G., Bergeron, C., Burns, A., Miller, B. L., Lovestone, S., Collerton, D., Jansen, E. N., Ballard, C., de Vos, R. A., Wilcock, G. K., Jellinger, K. A., & Perry, R. H. (1996). Consensus guidelines for the clinical and pathologic diagnosis of dementia with Lewy bodies (DLB): Report of the consortium on DLB international workshop. *Neurology, 47,* 1113–1124.
15. Yamada, T., McGeer, P. L., & McGeer, E. G. (1992). Lewy bodies in Parkinson's disease are recognized by antibodies to complement proteins. *Acta Neuropathol, 84,* 100–104.
16. Papolla, M. Z., Adorn, A. C., & Chou, S. M. (1964). Serum protein antigens with Lewy bodies of Parkinson's disease [Abstract]. *Ann. Neurol., 66,* 136.

17. Arai, H., Lee, V. M.Y., Hill, W. D., Greenberg, B. D., & Trojanowski, J. Q. (1992). Lewy bodies contain beta amyloid precursor proteins of Alzheimer's disease. *Brain Res., 585,* 386–390.

18. Gai, W. P., Blumbergs, P. C., & Blessing, W. W. (1996). Microtubule-associated protein 5 is a component of Lewy bodies and Lewy neurites in the brainstem and forebrain regions affected in Parkinson's disease. *Acta Neuropathol., 91,* 78–81.

19. Cummings, J. L. (1995). Lewy body disease with dementia: Pathophysiology and treatment *Brain Cognition, 28,* 266–280.

20. Hansen, L. A., Salmon, D., Galasko, D., Masliah, E., Katzaman, R., DeTeresa, R., Tahl, L., Pay, M. M., Hofstetter, R., Klauber, M., Rice, V., Butters, N., & Alford, M. (1990). The Lewy body variant of Alzheimer's disease: A clinical and pathologic entity. *Neurology, 40,* 1–8.

21. Katzman, R., Galasko, Saitoh, T., Thal, L. J., & Hansen, L. (1995). Genetic evidence that the Lewy body variant is indeed a phenotypic variant of Alzheimer's disease. *Brain Cognition, 28,* 259–265.

22. Kazee, A. M., & Han, L. Y. (1995). Cortical Lewy bodies in Alzheimer's disease. *Arch. Path. Lab. Med., 119,* 448–453.

23. Samuel, W., Galasko, D., Masliah, E., & Hansen, L. A. (1996). Neocortical Lewy body counts correlate with dementia in the Lewy body variant of Alzheimer's disease. *J. Neuropathol. Exp. Neurol., 55,* 44–52.

24. Smith, M. C., Mallory, M., Hansen, L. A., Ge, N., & Masliah, E. (1995). Fragmentation of the neuronal cytoskeleton in the Lewy body variant of Alzheimer's disease. *NeuroReport, 6,* 673–679.

25. Crystal, H. A., Dickson, D. W., Lizardi, J. E., Davies, P., & Wolfsen, L. I. (1990). Antemortem diagnosis of diffuse Lewy body disease. *Neurology, 40,* 1523–1528.

26. Dickson, D. W., Davies, P., Mayeux, R., Crystal, H., Horoupian, D. S., Thompson, A., & Goldman, J. E. (1987). Diffuse Lewy body disease: Neuropathological and biochemical studies of six patients. *Acta Neuropathol., 75,* 8–15.

27. Gibb, W. R. G., Luthert, P. J., Janota, I., & Lantos, P. L. (1989). Cortical Lewy body dementia: Clinical features and classification. *J. Neurol. Neurosurg. Psych., 52,* 185–192.

28. Ince, P., Irving, D., MacArthur, F., & Perry, R. H. (1991). Quantitative neuropathological study of Alzheimer-type pathology in the hippocampus: Comparison of senile dementia of Alzheimer type, senile dementia of Lewy body type, Parkinson's disease and non-demented elderly control patients. *J. Neurol. Sci., 106,* 142–152.

29. Lippa, C. F., Smith, T. W., & Swearer, J. M. (1994). Alzheimer's disease and Lewy body disease: A comparative clinicopathological study. *Ann. Neurol., 35,* 81–88.

30. Olichney, J. M., Galasko, D., Corey-Bloom, J., & Thal, L. J. (1995). The spectrum of disease with diffuse Lewy bodies. *Adv. Neurol., 65,* 159–170.

31. Perry, E. K., McKeith, I., Thompson, P., Marshall, E., Kerwin, J., Jabeen, S., Edwardson, J. A., Ince, P., Blessed, G., Irving, D., & Perry, R. H. (1991). Topography, extent and clinical relevance of neurochemical deficits in dementia of the Lewy body type, Parkinson's disease and Alzheimer's disease. *Ann. N. Y. Acad. Sci., 640,* 197–202.

32. Perry, R. H., Irving, D., Blessed, G., Fairbairn, A., & Perry, E. K. (1990). Senile dementia of the Lewy body type. A clinically and neuropathologically distinct form of Lewy body dementia in the elderly. *J. Neurol. Sci., 85,* 119–139.

33. Galvin, J. E., Lee, V. M. Y., Baba, M., Mann, D. M. A., Dickson, D. W., Yamaguchi, H., Schmidt, M. L., Iwatsubo, T., & Trojanowski, J. Q. (1997). Monoclonal antibodies to purified cortical Lewy bodies recognize the midsize neurofilament subunit. *Ann. Neurol., 42,* 595–603.

34. Iwatsubo, T., Yamaguchi, H., Fujimuro, M., Yokosawa, H., Ihara, Y., Trojanowski, J. Q., & Lee, V. M. Y. (1996). Purification and characterization of Lewy bodies from the brains of patients with diffuse Lewy body disease. *Am. J. Pathol., 148,* 1517–1529.

35. Nixon, R. A. (1993). The regulation of neurofilament protein dynamics by phosphorylation: Clues to neurofibrillary pathobiology. *Brain Pathol., 3,* 29–38.

36. Tu, P. H., & Lee, V. M. Y. (1998). Neurofilaments: Structure, function and involvement in neurodegeneration (in press). *Adv. Mol. Cell Biol.*

37. Tu, P. H., Robinson, K. A., de Snoo, F., Eyer, J., Peterson, A., Lee, V. M. Y., & Trojanowski, J. Q.

(1997). Selective degeneration of Purkinje cells with Lewy body-like inclusions in aged NFHLACZ transgenic mice. *J. Neurosci. 17*, 1064–1074.

38. Duffy, P. E., & Tennyson, V. M. (1965). Phase and electron microscopic observations of Lewy bodies and melanin granules in the substantia nigra and locus coeruleus in Parkinson's disease. *J. Neuropathol. Exp. Neurol., 24*, 398–414.

39. Trojanowski, J. Q., & Lee, V. M. Y. (1994). Phosphorylation of neuronal cytoskeletal proteins in Alzheimer's disease and Lewy body dementias. *Ann. N. Y. Acad. Sci., 747*, 92–109.

40. Balin, B. J., & Lee, V. M. Y. (1991). Individual neurofilament subunits reassembled in vitro exhibit unique biochemical, morphological and immunological properties. *Brain Res., 556*, 198–208.

41. Cole, J. S., Messing, A., Trojanowski, J. Q., & Lee, V. M. Y. (1994). Modulation of axonal diameter and neurofilament by hypomyelinating Schwann cells in transgenic mice. *J. Neurosci., 14*, 6956–6966.

42. Trojanowski, J. Q., Schmidt, M. L., Shin, R. W., Bramblett, G. T., Rao, D., & Lee, V. M. Y. (1993). Altered tau and neurofilament proteins in neurodegenerative disease: Diagnostic implication for Alzheimer's and Lewy body dementias. *Brain Pathol., 3*, 45–54.

43. Okazaki, H., Lipkin, L. E., & Aronson, S. M. (1961). Diffuse intracytoplasmic inclusions (Lewy type) associated with progressive dementia and quadraparesis in flexion. *J. Neuropathol. Exp. Neurol., 20*, 237–244.

44. Pollanen, M. S., Bergeron, C., & Weyer, L. (1992). Detergent-insoluble cortical Lewy body fibrils share epitopes with neurofilaments and tau. *J. Neurochem., 58*, 1953–1956.

45. Pollanen, M. S., Bergeron, C., & Weyer, L. (1993). Deposition of detergent-resistant neurofilaments into Lewy body fibrils. *Brain Res., 603*, 121–124.

46. Lee, V. M. Y., Carden, M. J., Schlaepfer, W. W., Trojanowski, J. Q., Otvos, L., Hollosi, M., Dietzschold, B., & Lazzarini, R. A. (1988). Identification of the major multiphosphorylation site in mammalian neurofilaments. *Proc. Natl. Acad. Sci. USA, 85*, 1998–2002.

47. Iwatsubo, T., Nakano, I., Fukunaga, K., & Miyamoto, E. (1991). Ca^{2+}/calmodulin-dependent protein kinase II immunoreactivity in Lewy bodies. *Acta Neuropathol., 82*, 159–163.

48. Brion, J. P., & Couck, A. M. (1995). Cortical and brainstem Lewy bodies are immunoreactive for cyclin dependent kinase 5. *Am. J. Pathol., 147*, 1465–1476.

49. Sternberger, L. A., & Sternberger, N. H. (1983). Monoclonal antibodies distinguish phosphorylated and nonphosphorylated forms of neurofilaments *in situ*. *Proc. Natl. Acad. Sci. USA, 80*, 6126–6130.

50. Tu, P. H., Elder, G., Lazzarini, R. A., Nelson, D., Trojanowski, J. Q., & Lee, V. M. (1995). Overexpression of the human NFM subunit in transgenic mice modifies the level of endogenous NFL and the phosphorylation states of NFH subunits. *J. Cell Biol., 129*, 1629–1640.

51. Lee, V. M. Y., Carden, M. J., Schlaepfer, W. W., & Trojanowski, J. Q. (1987). Monoclonal antibodies distinguish several differentially phosphorylated states of the two largest rat neurofilament subunits (NF-H and NF-M) and demonstrate their existence in the normal nervous system of adult rats. *J. Neurosci., 7*, 3474–3488.

52. Wisniewski, T., Haltia, M., Ghiso, J., & Frangione, B. (1991). Lewy bodies are immunoreactive with antibodies to gelsolin related amyloid-Finnish type. *Am. J. Pathol., 138*, 1077–1083.

53. Masaki, T., Ishiura, S., Sugita, H., & Kwak, S. (1994). Multicatalytic proteinase is associated with characteristic oval structures in cortical Lewy bodies: An immunocytochemical study with light and electron microscopy. *J. Neurol. Sci., 122*, 127–134.

54. Yamada, T., McGeer, P. L., Bainbridge, K. G., & McGeer, E. G. (1990). Relative sparing in Parkinson's disease of substantia nigra neurons containing calbindin-D 28K. *Brain Res., 526*, 303–307.

55. Lowe, J., Landon, M., Pike, I., Spendlove, I., McDermott, H., & Mayer, R. J. (1990). Dementia with β-amyloid deposition: Involvement of αβ-crystallin supports two main diseases. *Lancet, 336*, 515–516.

56. Mather, K., Watts, F. Z., Carroll, M., Whitehead, P., Swash, M., Cairns, N., & Burke, J. (1993). Antibody to an abnormal protein in amyotrophic lateral sclerosis identifies Lewy body-like inclusions in ALS and Lewy bodies in Parkinson's disease. *Neurosci. Lett., 160*, 13–16.

57. Nishimura, M., Tomimoto, H., Suenaga, T., Nakamura, S., Namba, Y., Ikeda, K., Akiguchi, I., & Kimura, J. (1994). Synaptophysin and chromogranin A immunoreactivities of Lewy bodies in Parkinson's disease brains. *Brain Res., 634,* 339–344.

58. Nishiyama, K., Murayama, S., Shimizu, J., Ohya, Y., Kwak, S., Asayama, K., & Kanazawa, I. (1995). Cu/Zn superoxide dismutase-like immunoreactivity is present in Lewy bodies from Parkinson's disease: A light and electron microscopic immunocytochemical study. *Acta Neuropathol., 89,* 471–474.

59. Shibata, N., Hirano, A., Kobayashi, M., Sasaki, S., Kato, T., Matsumato, S., Shiozawa, Z., Komori, T., Ikemoto, A., & Umahara, T. (1994). Cu/Zn superoxide dismutase-like immunoreactivity in Lewy body-like inclusions of sporadic amyotrophic lateral sclerosis. *Neurosci. Lett., 179,* 149–152.

60. Hirsch, E., Ruberg, M., Portier, M. M., Dardenne, M., & Agid, Y. (1988). Characterization of two antigens in parkinsonian Lewy bodies. *Brain Res., 441,* 139–144.

61. Yamada, T., Akiyama, H., & McGeer, P. L. (1991). Two types of spheroid bodies in the nigral neurons in Parkinson's disease. *Can. J. Neurol. Sci., 18,* 287–294.

62. Yamada, T. (1995). Further observations on MxA-positive Lewy bodies in Parkinson's disease brain tissues. *Neurosci. Lett., 195,* 41–44.

63. Wakabayashi, K., Takahashi, H., Obata, K., & Ikuta, F. (1992). Immunocytochemical localization of synaptic vesicle-specific protein in Lewy body–containing neurons in Parkinson's disease. *Neurosci. Lett., 138,* 237–240.

64. Nakashima, S., & Ikuta, F. (1984). Tyrosine hydroxylase proteins in Lewy bodies of parkinsonian and senile brains. *J. Neurosci. Res., 64,* 91–96.

65. Lowe, J., McDermott, H., Landon, M., Mayer, R. J., & Wilkinson, K. D. (1990). Ubiquitin carboxy-terminal hydroxylase (PGP 9.5) is selectively present in ubiquitinated inclusion bodies characteristic of human neurodegenerative disease. *J. Pathol., 161,* 153–160.

66. Selkoe, D. J. (1991). The molecular neuropathology of Alzheimer's disease. *Neuron, 6,* 487–498.

67. Schmidt, M. L., Martin, J. A., Lee, V. M. Y., & Trojanowski, J. Q. (1996). Convergence of Lewy bodies and neurofibrillary tangles in amygdala neurons of Alzheimer's disease and Lewy body disorders. *Acta Neuropathol., 91,* 475–481.

68. Schmidt, M. L., Carden, M. J., Lee, V. M. Y., & Trojanowski, J. Q. (1987). Phosphate dependent and independent neurofilament epitopes in the axonal swellings of patients with motor neuron disease and controls. *Lab. Invest., 56,* 282–294.

69. Collard, J. F., Cote, F., & Julien, J. P. (1995). Defective axonal transport in a transgenic mouse model of amyotrophic lateral sclerosis. *Nature, 375,* 61–64.

70. Cote, F., Collard, J. F., & Julien, J. P. (1993). Progressive neuronopathy in transgenic mice expressing the human neurofilament heavy gene: A mouse model of amyotrophic lateral sclerosis. *Cell, 73,* 35–46.

71. Julien, J. P., Cote, F., & Collard, J. F. (1995). Mice overexpressing the human neurofilament heavy gene as a model of ALS. *Neurobiol. Aging, 16,* 487–490; discussion 490–492.

72. Wong, P. C., Marszalek, J., Crawford, T. O., Xu, Z., Hsieh, S. T., Griffin, J. W., & Cleveland, D. W. (1995). Increasing neurofilament subunit NF-M expression reduces axonal NF-H, inhibits radial growth, and results in neurofilamentous accumulation in motor neurons. *J. Cell Biol. 130,* 1413–1422.

73. Xu, Z., Cork, L. C., Griffin, J. W., & Cleveland, D. W. (1993). Increased expression of neurofilament subunit NF-L produces morphological alterations that resemble the pathology of human motor neuron disease. *Cell, 73,* 23–33.

74. Tu, P. H., Raju, P., Robinson, K. A., Gurney, M. E., Trojanowski, J. Q., & Lee, V. M. Y. (1996). Transgenic mice carrying a human mutant superoxide dismutase transgene develop neuronal cytoskeletal pathology resembling human amyotrophic lateral sclerosis lesions. *Proc. Natl. Acad. Sci. USA, 93,* 3155–3160.

75. Lee, M. K., Marszalek, J. R., & Cleveland, D. W. (1994). A mutant neurofilament subunit causes massive, selective motor neuron death: Implications for the pathogenesis of human motor neuron disease. *Neuron, 13,* 975–988.

76. Eyer, J., & Peterson, A. (1994). Neurofilament-deficient axons and perikaryal aggregates in viable transgenic mice expressing a neurofilament-beta-galactosidase fusion protein. *Neuron, 12,* 389–405.

77. Tu, P. H., Gurney, M. E., Julien, J. P., Lee, V. M. Y., & Trojanowski, J. Q. (1996). Oxidative stress, mutant SOD1, and neurofilament pathology in transgenic mouse models of human motor neuron disease. *Lab. Invest. 76,* 1–16.

78. Polymeropoulos, M. H., Lavedan, C., Leroy, E., Ide, S. E., Dehejia, A., Dutra, A., Pike, B., Root, H., Rubenstein, J., Boyer, R., Stenroos, E. S., Chandrasekharappa, S., Athanassiadou, A., Papapetropoulos, T., Johnson, W. G., Lazzarini, A. M., Duvoisin, R. C., DiIorio, G., Golbe, L. I., & Nusbaum, R. L. (1997). Mutation in the α-synuclein gene identified in families with Parkinson's disease. *Science, 274,* 1197–1199.

79. Iwai, A., Masliah, E., Yoshimoto, M., Ge, N., Flanagan, L., deSilva, H. A., Kittel, A., & Saitoh, T. (1995). The precursor protein of the non-Aβ component of Alzheimer's disease amyloid is a presnyaptic protein of the central nervous system. *Neuron, 14,* 467–475.

NOTE ADDED IN PROOF

Since the submission of this chapter, it has been shown that NACP/α-synuclein is a component of the cortical and subcortical Lewy bodies in sporadic Parkinson's disease and dementia with Lewy bodies [Baba et al. (1998). In press. *Am. J. Pathol.*; Spillantini et al. (1997). *Nature, 388,* 839–840; Takeda, et al. (1998). *Am. J. Pathol., 152,* 367–372] which suggests that NACP/α-synuclein plays a mechanistic role in the pathogenesis of both sporadic and familial Parkinson's disease as well as in the pathogenesis of dementia with Lewy bodies.

11

MICROTUBULE-ASSOCIATED PROTEIN TAU

BIOCHEMICAL MODIFICATIONS, DEGRADATION, AND ALZHEIMER'S DISEASE

SHU-HUI YEN, PARIMALA NACHARAJU, LI-WEN KO, AGNES KENESSEY,* AND WAN-KYNG LIU

Department of Research, Mayo Clinic Jacksonville
Jacksonville, Florida 32224

Brain tissue of subjects with Alzheimer's disease (AD) differs from normal brain tissue by accumulation of abnormal filamentous elements in neurons to form neurofibrillary tangles (NFTs) and extracellularly to form amyloid in senile plaques.[1, 2] The ultrastructural and biochemical properties of filaments in these two types of structure are very different. NFTs consist of paired helical filaments (PHFs) and occasionally straight filaments,[3] which contain mainly a hyperphosphorylated form of microtubule-associated protein tau, referred to as PHF-tau.[4–6] PHF is considered to be a polymerized form of PHF-tau, whereas amyloid fibers are assembled from small peptides of 1–40/1–42 amino acid derived from a precursor membrane protein, referred to as APP.[7–9] PHF-tau contains mainly three isoforms of 60–70 kDa in apparent molecular weight.[4–6] The size of PHF-tau is larger than that of tau from normal adult brain (N-tau) or fetal brain (F-tau), and the isoelectrical charge is more acidic for PHF-tau than other tau proteins.[6] Figure 11.1 illustrates the domain structure of different isoforms of tau. The dif-

*Present address: *Department of Pathology, Albert Einstein College of Medicine, Bronx, New York 10461*

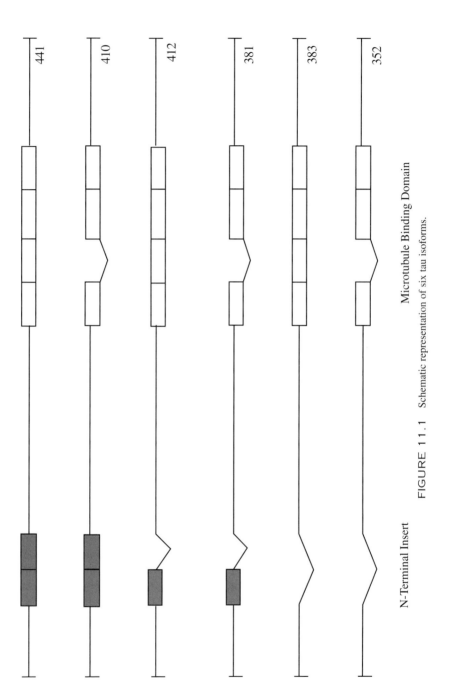

N-Terminal Insert

Microtubule Binding Domain

FIGURE 11.1 Schematic representation of six tau isoforms.

ference among these isoforms is due to alternative splicing of mRNA and post-translational modifications. F-tau is devoid of 58 amino acid residues at its amino terminus, and its microtubule binding domain has three rather than four imperfect tandem repeats.[10] In comparison, N-tau and PHF-tau contain a mixture of six isoforms. However, the proportion of different isofoms in N-tau and PHF-tau are not the same.[11, 12] Deletion of microtubule binding domain from tau removes its ability to bind tubulin.[13] Reduction of the number of tandem repeats leads to a decrease of tau–tubulin interactions.[14, 15] Therefore in order for a tau molecule to function normally, it is essential to have an intact microtubule binding domain.

Based on studies of tau preparations from AD and normal tissues, five types of modifications have been demonstrated to distinguish PHF-tau or PHF-enriched preparations from normal tau. These include phosphorylation, glycation, glycosylation, ubiquitination, and racemization.[16–22] The following discussions will focus on two of these modifications, namely phosphorylation and glycation, and on degradation of tau protein based on our recent studies.

PHOSPHORYLATION

According to our phosphate analysis,[22] N-tau contains 3–4 times less phosphate per mole protein than PHF-tau from autopsy brains with comparable postmortem intervals. N-tau prepared from biopsy or short postmortem brains are slightly more phosphorylated than those derived from autopsy brains.[23–25] And F-tau is similar but not identical to PHF-tau in phosphate content (7 mole phosphate/mole protein for F-tau versus 8 mol phosphate/ml protein for PHF-tau).[26] At least 20 sites are found to be phosphorylated in PHF-tau.[10, 27–34] Some of these sites are fully phosphorylated in PHF-tau, and others are incompletely phosphorylated, indicating the heterogeneous nature of PHF-tau and the continuous phosphorylation and dephosphorylation of tau in AD brains.[10, 34] Based on the work of many investigators, most (if not all) phosphorylated sites in PHF-tau are at regions flanking the microtubule binding domain (Fig. 11.2, *A*). A major consequence of tau phosphorylation in vitro, in situ, or in vivo (as in the case of PHF-tau) is the impairment of its binding to microtubules and ability to promote microtubule assembly.[35–37] Such an event is believed to take place at early stages of neurofibrillary degeneration[38] and to promote tau proteins to enter a state more favorable for tau self-interaction than tau–tubulin interaction. In unphosphorylated tau, the microtubule binding domain is positively charged while both the amino and carboxy ends are negatively charged. Phosphorylation of regions flanking the binding domain alters the charge of tau molecule and its accessibility to interact with other proteins, probably due to changing of conformation. Thus, beside the number of tandem repeats in microtubule binding domain, altering the charge of subregions exerts a marked effect on tau's normal function.

Many well-characterized protein kinases are shown to be capable of phosphorylating tau in vitro and in situ. These kinases can be classified into three

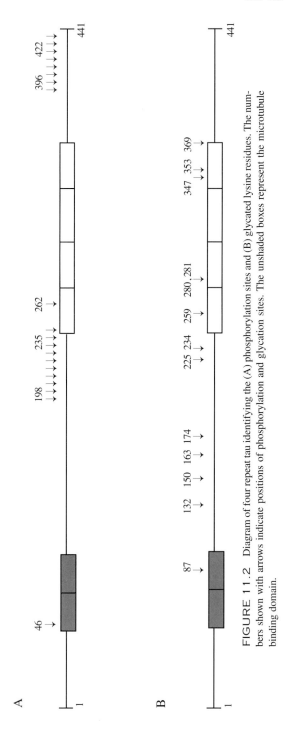

FIGURE 11.2 Diagram of four repeat tau identifying the (A) phosphorylation sites and (B) glycated lysine residues. The numbers shown with arrows indicate positions of phosphorylation and glycation sites. The unshaded boxes represent the microtubule binding domain.

groups: (1) second messenger kinase (protein kinase A, protein kinase C, calmodulin-dependent kinase II),[39–42] (2) proline-directed kinases [mitogen-activated protein kinases, glycogen synthase kinase 3, cyclin-dependent kinase 2 (cdc2), and cdk5],[33, 43–53] and (3) other kinases, which include casein kinase I, casein kinase II, DNA-dependent kinase, p110mapk, and brain kinase.[54–57] Although many of the phosphorylated sites identified in vitro are phosphorylated in PHF-tau, the significance of these known kinases in AD pathogenesis is not fully understood. PHF samples prepared from AD brains were found to display kinase activity.[58, 59] This type of kinase associated tightly with PHF and is believed to be important for PHF-tau phosphorylation. Some PHF-associated kinases appeared to be novel kinases belonging to the family of proline-directed Ser/Thr kinases, especially the cdc2-related kinase. We have identified and characterized such a kinase, which has a molecular weight of 35 kDa.[59] It shares epitopes with the authentic cdc2 kinase, regarded as p34cdc2 kinase. This kinase is capable of phosphorylating synthetic tau peptides, PHF-tau, and histone, but differs from the authentic kinase by the absence of immunoreactivity with antibodies to subdomain III that contains the PSTAIRE motif. Kinases with immunochemical properties comparable to the PHF-associated, cdc2-related kinase were detected in normal human and animal brains, indicating these kinases are not unique to AD brains. The cdc2-related kinase from bovine brain behaved the same as that isolated from autopsied human brain tissue. The electrophoretic mobility of human kinase, however, is not identical to that of nonhuman kinase. Additional studies are being conducted to determine the extent of homology between the human versus nonhuman kinases in amino acid sequences. We are purifying this kinase from bovine brain to high purity for structural studies. As revealed by immunoblotting analyses, the cdc2-related kinase is developmentally regulated. It was not found in fetal or newborn tissue, but appeared postnatally around Day 12 in rat brain. The appearance of this kinase coincided with the disappearance of the p34cdc2 kinase. We have yet to determine if a cyclin-like molecule is required for the activation of this kinase. The kinase immunoreactivity was detected in both the gray and white matter. In comparison to normal brains, AD brains proportionately contain more of this kinase in gray matter than in white matter. Such a difference in kinase distribution is consistent with the observations of more NFTs–PHFs in gray than in white matter. The results suggest that phosphorylation of PHF-tau in AD brains may be due to the abnormal expression and distribution of this cdc2-related kinase.

DEGRADATION OF TAU

In addition to altering tau's ability to interact with tubulin, phosphorylation can lead to a decrease in the susceptibility of tau to proteolysis. For example, exposure of cultured neuroblastoma cells to okadaic acid, a phosphatase inhibitor, was found to increase tau phosphorylation and reduce its turnover.[60–61]

Degradation of tau has been considered to involve proteases belonging to the family of calcium-activated neutral proteases, regarded as calpains. Depending on the concentration of calcium required for activation, calpains can be classified into two major groups. Calpain I requires μM calcium, while calpain II requires mM.[62, 63] Calpains could have a role in AD pathogenesis, since AD brains contain an increased amount of the enzymatically active form of calpain I.[64] In vitro studies showed that phosphorylation of bovine tau by protein kinase A led to an incorporation of 1 mol phosphate/mol protein and a decrease in the susceptibility of tau to calpain II degradation.[39] However, purified N-tau and F-tau displayed similar susceptibility to calpain II,[65] even though F-tau contains far more phosphate than N-tau. In contrast to the soluble form of tau, purified PHFs (containing hyperphosphorylated tau) were highly resistant to calpain II,[66] indicating that formation of tau polymers may decrease the accessibility of calpain cleavage sites to the protease. Comparable to this is our finding that PHF-tau and N-tau were similar in their susceptibility to calpain I (Fig. 11.3). A difference in resistance to calpain I digestion was observed only when unphosphorylated recombinant tau was compared to N-tau. Dephosphorylation of N-tau or PHF-tau by *Escherichia coli* alkaline phosphatase increased the electrophoretic mobility of tau, but had very little effect on the degradation of these proteins by calpain I. The results showed that the effect of phosphorylation on the degradation of tau by calpains I and II is similar and that phosphorylation of normal phosophory-lated tau has very little consequence on the breakdown rate of this protein. The location of calpain cleavage sites has not been identified definitely. However, based on the general rule that calpains favor cleaving at sites that display amino acid residues Lys/Tyr/Arg or Met preceded by Leu or Val,[67] several potential sites have been identified on tau molecule.[68] The majority of these potential sites with such structural motifs are located within the microtubule binding domain, which forms the core of PHF. Based on the size and immunoreactivity of peptide fragments generated from the cleavage of PHF by other proteases (for example, trypsin, chymotrypsin, and pronase) the microtubule binding domain of tau is also more resistant to degradation than other regions of the molecule.[68, 69]

Beside calpain, lysosomal enzymes are important for protein turnover. Brain tissues express a high level of cathepsin D.[70–72] Although lysosomal enzymes reside in the membranous compartment with acidic pH, in normal conditions, these enzymes could be involved in the degradation of tau through a process regarded as autophagy.[73] In brains from aged human or rat, the activity of cathepsin D was reported to increase,[74–78] and proportionally more cathepsin D was found to redistribute in the soluble fraction of brain extracts.[77] In AD brains the lysosomal function of neurons is augmented, and the distribution of lysosomal hydrolases are abnormal.[79, 80] The cytoplasm of tangle-bearing or tangle-free neurons displays more intense cathepsin D immunoreactivity than that of control brains, thus suggesting that this enzyme plays a role in the pathogenesis of AD. It is conceivable that neuronal lysosomes in AD may be leaky and lead to release of lysosomal proteases into cytoplasm to degrade proteins at a pH less acidic than

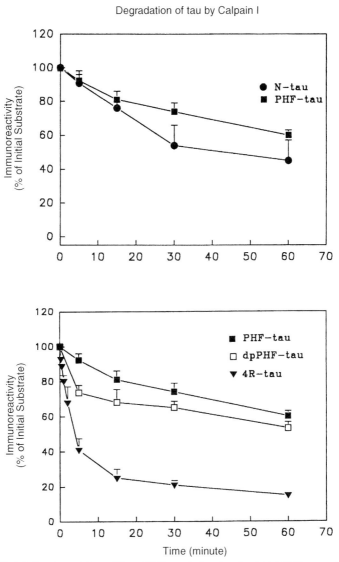

FIGURE 11.3 Degradation of N-tau, PHF-tau, dephosphorylated (dp) PHF-tau, and recombinant tau (4R-tau, 441 amino acid residues). Degradation of tau was analyzed by immunoblotting using Ab71, a monoclonal antibody reactive with an epitope resides at the carboxy end of tau molecule.

that of lysosomal or related vacuoles. Our studies showed that cathepsin D is capable of degrading tau proteins in both acidic (pH 3.5) and neutral pH (pH 7.0).[81] (Figure 11.4, schematic drawing, illustrates the cleavage sites in relation to the domain structure of tau.) However, a much higher amount of the enzyme was required for digestion of tau at neutral pH. Based on immunoblotting analyses using antibodies raised to defined tau epitopes, both the amino and the carboxy termini of tau were the first to be cleaved at acidic pH, whereas the amino end was more rapidly cleaved than the carboxy end at neutral pH. This property was shared by recombinant tau, F-tau, and PHF-tau at acidic pH. The regions susceptible to cathepsin D cleavage appeared to be similar in different tau preparations, but the rate of cleavage was faster in case of phosphorylated tau, especially at the amino end. Amino acid sequencing and mass spectroscopic analyses of peptide fragments generated from recombinant tau revealed several cathepsin D cleavage sites at the amino and carboxy ends of tau. These sites are at amino acid residues corresponding to Phe8-Glu9, Met419-Val420, Thr427-Leu428-Ala429, and Leu436-Ala437 of the largest isoform of brain tau. Additional sites at other regions of tau molecule were also cleaved when higher concentrations of this protease was used. At acidic pH (pH 3.5), cleavage at additional sites would initially generate large tau fragments with intact microtubule binding domain, but eventually these fragments were further degraded to smaller fragments containing only a portion of the binding domain. Cleavage sites located in the microtubule binding domain were less accessible to cathepsin D in neutral pH. Tau fragments with intact microtubule binding domain were detected in neutral digests even after prolonged incubation with the protease. Furthermore, fragments generated at acidic pH were different in size and immunoreactivity when compared with those obtained at neutral pH, suggesting cleavage at different sites. These differences may be due to the change of tau conformation at different pH, in association with the middle region of tau being basic charged versus the acidic nature of both the amino and carboxy ends. In comparison to N-tau or F- tau, PHF-tau was more susceptible to cathepsin D proteolysis. It has been demonstrated that in vitro incubation of truncated recombinant tau containing the microtubule binding domain could generate PHF-like filament. Together these results indicate that altering the pH of digestion mixture could change the preferential sites(s) of cleavage, that changing the state of phosphorylation could render tau more vulnerable to cathepsin D degradation, and that assembly-competent tau fragments could be generated by digestion with cathepsin D at neutral pH. It is possible that cathepsin D is involved in the pathogenesis of AD at more than one stage. During the early stage this enzyme may play a role in generation of fragments that are more capable of self-interaction than undegraded tau. At the late stage, when NFT-bearing neurons die, the enzyme could be involved in PHF metabolism. Consistent with this view are the finding that the amino and carboxy ends of tau protein are more rapidly cleaved by cathepsin D than other regions of the molecule at acidic pH and the fact that this is the phenotype of extracellular NFTs.[27, 82–84]

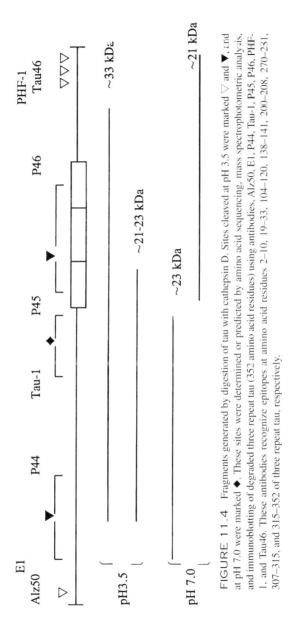

FIGURE 11.4 Fragments generated by digestion of tau with cathepsin D. Sites cleaved at pH 3.5 were marked ▽ and ▼; and at pH 7.0 were marked ◆. These sites were determined or predicted by amino acid sequencing, mass spectrophotometric analysis, and immunoblotting of degraded three repeat tau (352 amino acid residues) using antibodies, Alz50, E1, P44, Tau-1, P45, P46, PHF-1, and Tau46. These antibodies recognize epitopes at amino acid residues 2–10, 19–33, 104–120, 138–141, 200–208, 270–231, 307–315, and 315–352 of three repeat tau, respectively.

In discussing the formation of tau polymers it should be noted that the mere presence of assembly-competent tau fragments may not be sufficient, since filamentous structures were not detected in samples containing calpain I–degraded fragments of N-tau, F-tau, or recombinant tau. Tau polymers were also not detected in culture cells that were transfected with truncated tau cDNA containing microtubule binding domain. These findings indicate that other factors, which are either enriched or produced only under pathological conditions, may be required for the generation of tau polymers characteristic of AD.

What are the mechanisms involved in altering the activity and compartmentalization of lysosomal enzymes in aging or AD? Increase of lysosomal enzyme activity in the cytosol could be a result of changes in the integrity of lysosomal membrane as suggested previously by others.[78] Studies of the phospholipid composition of plasma membrane prepared from AD brains have shown a decrease in the content of ethanolamine glycerophospholipids and an increase in serine glycerophospholipids when compared to controls.[85] It remains to be determined if the compositions of lysosomal and endosomal membranes are altered in aging or AD and if components other than enzymes leak from these compartments as well. In this regard, various proteoglycan immunoreactivities have been demonstrated in NFTs in AD and Down syndrome brains.[86–88] Normally, proteoglycans are detected in extracellular matrix and in extracellular loci adjacent to plasma membranes. In AD and related neurodegenerative disorders, proteoglycan immunoreactivities are detected in neuronal cytoplasm and senile plaques. Electron microscopic studies revealed the presence of proteoglycan immunoreactivity at the periphery of NFTs.[86] It is conceivable that the abnormal distribution of proteoglycans is due to changes in membrane integrity and that glycans located near AD tangles play a role in initiation or facilitation of tau polymerization. These possibilities deserve consideration, since in vitro studies have shown polymerization of full-length recombinant tau in the presence of glycosaminoglycan and independent of phosphorylation.[88, 89] The ability of glycosaminogycans in promoting tau polymerization is likely due to its polyanionic property, since anionic charged molecules such as polyglutamic acid, RNA, and unsaturated fatty acids are effective in facilitating tau filaments assembly in vitro.[90, 91] According to Goedert et al.,[88] heparin interacts with the microtubule binding domain, as well as other regions of the tau molecule. Such interaction is believed to alter the conformation of tau by repelling the acidic segments at both ends from the basic region of tau containing the binding domain. This view is supported by our findings that the presence of heparin (Fig. 11.5), RNA, and polyglutamic acid (data not shown) retarded the degradation of tau by calpain I. This effect is unlikely to be tau protein specific. In our studies the presence of anionic charged molecules were found to alter the degradation of bovine serum albumin by calpain as well. Further studies are essential to determine the significance of anionic charged molecules in tau polymerization in intact cells and for determining whether negatively charged molecules have any effect on the assembly of filamentous structure from microtubule-associated proteins other than tau. The latter question was

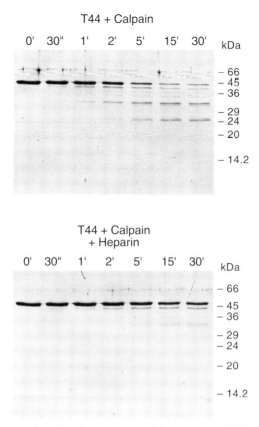

FIGURE 11.5 Heparin inhibits the degradation of three repeat tau (T44) by calpain I. Tau proteins digested for 30 sec to 30 min were subjected to gel electrophoresis and stained with Coomassie blue. Molecular weight standards were included as references.

raised because a microtubule binding domain comparable to that of tau is present in other microtubule-associated proteins such as MAP2 and MAP4.

GLYCATION

Exposure of protein to sugar can lead to modification of proteins by nonenzymatic glycosylation.[92–94] With time, the glycated protein would undergo complex chemical reactions including oxidation, fragmentation, and cross-linking, to become heterogeneous and insoluble products, referred to as advanced glycation end products (AGE).[95–97] A prime target for glycation is ε-amino groups of lysine residues.[98–101] Protein glycation has been demonstrated as a factor involved in complications of diseases such as diabetes and cataract.[102–105] Several proteins have long been established to be glycated in vivo. For example, hemoglobin, albumin, collagen, lens crystalline, and some enzymes.[98, 100, 106–110] More recently, NFTs, senile plaques, and abnormal neurites in AD brains were shown to contain glycated proteins as well.[111–112] These filamentous structures display immunoreactivity with antibodies to pentosidine, a defined AGE product, and to undefined AGE products derived from reaction of bovine serum albumin with sugar.[111] Modification of lysine residue in PHF protein was supported also by comparative analysis of the amino acid content of PHF-tau versus N-tau, in which PHF-tau was found to contain 35% less lysine.[113]

The role of glycation in AD pathogenesis is not fully understood. Evidences exist to suggest that glycation may increase the stability of tau polymers formed within affected neurons. Incubation of preparations containing dispersed PHF with sugar was found to lead to filament aggregation and to decrease the solubility of PHF-tau.[114] In comparison to nonglycated, soluble tau from AD brains, PHF-tau exhibited a higher ability for self-association. Such ability can be further enhanced by glycation of PHF-tau. In vitro studies also showed that glycation of phosphorylated tau facilitated the formation of tau polymer.[115] Cultured cells loaded with glycated tau display oxidant stress.[17] In vitro studies showed that glycation also increases the fibrinogenicity of β-amyloid peptides.[116] Glycated amyloid peptides bind to AGE receptors, and such interaction has been shown to elicit a cascade of reactions in microglia cells, leading to the release of cytokines[117] and possibly other factors, which in turn would affect neurons and glial cells in the vicinity. Protein glycation therefore could be a contributing factor to the irreversibility of neurofibrillary changes in AD or related disorders. If this view is correct, then the development of agents capable of preventing or reducing glycation would be of therapeutic value in treating AD. It is conceivable that protein glycation in AD brains may not be limited to PHF proteins and amyloid peptides. Identification and characterization of glycated proteins in AD versus age-matched, normal brains, as well as in early versus late stages of AD, therefore would be crucial to understanding the pathogenesis of such disorder.

Like phosphorylation, glycation reduced the ability of tau protein to bind tubulin and hence compromised the function of tau in promoting microtubule assembly. The approximate sites of glycation on tau molecule were determined earlier by probing of glycated tau with a polyclonal antibody raised to an in vitro glycated synthetic peptide, which contains 23 amino acid residues, and corresponds to the second repeat in microtubule binding domain.[118] The sites of tau glycation were determined by our analysis of recombinant tau glycated with radiolabeled glucose.[119] In this study, glycated tau was reduced with excess $NaBH_4$ and digested with trypsin. Glycated tryptic peptides were then purified by HPLC to homogeneity and subjected to amino acid analysis. With this experimental scheme, lysine residues modified by glycation would be resistant to tryptic digestion and avoid the detection by amino acid analysis. This approach has led us to identify multiple glycation sites, corresponding to lysine residues #87, 132, 150, 163, 174, 225, 234, 259, 280, 281, 347, 353, and 369 of the longest isoform of brain tau. Figure 11.2, *B* illustrates the sites of glycation in different structural domains of tau. Nearly half of the sites (aa #259–369) were on microtubule binding domain, two additional sites (Lys225, Lys234) were located adjacent to the binding domain, and the remaining glycation sites were at regions between the amino terminus and the binding domain. In this regard, it is interesting to note that amino terminal sequences spanning residues 154–173 were reported to be important for the microtubule nucleation activity of tau. The lost ability of tau to promote microtubule assembly by glycation therefore is directly due to the modification of lysine residues in domains crucial for microtubule nucleation and binding. Moreover, turning off the normal function of tau makes the molecule prone to form aggregates or polymers. An issue that requires further investigation is whether in AD brains, glycation of tau is an event that takes place prior or in parallel to phosphorylation.

SUMMARY

Alzheimer neurofibrillary degeneration involves posttranslational modification of tau proteins by different mechanisms. Phosphorylated sites in PHF-tau have been identified, most of which reside at regions flanking the microtubule binding domain. Multiple sites for glycation were identified in the tau molecule as well. Many of the glycation sites were distributed within the microtubule binding domain. These modifications alter the function of tau by decreasing its accessibility to tubulin binding, leading to a reduction in microtubule polymerization and stability. These posttranslational modifications also affect the degradation of tau by proteases. Depending on the acidity of the environment in which the tau molecule resides, proteolytic degradation of tau could generate fragments competent for the assembly of filamentous structures. Development of agents capable of preventing or reducing undesirable modification or degradation of tau could be of therapeutic value in treating AD.

ACKNOWLEDGMENTS

This work was supported by NIH grants AG01136 (S.-H. Y.), AG04145 (S.-H. Y.), Alzheimer Association (ZEN-94-012), T32 NS07098, and 5T32AG00194.

REFERENCES

1. Hirano, A., & Zimmerman, H. M. (1962). *Arch. Neurol., 7,* 73–88.
2. Wisniewski, H. M., Terry, R. D., & Hirano, A. (1970). *J. Neuropathol. Exp. Neurol., 29,* 163–176.
3. Kidd, M. (1964). *Brain, 87,* 307–320.
4. Greenberg, S., & Davies, P. (1990). *Proc. Natl. Acad. Sci. USA, 87,* 5827–5831.

5. Grundke-Iqbal, I., Iqbal, K., Quinlan, M., Tung, Y.-C., Zaidi, M. S., & Wisniewski, H. M. (1986). *J. Biol. Chem., 261,* 6084–6089.
6. Ksiezak-Reding, H., Binder, L. I., & Yen, S.-H. (1990). *J. Neurosci. Res., 25,* 420–430.
7. Gravina, S. A., Ho, L., Eckman, C. B., Long, K. E., Otvox, L., Jr., Younkin, L. H., Suzuki, N., & Younkin, S. G. (1995). *J. Biol. Chem., 270,* 7013–7016.
8. Miller, D. L., Papadyannopulos, I. A., Styles, J., Robin, S. A., Lin, V. V., Biemann, K., & Iqbal, K. (1993). *Arch. Biochem., 301,* 41–51.
9. Kang, J., Lemaire, H. G., Unterbeck, A., Salbqum, J. M., Master, C. L., Grzeschik, K. M., Multhaup, G., Beyreuther, K., & Muller-Hill, B. *Nature, 325,* 733–736.
10. Himmler, A. (1989). *Mol. Cell. Biol., 9,* 1381–1388.
11. Ksiezak-Reding, H., Shafit-Zagardo, B., & Yen, S.-H. (1995). *J. Neurosci. Res., 41,* 583–593.
12. Liu, W.-K., Dickson, D. W., & Yen, S.-H. (1993). *Am. J. Pathol., 142,* 387–394.
13. Lee, G., Neve, R. L., & Kosik, K. S. (1989). *Neuron, 2,* 1615–1624.
14. Ennulat, D. J., Liem, R. K., Hasim, G. A., & Shelanski, M. L. (1989). *J. Biol. Chem., 264,* 5327–5330.
15. Goedert, M., & Jakes, R. (1990). *EMBO J., 9,* 4225–4230.
16. Kenessey, A., Yen, S.-H., Liu, W.-K., Yang, X.-R., & Dunlop, D. S. (1995). *Brain Res., 675,* 183–189.
17. Yan, S. D., Chen, X., Schmidt, A.-M., Brett, J., Godman, G., Zou, Y.-S., Scot, C. W., Caputo, C., Grappier, T., Smith, M. A., Perry, G., Yen, S.-H., & Stern, D. (1994). *Proc. Natl. Acad. Sci. USA, 92,* 7787–7791.
18. Ledesma, M. D., Bonay, P., Colaco, C., & Avila, J. (1994). *J. Biol. Chem., 269,* 21614–21619.
19. Wan, J. Z., Grundke-Iqbal, I., & Iqbal, K. (1996). *Nature Med., 2,* 871–875.
20. Mori, H., Kondo, J., & Ihara, Y. (1987). *Science, 235,* 1641–1644.
21. Perry, G., Friedman, R., Shaw, G., & Chau, V. (1987). *Proc. Natl. Acad. Sci. USA, 84,* 3033–3036.
22. Ksiezak-Reding, H., Liu, W.-K., & Yen, S.-H. (1992). *Brain Res., 597,* 209–219.
23. Matsuo, E. S., Shin, R.-Y., Billingsley, M. L., Van de Voorde, A., O'Connor, M., & Lee, V. M.-Y. (1994). *Neuron, 13,* 989–1002.
24. Liu, W.-K., & Yen, S.-H. (1996). *J. Neurochem., 66,* 1131–1139.
25. Sergeant, N., Bissoere, T., Vermersch, P., & Delacourte, A. (1995). *NeuroReport, 6,* 2217–2220.
26. Kenessey, A., & Yen, S.-H. (1993). *Brain Res., 629,* 40–46.
27. Brion, J. P., Hanger, M. Y., Bruce, M. Y., Couck, A. M., Flament-Durand, J., & Anderton, B. H. (1991). *Biochem. J., 273,* 127–133.
28. Goedert, M., Jakes, R., Browther, R. A., Six, J., Lubke, U., Vandermeeren, M., Cras, P., Trojanowski, J. Q., & Lee, V. M.-Y. (1993). *Proc. Natl. Acad. Sci. USA, 90,* 5066–5070.
29. Hasegawa, M., Morishima-Kawashima, M., Takio, K., Suzuki, M., Titani, K., & Ihara, Y. (1992). *J. Biol. Chem., 267,* 17047–17054.
30. Iqbal, K., Grundke-Iqbal, I., Smith, A. J., George, L., Tung, Y.-C., & Zaidi, T. (1989). *Proc. Natl. Acad. Sci. USA, 86,* 5646–5650.
31. Kanemaru, M., Miura, R., Hasegawa, M., Kuzuhara, A., & Ihara, Y. (1992). *J. Neurochem., 58,* 1667–1675.
32. Lang, E., Szendrei, G. I., Lee, V. M.-Y., & Otvos, L. (1992). *Biochem. Biophys. Res. Commun., 187,* 783–790.
33. Liu, W.-K., Moore, W. T., Williams, R. T., Hall, F. L., & Yen, S.-H. (1993). *J. Neurosci. Res., 34,* 371–376.
34. Liu, W.-K., Dickson, D. W., & Yen, S.-H. (1994). *J. Neurochem., 62,* 1055–1061.
35. Biernat, J., Gustke, N., Drewes, G., Mandelkow, E.-M., & Mandelkow, E. (1993). *Neuron, 11,* 153–163.
36. Lindwall, G., & Cole, R. D. (1984). *J. Biol. Chem., 259,* 5301–5305.
37. Yoshida, H., & Ihara, Y. (1993). *J. Neurochem., 61,* 1183–1186.
38. Bancher, C., Brunner, C., Lassmann, H., Budk, H., Jellinger, J., Wiche, G., Seitelberger, F., Grundke-Iqbal, I., Iqbal, K., & Iwsniewski, H. M. (1991). *Brain Res., 539,* 11–18.

39. Litersky, J. M., & Johnson, G. V. W. (1992). *J. Biol. Chem., 267,* 1563–1568.
40. Baudier, J., Lee, S.-H., & Cole, R. D. (1987). *J. Biol. Chem., 262,* 17584–17590.
41. Correas, I., Diaz-Nindo, J., & Avila, J. (1992). *J. Biol. Chem., 267,* 15721–15728.
42. Steiner, B., Mandelkow, E.-M., Biernat, J., Gustke, N., & Mandelkow, E. (1990). *EMBO J., 11,* 3539 3544.
43. Roder, H. M., Eden, P. A., & Ingram, V. M. (1993). *Biochem. Biophys. Res. Commun., 193,*639–647.
44. Mandelkow, E.-M., Drewes, G., Biernat, J., Gustke, N., Vanlint, J., Vandenheede, J. R., & Mandelkow, E. (1992). *FEBS Lett , 314,* 315–321.
45. Hanger, D. P., Hughes, K., Woodgett, R., Brion, J.-P., & Anderton, B. H. (1992). *Neurosci. Lett., 147,* 58–62.
46. Yang, S.-D., Song, J.-S., Yu, J.-S., & Shiah, S.-G. (1993). *J. Neurochem., 61,* 1742–1747.
47. Ishiguro, K., Shiratsuchi, A., Sato, S., Omori, A., Arioka, M., Kobayashi, S., Uchida, T., & Imahori, K. (1993). *FEBS Lett., 325,* 167–172.
48. Baumann, K., Mandelkow, E.-M., Biernat, J., Piwnica-Worms, H., & Mandelkow, E. (1993). *FEBS Lett., 336,* 417–424.
49. Morishima-Kawashima, M., Hasegawa, M., Takio, K., Suzuki, M., Yoshidq, H., Titani, K., & Ihara, Y. (1995). *J. Biol. Chem., 270,* 823–829.
50. Hall, F. L., Braun, R., Mitchell, J. P., & Vulliet, P. R. (1990). *Proc. West Pharmacol. Soc., 33,* 213–217.
51. Mawal-Dewan, M., Parimal, C. S., Abdel-Ghany, M., Shalloway, D., & Racker, E. (1992). *J. Biol. Chem., 267,* 19705–19709.
52. Scott, C. W., Vulliet, P. R., & Caputo, C. B. (1993). *Brain Res., 611,* 237–242.
53. Vulliet, R., Halloran, S. M., Braun, R. K., Smith, A. L., & Lee, G. (1992). *J. Biol. Chem., 267,* 22570–22574.
54. Sing, T. J., Grunkde-Iqbal, I., & Iqbal, K. (1995). *J. Neurochem., 64,* 1420–1423.
55. Greenwood, J. A., Scott, C. W., Spreen, R. C., Caputo, C. B., & Johnson, G. V. (1994). *J. Biol. Chem., 269,* 4373–4380.
56. Drewes, G., Trinczek, B., Illenberger, S., Biernat, J., Schmit-Ulms, G., Meyer, H. D., Mandelkow, E. M., & Mandelkow, E. J. (1995). *J. Biol., 270,* 7679–7688.
57. Wu, J. M., Chen, Y., Hsieh, T. C., Brandt, R., & Lee, G. (1996). *Biochem. Mol. Biol. Int., 40,* 21–31.
58. Vincent, I., & Davies, P. (1992). *Proc. Natl. Acad. Sci. USA, 89,* 2827–2882.
59. Liu, W.-K., Williams, R., Hal, F. L., Dickson, D. W., & Yen, S.-H. (1995). *Am. J. Pathol., 146,* 228–238.
60. Vincent, I., Rosado, M., Kim, E., & Davies, P. (1994). *J. Neurochem., 62,* 715–723.
61. Litersky, J. M., & Johnson, G. V. W. (1995). *J. Neurochem., 65,* 903–911.
62. Croall, E. E., & DeMartino, G. N. (1991). *Physiol. Rev., 71,* 813–847.
63. Saido, T. C., Sorimachi, H., & Suzuki, K. (1994). *FASEB J., 8,* 814–822.
64. Saito, K.-I., Elce, J. S., Hamon, J. E., & Nixon, R. A. (1993). *Proc. Natl. Acad. Sci. USA, 90,* 2628–2632.
65. Yang, L.-S., & Ksiezak-Reding, H. (1995). *Eur. J. Biochem., 233,* 9–19.
66. Yang, L. S., & Ksiezak-Reding, H. (1995). *J. Neurochem., 64,* S54B.
67. Sasaki, T., Kikuchi, T., Yumoto, N., Yoshimura, N., & Murachi, T. (1984). *J. Biol. Chem., 259,* 12489–12494.
68. Ksiezak-Reding, H., & Yen, S.-H. (1991). *Neuron, 6,* 717–728.
69. Wischik, D. M., Novak, M., Thogersen, H. C., Edwards, P. C., Runswick, M. J. R., Walke, J. E., Milstein, C., Roth, M., & Klug, A. (1988). *Proc. Natl. Acad. Sci. USA, 85,* 4506–4510.
70. Marks, N., & Lajtha, A. (1965). *Biochem. J., 97,* 74–83.
71. Whitaker, J. N., Terry, L. C., & Whetsell, W. O., Jr. (1986). *Brain Res., 216,* 109–124.
72. Berstein, H.-G., Weideranders, B., Rinne, A., & Dorn, A. (1985). *Acta Histochem., 77,* 139–142.
73. Seglen, P. O., & Bohley, P. (1992). *Experentia, 48,* 158–172.

74. Kenessey, A., Banay-Schwartz, M., DeGuzman, T., & Lajtha, A. (1989). *J. Neurosci. Res., 23,* 454–456.
75. Nakamura, Y., Takeda, M., Suzuki, H., Morita, H., Tada, K., Hariguchi, S., & Nishimura, T. (1989). *Mech. Ageing Dev., 50,* 215–225.
76. Banay-Schwartz, M., DeGusman, T., Kenessey, A., Palkovits, M., & Lajtha, A. (1992). *J. Neurochem., 58,* 2207–2211.
77. Nakinishi, H., Tominaga, K., Amano, T., Hirotsu, I., & Yamamato, K. (1994). *Exp. Neurol., 126,* 119–128.
78. Nakamura, Y., Takeda, M., Suzuki, H., Morita, H., Tada, K., Hariguchi, S., & Nishimura, T. (1986). *Neurosci. Lett., 97,* 215–220.
79. Cataldo, A. M., Thayer, C. Y., Bird, E. D., Wheelock, R. R., & Nixon, R. A. (1990). *Brain Res., 513,* 181–192.
80. Nakamura, Y., Takeda, M., Suzuki, H., Hattori, H., Tada, K., Hariguchi, S., & Nishimura, T. (1991). *Neurosci. Lett., 130,* 195–198.
81. Kenessey, A., Nacharaju, P., Ko, L., & Yen, S.-H. (1997). *J. Neurochem., 69,* 2026–2038.
82. Bondareff, W., Wischik, C. M., Novak, M., Amos, W. B. A., & Roth, M. (1990). *Am. J. Pathol., 137,* 711–723.
83. Dickson, D. W., Ksiezak-Reding, H., Liu, W.-K., Davies, P., Crowe, A., & Yen, S.-H. (1992). *Acta Neuropathol., 84,* 596–605.
84. Endoh, R., Ogawara, M., Iwatsubo, T., Nakano, I., & Mori, H. (1993). *Brain Res., 601,* 164–172.
85. Farooqui, A. A., Wells, K., & Horrocks, L. A. (1995). *Mol. Chem. Neuropathol., 25,* 155–173.
86. Snow, A. D., Mar, H., Nochlin, D., Sekiguchi, R. T., Kimata, K., Koike, Y., & Wight, T. N. (1990). *Am. J. Pathol., 137,* 1253–1270.
87. Perry, G., Siedlak, S. L., Richey, P., Kawai, M., Crass, P., Kalaria, R. N., Galloway, P. G., Scardina, J. M., Cordell, B., Greenberg, B. D., Ledbetter, S. R., & Gambetti, P. (1991). *J. Neurosci., 11,* 3679–3683.
88. Goedert, M., Jakes, R., Spillantini, M. G. M., Smith M. J., & Crowther, R. A. (1996). *Nature (London), 383,* 550–553.
89. Perez, M., Valpuesta, J. M., Medina, M., Garcini, E. M., & Avila, J. J. (1996). *J. Neurochem., 67,* 1183–1190.
90. Kampers, T., Friedhoff, P., Biernat, J., Mandelkow, E. M., & Mandelkow, E. (1996). *FEBS Lett., 399,* 344–439.
91. Wilson, D. M., & Binder, L. I. (1997). *Am. J. Pathol., 150,* 2181–2195.
92. Baynes, J. W. (1991). *Diabetes, 40,* 405–412.
93. Brownlee, M., Cerami, A., & Vlassara, H. (1988). *N. Engl. J. Med., 318,* 1315–1321.
94. Sell, D. R., & Monnier, V. M. (1989). *J. Biol. Chem., 264,* 21597–21602.
95. Dunn, J. A., Patrick, J. S., Thorpe, S. R., & Baynes, J. W. (1989). *Biochemistry, 28,* 9464–9468.
96. Grandhee, S. K., & Monnier, V. M. (1991). *J. Biol. Chem., 266,* 11649–11653.
97. Fu, M., Wells-Knech, K. J., Blackledge, J. A., Lyon, T. J., Thorpe, S. R., & Baynes, J. W. (1994). *Diabetes, 43,* 676–683.
98. Shapiro, R., McManus, M. J., Zalut, C., & Bunn, H. F. (1980). *J. Biol. Chem., 255,* 3120–3127.
99. Watkins, N. G., Thorpe, S. R., & Baynes, J. W. (1985). *J. Biol. Chem., 260,* 10629–10636.
100. Iberg, N., & Fluckiger, R. (1986). *J. Biol. Chem., 261,* 13542–13545.
101. Shilton, B. H., & Walton, D. J. (1991). *J. Biol. Chem., 266,* 5587–5595.
102. Singer, E. E., Nathan, D. M., Anderson, K. M., Wilson, P. W. F., & Evans, J. C. (1992). *Diabetes, 41,* 202–208.
103. Brownlee, M. (1992). *Diabetes Care, 15,* 1835–1843.
104. Monnier, V. M., Sell, D. R., Nagaraj, R. H., Miayata, S., Grandhee, S., Odetti, P., & Ibrahim, S. A. (1992). *Diabetes, 41,* 36–41.
105. Palinski, W., Koschinsky, T., Butler, S. W., Miller, E., Vlassara, H., Cerami, A., & Witztum, J. L. (1995). *Atheroscler. Thromb. Vasc. Biol., 15,* 571–582.
106. Day, J. F., Thorpe, S. R., & Baynes, J. W. (1979). *J. Biol. Chem., 254,* 595–597.

107. Ponde, A., Garner, W. H., & Spector, A. (1979). *Biochem. Biophys. Res. Commun., 89,* 1260–1266.
108. Agarwal, K. C., Parks, R. E., Widness, J. A., & Schwartz, R. (1985). *Diabetes, 34,* 251–255.
109. Arai, K., Iizuka, S., Tada, Y., Oikawa, K., & Taniguchi, N. (1987). *Biochem. Biophys. Acta, 924,* 292–296.
110. McPherson, J. D., Shilton, B. H., & Walton, D. J. (1988). *Biochem. Biophys. Res. Commun., 152,* 711–716.
111. Smith, M. A., Taneda, S., Richey, P. L., Miyata, S., Yan, S.-D., Sayre, L. M., Monnier, V. M., & Perry, G. (1994). *Proc. Natl. Acad. Sci. USA, 91,* 5710–5714.
112. Dickson, D. W., Sinicropi, S., Yen, S.-H., Ko, L., Bucala, R., & Vlassara, H. (1996). *Neurobiol. Aging, 17,* 733–743.
113. Liu, W.-K., Ksiezak-Reding, H., & Yen, S.-H. (1991). *J. Biol. Chem., 266,* 21723–21727.
114. Ledesma, M. D., Bonay, P., Colaco, C., & Avila, J. (1994). *J. Biol. Chem., 269,* 21614–21619.
115. Ledesma, M. D., Medina, M., & Avila, J. (1996). *Mol. Chem. Neuropathol., 27,* 249–258.
116. Vitek, M. P., Bhattacharya, K., Glendening, J. M., Stopa, E., Vlassara, H., Bucala, R., Manogue, K., & Cerami, A. (1994). *Proc. Natl. Acad. Sci. USA, 91,* 4766–4770.
117. Yan, S. D., Yan, S. F., Chen, Z., Fu, J., Chen, M., Kuppusamy, P., Smith, M. A., Perry, G., Godman, G. C., Nawroth, P., Zweier, J. L., & Stern, D. (1995). *Nature Med., 1,* 693–699.
118. Ledesma, M. D., Bonay, P., Colaco, C., & Avila, J. (1995). *J. Neurochem., 65,* 1658–1664.
119. Nacharaju, P., Ko, L., & Yen, S.-H. (in press). *J. Neurochem.*

12

A NOVEL GENE IN THE ARMADILLO FAMILY INTERACTS WITH PRESENILIN 1

KENNETH S. KOSIK, CAROLE HO,
UDAYA LIYANGE, CYNTHIA LEMERE,
MIGUEL MEDINA, AND JIANHUA ZHOU

Center for Neurologic Diseases
Brigham and Women's Hospital and Harvard Medical School
Boston, Massachusetts 02115

Like a child with a new toy, Alzheimer scientists have flocked to presenilin as a gene that promises new revelations about the disease. Although the genes that encode the components of plaques and tangles have been known for some time, the fundamental disease process has remained elusive. Clearly, a number of processing and regulatory steps in the metabolism of the amyloid precursor protein (APP) are missing. So it was no surprise that interest in the presenilins soared when this previously unsuspected player appeared on the Alzheimer stage. Part of the excitement was due to presenilin's dramatic entrance through the door of genetics. The gene was discovered not by purifying components of the pathological structures, but by the use of positional cloning techniques in families with an inherited form of the disease. As such, presenilin was not pinned to any one camp among the vying factions who place their chips in the various Alzheimer bins of amyloid, tau, or apoE.

Presenilin 1 (PS1) is the most commonly mutated gene in early-onset Alzheimer's disease (AD); there are more than 40 different reported mutations.[1, 2] Mutations in this gene result in a clinical phenotype that is nearly identical to sporadic AD; the most prominent distinction is the early age of onset. The most comprehensive clinical description of patients with a presenilin mutation is that of Lopera et al.[3] These patients from the largest kindred in the world with familial AD harbor a glutamic acid to alanine mutation at position 280 of PS1.[4] The families live in Antioquia, Colombia and consist of five extended pedigrees, all of whom probably share a common founder. From a total of more than 3000 individuals described in the family trees, a case series of 128 individuals was identified of which 6 had autopsy-proven early-onset AD, 93 had probable early onset AD, and 29 had possible early onset AD. The disease had a mean age of onset of 46.8 years with a range of 34 to 62 years. The average interval until death was 8 years. The most frequent presentation was memory loss followed by behavior and personality changes and progressive loss of language ability. In the final stages gait disturbances, seizures, and myoclonus were frequent. Other than the early onset, the clinical phenotype was indistinguishable from sporadic AD. Despite the uniform genetic basis for the disease, there was significant variability in the age of onset, suggesting an important role for environmental factors or genetic modifiers in determining the age of disease onset.

Perhaps the most compelling datum that links the presenilins to the disease process is the unanimity among investigators that presenilin mutations increase $A\beta_{42}$ secretion.[5-9] These observations performed in many independent systems by different investigators strengthen the idea that $A\beta$ secretion represents a key step in the pathogenesis of AD, and increased secretion represents one pathogenetic mechanism. Compared to the shorter $A\beta_{40}$ form of the peptide, $A\beta_{42}$ shows enhanced fibrillogenic properties in vitro[10] and deposits earlier in the disease course.[11] In AD caused by presenilin mutations, not only is $A\beta_{42}$ increased when measured by enzyme-linked immunosorbent assay (ELISA), but $A\beta_{42}$ deposits are increased.[8] In this study, four autopsied patients with the E280A mutation were analyzed by computer-assisted quantification of the brain sections stained with antibodies specific for the alternative carboxy termini of $A\beta$. There was an increased burden of $A\beta_{42}$-immunoreactivity but not of $A\beta_{40}$ compared to 12 sporadic control cases. Significant cerebellar pathology that included $A\beta_{42}$-reactive plaques often bearing dystrophic neurites and reactive glia was also noted. Together these results suggest that PS1 mutations alter APP processing in a way that favors the generation and deposition of $A\beta_{42}$.

Few clues regarding the function of PS1 are known. It is a polytopic protein resident in the endoplasmic reticulum[12] that generates two stable endoproteolytic fragments of ~27–28 kDa and ~17–18 kDa.[13] Both of the fragments remain in the endoplasmic reticulum, and the amounts of these fragments are highly regulated in that increased expression of the full-length PS1 does not increase the generation of fragments; instead overexpression results in increased uncleaved PS1.[6] Its amino and carboxy termini, as well as a hydrophilic "loop" region, lie within the cytoplasm.[14, 15] PS1 is expressed in many tissues, but is enriched in the brain and in neurons as evidenced by in situ hybridization, immunocytochemistry, and Northern blots.[12, 16, 17]

Despite its name, PS1 plays a key role early in development. Mice with a targeted null mutation in the murine homolog of PS1 die shortly after natural birth.[18, 19] Defects in the axial skeleton and somite organization similar to those in a Notch1 deleted animal[20, 21] were present. Remarkably, within the brain there was a symmetrical cavitation of the ventrolateral region of the ventricular zone in the posterior portion of the mutant brain. Additionally, there was degeneration in the subcortical region of the temporal lobe. Of particular interest was the failure to detect Notch expression or Dll1/Notch ligand expression in presomitic tissue caudally.[19] Within this caudal region, both Notch$^{-/-}$and PS$^{-/-}$ animals show blurring of the intersomite boundaries. The finding of a functional interaction between Notch and presenilin in development also came from a suppressor screen in *Caenorhabditis elegans* to detect genes that rescue the effects of mutations in lin-12, a Notch1 homolog.[22] An allele of a gene called sel-12 suppressed the gain-of-function mutation, and remarkably this gene shared a 50% identity with PS1. Furthermore, normal human presenilin genes can rescue the mutant lin-12 phenotype.[23] Of course, none of these studies preclude a role for PS1 in the mature brain, they simply reflect the bias of genetic techniques to detect developmental defects. Nevertheless, PS1 Alzheimer mutations may begin to have consequences even during development because PS1 mutations only partially rescue the sel-12 mutant phenotype.[23] On the other hand, patients with PS1 mutations have no detectable abnormalities before the onset of AD.

Given the strong evidence in favor of a genetic interaction between sel-12/presenilin and the Notch pathway, other intermediates in related signaling cascades seemed likely to appear. One related signaling pathway is Wingless (Wg)—a downstream effector in the Wingless pathway called *disheveled* gene (Dsh) can inhibit Notch.[24] Therefore it was particularly exciting to discover in our initial two-hybrid screen using the loop region of PS1 as the bait, a novel gene that had homology with Armadillo.[25] These data will be discussed in the following; however, first we present some background on the key molecules involved in the Wg signaling pathway.

Wg and its vertebrate homologs, the Wnt genes, represent a striking connection between development and oncogenesis.[26] Ectopic expression of Wnt induces axis duplication in frog embryos[27] and mammary cancer in mice; Wnt deficiency results in impaired brain development in mammals[28, 29] and impaired segmentation in insects.[30] Wnt is a secreted glycoprotein that can cause cells to proliferate, differentiate, or simply survive by signaling through autocrine and paracrine routes.

Armadillo and its vertebrate homolog, β-catenin, lie downstream in the Wg/Wnt pathway. Mutations in either Armadillo or Wg cause similar phenotypic changes in segment polarity when mutated in *Drosophila*,[31] and are required for the correct expression of a large set of genes including *engrailed* in the ectoderm and other homeobox genes in the endoderm and mesoderm [reviewed in Klingensmith and Nusse[32]]. Wg is believed to signal through the frizzled (Dfz2) receptor[33] and the Dsh to inactivate the kinase zeste-white 3 (the *Drosophila*

homolog of glycogen synthetase kinase-3β [GSK-3β]).[34] GSK-3β is a kinase that
is active under basal conditions and turned off when the Wg/Wnt pathway is
stimulated. GSK-3β phosphorylation of Arm/β-catenin destabilizes the protein
and keeps the levels of this protein low in unstimulated cells. Arm/β-catenin levels are also kept low by binding to the tumor suppressor gene, adenomatous polyposis coli (APC), which is also a substrate for GSK-3β.[35] Why must β-catenin
levels be kept low? One possible answer is the effects of β-catenin on transcription. β-Catenin interacts with Lef-1/TCF DNA-binding proteins, travels to the
nucleus, and forms an active transcriptional complex.[36–39]

β-Catenin selectively interacts with various binding partners via discrete signaling motifs within the protein. These so-called arm motifs are imperfectly
repeated stretches of 42 amino acids that appear to mediate specific protein–protein interactions similar to SH2 or ankyrin domains.[40] However, in contrast to
some binding motifs, such as SH2 or SH3, interactions with arm domains often
require several domains for a successful interaction. Proteins containing arm
motifs are a growing and diverse group that contain 7–13 copies of the arm motif
arranged in tandem with little or no intervening sequence. In addition to β-
catenin the group consists of γ-catenin,[41] p120,[42] p0071,[43, 44] SRP1,[43] smgGDS,[45]
band 6 protein, and APC.[46, 47] The arm domains of these proteins are believed to
serve as distinct signaling motifs.

β-Catenin not only has a signaling function but is also present in the adherens
junction, where it forms part of a complex that links the extracellular domains of
adhesive molecules to the cytoskeleton. Adherens junctions are associated with
homotypic cell adhesion and consist of a cadherin receptor linked via α-catenin
to the actin cytoskeleton [reviewed in Gumbiner[48]]. Desmosomes, which are
prominent in epithelial cells, are related structures linked to intermediate filaments and contain plakoglobin, a close relative of β-catenin. Cadherins from two
different cells form tight antiparallel homotypic dimers by binding at their N-ter-
mini. The cytoplasmic portion of the cadherins are bound to complexes which, in
addition to α-catenin, include the arm motif proteins β-catenin, plakoglobin (also
known as γ-catenin), and p120. The ability of β-catenin to function in cell–cell
adhesive interactions and to regulate the transcription of specific genes suggests
a key role for this molecule and its family members in the assembly and maintenance of three-dimensional tissue architecture. Although much of the discussion
to this point has centered on development, it is useful to point out here, because
of our interest in the mature brain, that adhesive interactions may become more
stable in the adult, but they are not static. In fact, the maintenance of such interactions probably requires active cellular mechanisms.

In brain, catenins are found within a subdomain of the synaptic junction bordering the transmitter release zone in a structure referred to as the synaptic
adherens junction.[49, 50] Furthermore, Dsh, which is upstream of Arm/β-catenin in
the Wg/Wnt signaling pathway, contains a PDZ domain, which is a sequence
motif that may control protein–protein interactions adjacent to cell membranes
and the formation of macromolecular signal transduction complexes [reviewed in

Ponting and Phillips [51]. PDZ domains are present in other proteins including the brain-specific protein PSD-95. Thus, the synaptic junction contains the structural elements of the adherens junction and may contain related signaling elements of developmental pathways possibly co-opted for plasticity functions. Interestingly, synapses represent a target site of the AD process and plasma membranous APP is localized to sites of cell–cell contact.[52]

DISCOVERY OF δ-CATENIN

We have discovered two interactors with PS1: One is a novel member of the Armadillo family called δ-catenin and the other is β-catenin. The interest in this interaction arises from the strong evidence that the Wg pathways can exert an inhibitory influence on Notch. The Dsh protein, an effector required for cells to respond to Wg, can form an inhibitory complex with the intracellular domain of Notch.[24] Overexpression of dsh in the *Drosophila* wing attenuates Notch signaling in a Wg-dependent manner. In vertebrate organisms there are a plethora of Wnt genes expressed in the embryo. The overlap of their expression domains with Notch expression domains suggests that interactions between these signaling pathways extend well beyond the fly wing. The known link between the Notch and Wingless pathways provides a possible biological rationale for our discovery of an arm protein interacting with PS1.

The bait used for the two-hybrid screen was amino acids 263–407 of PS1 fused to the DNA-binding domain of GAL4. Approximately 4 million library transformants were screened from which 8 clones that co-activate the GAL4-responsive HIS3 and lacZ reporter genes of Y190 were isolated. Two of the isolates interacted specifically in a yeast two-hybrid mating assay that utilized the irrelevant proteins, murine p53 as a positive control and pLAM 5'-1 human lamin C as a negative control. A nucleotide sequence analysis of these two independent isolates revealed a novel gene with homology to Armadillo. One of these two clones was designated A1. Portions of the 3′ end of the A1 sequence were present in the database in a clone designated NIBA2. This clone was obtained from two YACs that spanned the ~2Mb cri-du-chat critical region (CDCCR) of chromosome 5p[53] and were used to screen high-density arrays of clones directly from a normalized infant brain (NIB) cDNA library.[54] This led to the identification of a ~2.0 kB cDNA, designated NIBA2, that mapped back to YAC 938G6 but beyond the boundaries of the CDCCR.[55] To extend A1 in the 5′ direction we utilized human adult brain Marathon cDNA with Taq polymerase and 5′ Race PCR and assembled the entire sequence.

The full-length sequence showed greatest homology to p0071 (69.3% identical, EMBL GenBank accession number X81889) and somewhat less homology to p120 (48.0% identical, accession number Z17804), two proteins along with B6P/plakophilin 1 (accession number X79293), that are considered a subfamily of the catenins. We termed the novel gene δ-catenin. In contrast to β-catenin, these proteins have 10 repeat arm motifs that span 482 amino acids for δ-catenin, 479 amino acids for p0071, and 554 amino acids for p120. Very little is known about p0071, whereas somewhat more is known about p120. This protein was initially identified as one of several substrates of pp60^{src}.[56, 57] In fact, a mutational analysis suggested that tyrosine phosphorylation of p120 was necessary for pp60^{src}-mediated transformation.[56, 58] It is membrane associated and can be myristoylated, but is not glycosylated.

To confirm the validity of the interaction, PS1 and δ-catenin were shown to co-immunoprecipitate in vitro. cDNAs encoding A1 (the carboxy terminus of δ-catenin), full-length PS1, or the loop region of PS1 were subcloned into the PSV-K3 expression vector, and these plasmids were used for coupled in vitro transcription/translation in rabbit reticulocyte lysates. Translation of the plasmids in the presence of [^{35}S] methionine generated a radiolabeled A1 polypeptide of ~45 kDa, a PS1 band of ~43 kDa, and a PS1 loop peptide of 16 kDa. Various combinations of these in vitro translated proteins were mixed in binding buffer and immunoprecipitated. The antibody immunoprecipitated the PS1 loop peptide with A1 and full-length PS1 with A1. An in vitro interaction was also demonstrated by mixing equivalent aliquots of the radiolabeled protein with purified glutathione S-transferase (GST) or with purified GST-fusion proteins containing either A1, an amino terminal PS1 construct from amino acid 1–80, or residues 263–407 of PS1. Each mixture of GST proteins was adsorbed to glutathione-agarose beads. In vitro translated A1 was specifically retained on the beads by the PS1 loop–GST fusion protein, but not by the parental GST polypeptide or by GST fused to the amino terminus of PS1. The reverse experiment demonstrated that in vitro translated PS1 residues 263–407 specifically bound to the A1–GST fusion protein. A more definitive proof for an interaction with PS1 was obtained using CHO cells stably transfected with wild-type PS1. Full-length δ-catenin was fused in frame to an HA tag and transfected into the CHO cells. The cells were homogenized and immunoprecipitated with our antibody prepared against the N-terminus of PS1. The blot was then stained with an HA antibody, and a band at the appropriate molecular weight was observed in the immunoprecipitate. This experiment provides additional evidence that PS1 and δ-catenin interact in living cells.

At this point we sought other members of the Armadillo family that interact with PS1. We utilized 293 cells for co-immunoprecipitation because they have relatively high levels of catenins and nondetectable levels of δ-catenin. Two different PS1 antibodies precipitated β-catenin, but failed to co-immunoprecipitate α-catenin or γ-catenin. In the reverse experiment, an antibody to β-catenin co-immunoprecipitated PS1. We concluded that PS1 also forms stable complexes in vivo with β-catenin and that the specific binding partner for PS1 may vary depending upon the cell type.

The demonstration of an interaction between PS1 and at least two members of the Armadillo family, β-catenin and δ-catenin, raises the possibility that the interaction is regulated by signals known to affect β-catenin stability, nuclear translocation, and ultimately gene transcription. The upstream signals known to regulate this pathway are probably initiated by binding of the Wg/Wnt ligand to the frizzled (Fz) receptor [reviewed in Perrimon[59]]. Both the ligand and receptor are members of larger families.[33] An additional level of regulation in the stimulation of this pathway occurs via the Frzb gene which is structurally related to a putative Wnt binding domain of Fz.[60, 61] Frzb binds a Wnt and blocks its signaling. Because intermediates in the Wingless pathway can regulate Notch,[24] the finding of an interaction between PS1 and a signaling molecule in the Wnt/Wg pathway may explain the genetic observations in *C. elegans.* How early developmental genes are tied to processes that occur in the aging brain stands as an investigative arena of intense interest.

Within the cell, both presenilin[12] and members of the Wnt family[62] are retained in the endoplasmic reticulum. Although PS1 is thought to be resident in the endoplasmic reticulum, Wnt, as a secreted glycoprotein, probably lies in an export pathway. Also within the endoplasmic reticulum, APP begins to traverse its way to its destination in the plasma membrane, where it resides as a mature protein and is secreted. Because published data describe co-immunoprecipitation of APP and PS1,[63] a complex may exist that includes both δ-catenin and APP.

REFERENCES

1. Tanzi, R. E., Kovacs, D. M., Kim, T.-W., Moir, R. D., Guenette, S. Y., & Wasco, W. (1996). The presenilin genes and their role in early-onset familial Alzheimer's disease. *Alzheimer's Disease Review, 1,* 91–98.

2. Hardy, J. (1997). Amyloid, the presenilins and Alzheimer's disease. *Trends Neurosci., 20,* 154–159.

3. Lopera, F., Ardilla, A., Martinez, A., Madrigal, L., Arango-Viana, J. C., Lemere, C. A., Arango-Lasprilla, J. C., Hincapie, L., Arcos-Burgos, M., Ossa, J. E., Behrens, I. M., Norton, J., Lendon, C., Goate, A. M., Ruiz-Linares, A., Roselli, M., & Kosik, K. S. (1997). Clinical features of early-onset Alzheimer disease in a large kindred with an E280A presenilin-1 mutation. *JAMA, 277,* 793–799.

4. Clark, R. F., Hutton, M., Fuldner, R. A., Forelich, S., Karran, E., Talbot, C., Crook, R., Lendon, C., Prihar, G., He, C., Korenblat, K., Martinez, A., Wragg, M., Busfield, F., Behrens, M. I., Myers, A., Norton, J., Morris, J., Mehta, N., Pearson, C., Lincoln, S., Baker, M., Duff, K., Zehr, C., Perez-Tur, J., Houlden, H., Ruiz, A., Ossa, J., Lopera, F., Arcos, M., Madrigal, L., Collinge, J., Humphreys, C., Ashworth, A., Sarner, S., Fox, N., Harvey, R., Kennedy, A., Roques, P., Cline, R. T., Philips, C. A., Venter, J. C., Forsell, L., Axelman, K., Lilius, L., Johnston, J., Cowburn, R., Viitanen, M., Winblad, B., Kosik, K., Haltia, M., Poyhonen, M., Dickson, D., Mann, D., Neary, D., Snowden, J., Lantos, P., Lannfelt, L. M. R., Roberts, G. W., Adams, M. D., Hardy, J., & Goate, A. (1995). The structure of the presenilin 1 (S182) gene and identification of six novel mutations in early onset AD families. *Nature Genet., 11,* 219–222.

5. Scheuner, D., Eckman, C., Jensen, M., Song, X., Citron, M., Suzuki, N., Bird, T. D., Hardy, J., Hutton, M., Kukull, W., Larson, E., Levy-Lahad, E., Viitanen, M., Pesking, E., Poorkaj, P., Schellenberg, G., Tanzi, R., Wasco, W., Lannfelt, L., Selkoe, D., & Younkin, S. (1996). Secreted amyloid beta-protein similar to that in the senile plaques of Alzheimer's disease is increased in vivo by the presenilin 1 and 2 and APP mutations linked to familial Alzheimer's disease. *Nature Med., 2,* 848–870.

6. Borchelt, D. R., Thinakaran, G., Eckman, C. B., Lee, M. K., Davenport, F., Ratovitsky, T., Prada, C.-M., Kim, G., Seekins, S., Yager, D., Slunt, H. H., Wang, R., Seeger, M., Levey, A. I., Gandy,

S. E., Copeland, N. G., Jenkins, N. A., Price, D. L., Younkin, S. G., & Sisodia, S. S. (1996). Familial Alzheimer's disease-linked presenilin 1 variants elevate Aβ1-42/1-40 ratio in vitro and in vivo. *Neuron, 17,* 1005–1013.

7. Citron, M., Westaway, D., Xia, W., Carlson, G., Diehl, T., Levesque, G., Johnson-Wood, K., Lee, M., Seubert, P., Davis, A., Kholodenko, D., Motter, R., Sherrington, R., Perry, B., Yao, H., Strome, R., Lieberburg, I., Rommens, J., Kim, S., Schenk, D., Fraser, P., St. George-Hyslop, P., & Selkoe, D. (1997). Mutant presenilins of Alzheimer's disease increase production of 42-residue amyloid β-protein in both transfected cells and transgenic mice. *Nature Med., 3,* 67–72.

8. Lemere, C. A., Lopera, F., Kosik, K. S., Lendon, C. L., Ossa, J., Saido, T. C., Yamaguchi, H., Ruiz, A., Martinez, A., Madrigal, L., Hincapie, L., Arango, L. J. C., Anthony, D. C., Koo, E. H., Goate, A. M., Selkoe, D. J., & Arango, V. J. C. (1996). The E280A presenilin 1 Alzheimer mutation produces increased Aβ42 deposition and severe cerebellar pathology. *Nature Med., 2,* 1146–1148.

9. Duff, K., Eckman, C., Zehr, C., Yu, X., Prada, C.-M., Perez-Tur, M. J., Hutton, M., Buee, L., Harigaya, Y., Morgan, D., Gordon, M. N., Holcomb, L., Refoloe, L., Zenk, B., Hardy, J., & Younkin, S. (1996). Increased amyloid Aβ42(43) in brains of mice expressing mutant presenilin 1. *Nature, 383,* 710–713.

10. Jarrett, J. T., & Lansbury, J. P. T. (1993). Seeding "one-dimensional crystallization" of amyloid: A pathogenic mechanism in Alzheimer's disease and scrapie? *Cell, 73,* 1055–1058.

11. Iwatsubo, T., Mann, D. M., Odaka, A., Suzuki, N., & Ihara, Y. (1995). Amyloid β protein (Aβ) deposition: Aβ42(43) precedes Aβ40 in Down syndrome. *Ann. Neurol., 37,* 294–299.

12. Kovacs, D. M., Fausett, H. J., Page, K. J., Kim, T.-W., Moir, R. D., Merriam, D. E., Hollister, R. D., Hallmark, O. G., Mancini, R., Felsenstein, K. M., Hyman, B. T., Tanzi, R. E., & Wasco, W. (1996). Alzheimer-associated presenilins 1 and 2: Neuronal expression in brain and localization to intracellular membranes in mammalian cells. *Nature Med., 2,* 224–229.

13. Thinakaran, G., Borchelt, D. R., Lee, M. K., Slunt, H. H., Spitzer, L., Kim, G., Ratovitsky, T., Davenport, F., Nordstedt, C., Seeger, M., Hardy, J., Levey, A. I., Gandy, S. E., Jenkins, N. A., Copeland, N. G., Price, D. L., & Sisodia, S. S. (1996). Endoproteolysis of presenilin 1 and accumulation of processed derivatives in vivo. *Neuron, 17,* 181–190.

14. Doan, A., Thinakaran, G., Borchelt, D. R., Slunt, H. H., Ratovitsky, T., Podlisny, M., Selkoe, D. J., Seeger, M., Gandy, S. E., Price, D. L., & Sisodia, S. S. (1996). Protein topology of presenilin 1. *Neuron, 17,* 1023–1030.

15. Li, X., & Greenwald, I. (1996). Membrane topology of the *C. elegans* SEL-12 presenilin. *Neuron, 17,* 1015–1021.

16. Suzuki, T., Nishiyama, K., Murayama, S., Yamamoto, A., Sato, S., Kanazawa, I., & Sakaki, Y. (1996). Regional and cellular presenilin 1 gene expression in human and rat tissues. *Biochem. Biophys. Res. Commun., 219,* 708–713.

17. Cribbs, D. H., Chen, L., Bende, S. M., & LaFerla, F. M. (1996). Widespread neuronal expression of the presenilin-1 early-onset Alzheimer's disease gene in the murine brain. *Am. J. Pathol., 148,* 1797–1806.

18. Shen, J., Bronson, T., Chjen, D. F., Xia, W., Selkoe, D. J., & Tonegawa, S. (1997). Skeletal and CNS defects in *presenilin-1*-deficient mice. *Cell, 89,* 629–639.

19. Wong, P., Zheng, H., Chen, H., Becher, M., Sirinathsinghji, D., Trumbauer, M., Chen, H., Price, D., Van fer Ploeg, L., & Sisodia, S. (1997). Presenilin 1 is required for Notch 1 and Dll 1 expression in the paraxial mesoderm. *Nature, 387,* 288–292.

20. Conlon, R., Reaume, A., & Rossant, J. (1995). Notch 1 is required for the coordinate segmentation of somites. *Development, 121,* 1533–1545.

21. Swiatek, P., Lindsell, C., del Amo, F., Weinmaster, G., & Grindley, T. (1994). Notch 1 is essential for postimplantation development in mice. *Genes Develop., 8,* 707–719.

22. Levitan, D., & Greenwald, I. (1995). Facilitation of lin 12-mediated signalling by sel-12, a *Caenorhabiditis elegans* S182 Alzheimer's disease gene. *Nature, 377,* 351–354.

23. Levitan, D., Doyle, T. G., Brousseau, D., Lee, M. K., Thinakaran, G., Slunt, H. H., Sisodia, S. S., & Greenwald, I. (1996). Assessment of normal and mutant human presenilin function in *Caenorhabditis elegans*. *Proc. Natl. Acad. Sci. USA, 93,* 14940–14944.

24. Axelrod, J. D., Matsuno, K., Artavanis-Tsakonas, S., & Perrimon, N. (1996). Interaction between Wingless and Notch signalling pathways mediated by dishevelled. *Science, 271,* 1826–1832.

25. Zhou, J., Lyanage, U., Medina, M., Ho, C., Simmons, A. D., Lovett, M., & Kosik, K. S. (1997). Presenilin 1 interacts in brain with a novel member of the armadillo family. *NeuroReport, 8,* 1489–1494.

26. Nusse, R., & Varmus, H. (1992). Wnt genes. *Cell, 69,* 1073–1087.

27. McMahon, A. P., & Moon, R. T. (1989). Ectopic expression of the proto-oncogene int-1 in Xenopus embryos leads to duplication of the embryonic axis. *Cell, 58,* 1075–1084.

28. McMahon, A. P., & Bradley, A. (1990). The Wnt-1 (int-1) proto-oncogene is required for development of a large region of the mouse brain. *Cell, 62,* 1073–1085.

29. Thomas, K. R., & Capecchi, M. R. (1990). Targeted disruption of the murine int-1 proto-oncogene resulting in severe abnormalities in the midbrain and cerebellar development. *Nature, 346,* 847–850.

30. Rijsewijk, F., Schuermann, M., Wagenaara, E., Parren, P., Weigel, D., & Nusse, R. (1987). The drosophila homolog of the mouse mammary oncogene int-1 is identical to the segment polarity gene wingless. *Cell, 50,* 649–657.

31. Nusslein-Volhard, C., & Wieschaus, E. (1980). Mutations affecting segment number and polarity in Drosophila. *Nature, 287,* 795–801.

32. Klingensmith, J., & Nusse, R. (1994). Signaling by wingless in Drosophila. *Dev. Biol., 166,* 396–414.

33. Bhanot, P., Brink, M., Harryman Samos, C., Hsieh, J.-C., Wang, Y., Macke, J. P., Andrew, D., Nathans, J., & Nusse, R. (1996). A new member of the frizzled family from Drosophila functions as a Wingless receptor. *Nature, 382,* 225–230.

34. Noordermeer, J., Klingensmith, J., Perrimon, N., & Nusse, R. (1994). Disheveled and armadillo act in the wingless signaling pathway in Drosophila. *Nature, 367,* 80–83.

35. Rubinfeld, B., Albert, I., Porfiri, E., Fiol, C., Munemitsu, S., & Polakis, P. (1996). Binding of GSK3 beta to the APC-beta-catenin compolex and regulation of comples assembly. *Science, 272,* 1023–1026.

36. Behrens, J., Von Kries, J. P., Kuhl, M., Bruhn, L., Wedlich, D., Grosschedl, R., & Birchmeier, W. (1996). Functional interaction of beta-catenin with the transcription factor LEF-1. *Nature, 382,* 638–642.

37. Molenaar, M., van de Wetering, M., Oosterwegel, M., Peterson-Maduro, J., Godsave, S., Korinek, V., Roose, J., Destree, O., & Clevers, T. (1996). XTcf-3 transcription factor mediates beta-catenin-induced axis formation in Xenopus embryos. *Cell, 86,* 391–399.

38. Huber, O., Korn, R., McLaughlin, J., Ohsugi, M., Herrmann, B. G., & Kemler, R. (1996). Nuclear localization of beta-catenin by interaction with transcription factor LEF-1. *Mech. Dev., 59,* 3–10.

39. Brunner, E., Peter, O., Schweizer, L., & Basler, K. (1997). Pangolin encodes a LEF-1 homologue that acts downstream of Armadillo to transduce the Wingless signal in Drosophila. *Nature, 385,* 829–833.

40. Peifer, M., Berg, S., & Reynolds, A. B. (1994). A repeating amino acid motif shared by proteins with diverse cellular roles [Letter]. *Cell, 76,* 789–791.

41. Franke, W. W., Goldschmidt, M. D., Zimblemann, R., Mueller, H. M., Schiller, D. L., & Cowin, P. (1989). *Proc. Natl. Acad. Sci. USA, 86,* 4027–4031.

42. Reynolds, A. B., Herbert, L., Cleveland, J. L., Berg, S. T., & Gaut, J. R. (1992). p120, a novel substrate of protein tyrosine kinase receptors and of p60^{v-src}, is related to cadherin-binding factors β-catenin, plakoglobin and *armadillo. Oncogene, 7,* 2439–2445.

43. Yano, R., Oakes, M., Yamaghishi, M., Dodd, J. A., & Nomura, M. (1992). Cloning and characterization of SRP-1, a suppressor of temperature-sensitive RNA polymerase I mutations, in *Saccharomyces cerevisiae. Mol. Cell Biol., 12,* 5640–5651.

44. Hatzfeld, M., & Nachtsheim, C. (1996). Cloning and characterization of a new armadillo family member, P0071, associated with the junctional plaque-evidence for a subfamily of closely related proteins. *J. Cell Sci., 109,* 2767–2778.

45. Kikuchi, A., Kaibuchi, K., Hori, Y., Nonaka, H., Sakoda, T., Kawamura, M., Mizuno, T., & Takai,

Y. (1992). Molecular cloning of the human cDNA for a stimulatory GDP/GTP exchange protein for c-Ki-ras p21 and smg p21. *Oncogene, 7,* 289–293.

46. Kinzler, K. W., Nilbert, M. C., Su, L. K., Vogelstein, B., Bryan, T. M., Levy, D. B., Smith, K. J., Preisinger, A. C., Hedge, P., McKechnie, D., Finnier, R., Markham, A., Groffen, J., Boguski, M. S., Altschul, S. F., Horii, A., Ando, H., Miyoshi, Y., Miki, Y., Nishisho, I., & Nakamura, Y. (1991). Identification of FAP locus genes from chromosome 5Q21. *Science, 253,* 661–665.

47. Groden, J., Thliveris, A., Samowitz, W., Carlson, M., Gelbert, L., Albertsen, H., Joslyn, G., Stevens, J., Spirio, L., Robertson, M., Sargeant, L., Krapcho, K., Wolff, E., Burt, R., Hughes, J. P., Warrington, J., McPherson, J., Wasmuth, J., Le Paslier, D., Abderrahim, H., Cohen, D., Leppert, M., & White, R. (1991). Identification and characterization of the familial adenomatous polyposis coli gene. *Cell, 66,* 589–600.

48. Gumbiner, B. M. (1996). Cell adhesion: The molecular basis of tissue architecture and morphogenesis. *Cell, 84,* 345–357.

49. Uchida, N., Honjo, Y., Johnson, K. R., Wheelock, M. J., & Takeichi, M. (1996). The catenin/cadherin adhesion system is localized in synaptic junctions bordering transmitter release zones. *J. Cell Biol., 135,* 767–779.

50. Fannon, A. M., & Colman, D. R. (1996). A model for central synaptic junctional complex formation based on the differential adhesive specificities of the cadherins. *Neuron, 17,* 423–434.

51. Ponting, C. P., & Phillips, C. (1995). DHR domains in syntrophins, neuronal NO synthases and other intracellular proteins. *Trends Biochem. Sci., 20,* 102–103.

52. Shivers, B. D., Hilbich, C., Multhaup, G., Salbaum, M., Beyreuther, K., & Seeburg, P. H. (1988). Alzheimer's disease amyloidogenic glycoprotein: Expression pattern in rat brain suggests a role in cell contact. *EMBO J., 7,* 1365–1370.

53. Goodart, S. A., Simmons, A. D., Grady, D., Rojas, K., Moyzis, R. K., Lovett, M., & Overhauser, J. (1994). A yeast artificial chromosome contig of the critical region for cri-du-chat syndrome. *Genomics, 24,* 63–68.

54. Soares, M. B., Bonaldo, M. F., Jelene, P., Su, L., Lawton, L., & Efstratiadis, A. (1994). Construction and characterization of a normalized cDNA library. *Proc. Natl. Acad. Sci. USA, 91,* 9228–9232.

55. Simmons, A. D., Overhauser, J., & Lovett, M. (1997). Isolation of cDNAs from the cri-du-chat critical region by direct screening of a chromosome 5-specific cDNA library. *Genome Res., 7(2),* 118–127.

56. Reynolds, A. B., Roesel, D. J., Kanner, S. B., & Parsons, J. T., (1989). Transformation-specific tyrosine phosphorylation of a novel cellular protein in chicken cells expressing oncogenic variants of the avian cellular src gene. *Mol. Cell Biol., 9,* 629–638.

57. Kanner, S. B., Reynolds, A. B., Vines, R. R., & Parson, J. T. (1990). Monoclonal antibodies to individual tyrosine-phosphorylated protein substrataes of oncogene-encoded tyrosine kinases. *Proc. Natl. Acad. Sci. USA, 87,* 3328–3332.

58. Linder, M. E., & Burr, J. G. (1988). Nonmyristoylated p60[v-src] fails to phosphorylate proteins of 115-120 kDa in chicken embryo fibroblasts. *Proc. Natl. Acad. Sci. USA, 85,* 2608–2612.

59. Perrimon, N. (1996). Serpentine proteins slither into the Wingless and Hedgehog fields. *Cell, 86,* 513–516.

60. Leyns, L., Bouwmeester, T., Kim, S. H., Piccolo, S., & De Robertis, E. M. (1997). Frzb-1 is a secreted antagonist of Wnt signaling expressed in the Spemann organizer, *Cell, 88,* 747–756.

61. Wang, S., Krinks, M., Lin, K., Luyten, F. P., & Moos, M. J. (1997). Frzb, a secreted protein expressed in the Spemann organizer, binds and inhibits Wnt-8. *Cell, 88,* 757–766.

62. Burrus, L. W., & McMahon, A. P. (1995). Biochemical analysis of murine Wnt proteins reveals both shared distinct properties. *Exp. Cell Res., 220,* 363–373.

63. Weidemann, A., Paliga, K., Durrwang, U., Czech, C., Evin, G., Masters, C. L., & Beyreuther, K. (1997). Formation of stable complexes between two Alzheimer's disease gene products: Presenilin-2 and β-amyloid precursor protein. *Nature Med., 3,* 328–332.

13

PUTATIVE LINKS BETWEEN SOME OF THE KEY PATHOLOGICAL FEATURES OF THE ALZHEIMER'S BRAIN

RÉMI QUIRION, DANIEL AULD, UWE BEFFERT, JUDES POIRIER, AND SATYABRATA KAR

Douglas Hospital Research Centre
McGill Center for Studies on Aging
and Department of Psychiatry, McGill University
Montréal, Québec, Canada H4H 1R3

Life expectancy has dramatically increased over the past century. One of the impacts of this increment is the greater incidence of diseases associated with old age such as diabetes, hypertension, and disorders of the aged brain, including Alzheimer's disease (AD). Although effective therapies have been developed for diseases of peripheral origin, limited progress has been made regarding chronic neurodegenerative disorders. In that context, the development of effective therapeutic approaches for diseases such as AD certainly represents a major challenge of the 21st century.

AD is characterized by the progressive impairment of memory and language abilities.[1] Among the various neuropathological features of the AD brain possibly leading to these deficits, much research has focused on the role of (1) senile plaques enriched in β-amyloid (Aβ) deposits; (2) neurofibrillary tangles composed of phosphorylated tau proteins; (3) the apolipoprotein E_4 gene; (4) the two presenilin genes; (5) components of the immune defense system; and (6) deficits in cholinergic innervation, acetylcholine being a key neurotransmitter

181

involved in cognitive processes (for review see Gauthier[1]). While significant advances have been made over the past decade on the respective relevance of each of these systems in the symptomatology of AD, very little attention has been given to their possible interrelationships in the etiology of this disorder, most scientists focusing on a single, unitary hypothesis often without attempting to come up with a more unifying theory that takes into account key neuropathological features of the AD brain.

As an effort toward that goal, we recently focused our attention on the existence of links between some of the alterations seen in the AD brain, and their possible relevance to the etiology and treatment of this disease. Although the picture is still far from complete, our data already suggest important interrelationships between the cholinergic system, $A\beta$ proteins, and the apolipoprotein E_4 (ApoE$_4$) gene.

CHOLINERGIC DEFICITS AND
$A\beta$ PEPTIDES

Various groups, including ours, have recently shown that very low concentrations of $A\beta$ peptides can directly regulate cholinergic functions (Table 13.1), suggesting a possible link between the well-established amyloid burden and choli-

TABLE 13.1 Major Steps in Cholinergic Neurotransmission on Which Low Concentrations (pM–nM) of Solubilized $A\beta$ Induce Hypofunction without Apparent Neurotoxicity[a]

Fig. 13.2 location	Peptide fragment	Effect	Concentration	Model/phenotype	Exposure
1.	1–42, 1–40 1–28, 25–35	ACh release inhibition	pM–nM	Cortical and hippocampal brain slices	1 hr
2.	1–42, 1–28 25–35, 25–28	Intracellular ACh concentration and ChAT activity reduced	pM–nM	SN56 cell line culture	48 hr
3.	1–42	Intracellular ACh concentration and activity reduced	nM	Primary septal culture	12 hr
4.	1–40, 25–35	Impaired muscarinic M$_1$-like receptor signaling	nM–μM	Primary cortical culture	4 hr

[a]Modified from Auld, D. S., Kar, S., & Quirion, R. (1998). β-Amyloid peptides as direct cholinergic neuromodulators: A missing link? *Trends Neurosci, 21,* 43–49. Copyright (1998), with permission from Elsevier Science.

Note: ACh, acetylcholine; *ChAT,* choline acetyltransferase.

nergic hypofunction in AD. Additionally, the very high potency of various Aβ derivatives in modulating various cholinergic responses, the apparent lack of acute toxicity, and recent evidence showing that Aβ peptides are produced in the normal brain suggest that APP–amyloid derivatives may act as physiologically relevant neuromodulators.[2]

Aβ PEPTIDES ACUTELY MODULATE ACETYLCHOLINE RELEASE

Solubilized Aβ inhibits several steps in acetylcholine (ACh) synthesis and release without inducing any apparent neurotoxicity. K^+-stimulated release of ACh from rat hippocampal and cortical slices is acutely reduced by pM–nM concentrations of Aβ. This effect is tetrodotoxin-insensitive, suggesting that Aβ acts at the level of cholinergic nerve terminals. Several Aβ fragments including $Aβ_{25-35}$, $Aβ_{1-28}$, $Aβ_{1-40}$, and $Aβ_{1-42}$ (but not the random sequence, all D-isomers or the reverse sequence control peptides) inhibit K^+-stimulated ACh release (Fig. 13.1).[3] Interestingly, striatal ACh release is insensitive to Aβ inhibition. The basis for this region-selective profile is unclear but most likely relates to distinct properties of different cholinergic populations. It is rather intriguing that most vulnerable brain regions in AD, which include the cortex and the hippocampus,

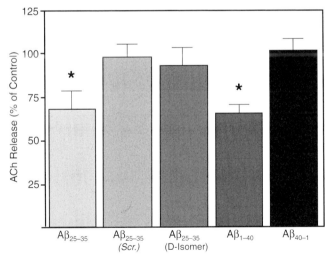

FIGURE 13.1   Comparative effects of $Aβ_{25-35}$, scrambled *(Scr.)* $Aβ_{25-35}$, all-D-isomer $Aβ_{25-35}$, $Aβ_{1-40}$, and reverse $Aβ_{1-40}$, (i.e., $Aβ_{40-1}$) on evoked hippocampal ACh release. Slices were depolarized with 25 mM K^+ buffer in the presence or absence (control) of 10^{-8} M of each of these peptides. Unlike for the regular peptides ($Aβ_{25-35}$ or $_{1-40}$), endogenous ACh release was not altered significantly in the presence of any of the control peptides. Data for each peptide are expressed as mean ± SEM ($n = 10-15$). *$p < 0.05$. [Figure 3 from Kar, S., Seto, D., Gaudreau, P., & Quirion, R. (1996). β-amyloid-related peptides inhibit potassium-evoked acetylcholine release from rat hippocampal slices. *J. Neurosci., 16* (3), 1038.]

are sensitive to Aβ neuromodulation in the rat brain, whereas striatal cholinergic interneurons are neither as affected in AD nor sensitive to Aβ neuromodulation.[2, 3] This functional interaction suggests that Aβ may directly contribute to cholinergic deficits in AD, provided that a similar modulation can be demonstrated in the human brain.

Several processes involved in ACh release could be disrupted by Aβ peptides. Aβ may act on more than one of these, with targets ranging from precursor availability to vesicular fusion (Fig. 13.2, see Table 13.1). The rate-limiting step for ACh synthesis is the availability of choline, and cholinergic terminals possess high-affinity choline transporters that internalize choline to ensure adequate supplies. It has been demonstrated that Aβ enhances choline conductance from PC12 cells, resulting in a leakage of choline from the cell.[4] Moreover, we observed that pM–nM concentrations of Aβ can reduce choline levels by inhibiting high-affin-

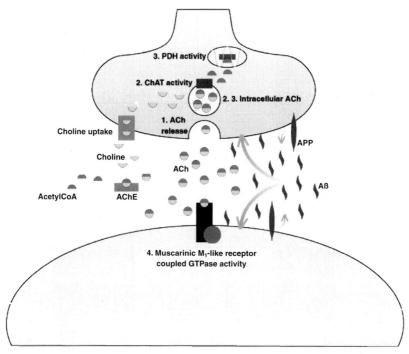

FIGURE 13.2   Major steps in cholinergic neurotransmission on which low concentrations (pM–nM) of solubilized Aβ have been indicated to induce hypofunction without apparent neurotoxicity. Numbers correspond to descriptions in Table 13.1. The cholinergic synapse could be modulated by Aβ derived from one or more sources including presynaptic and/or postsynaptic neurons, as well as extrasynaptic neurons and/or glial cells. *ACh,* acetylcholine; *AChE,* acetylcholinesterase; *acetylCoA,* acetyl-coenzyme A; *APP,* amyloid precursor protein; *Aβ,* β-amyloid; *ChAT,* choline acetyltransferase; *GTPase,* guanylyl triphosphatase; *PDH,* pyruvate dehydrogenase. [Reprinted from Auld, D. S., Kar, S., Quirion, R. (1998). β-Amyloid peptides as direct cholinergic neuromodulators: A missing link? *Trends Neurosci., 21,* 43–49. Copyright (1998), with permission from Elsevier Science.]

ity choline uptake.[5] These studies suggest that Aβ may disrupt the availability of choline and hence alter ACh synthesis.

Aβ PEPTIDES AND CHOLINERGIC NEURON AUTOCANNIBALISM

If Aβ indeed impairs the capacity of the nerve terminal for choline uptake, cholinergic neurons in AD might ultimately have a limited access to this precursor. Choline deprivation has already been postulated to contribute to the cholinergic vulnerability seen in AD.[6] During choline deficiency, autocannibalism may occur in cholinergic neurons to meet demands for ACh by breaking down membrane phosphatidylcholine. Phosphatidylcholine levels are known to be reduced in the AD cortex.[7] If Aβ-induced reduction of choline uptake relates to the inhibitory effect on ACh release, it could leave afflicted cholinergic populations more vulnerable to autocannibalism and consequent membrane compromise. Hence, in addition to any direct neurotoxic properties of Aβ seen at higher concentrations (μM), select populations of cholinergic neurons may suffer from an additional vulnerability because of their dependence on choline for both structural and neurotransmission purposes.

Aβ PEPTIDES MODULATE CHOLINERGIC ENZYME ACTIVITY

Aβ reduces intracellular concentrations of ACh following a 48-hr treatment with nM concentrations of $Aβ_{25-35}$, $Aβ_{1-28}$, and $Aβ_{1-42}$ in SN56 cells (a cell line expressing the cholinergic phenotype) without evidence of neurotoxicity.[8] Concomitant with these ACh decreases are reductions in choline acetyltransferase (ChAT) activity. Acetylcholinesterase (AChE) activity is unchanged following Aβ peptides treatment, suggesting that these peptides do not act as general modulators of cholinergic enzymatic activities. As ChAT activity deficits are a prominent feature of the AD brain,[1] the data obtained in SN56 immortalized cells are particularly interesting but need to be confirmed in primary cultured neurons.

Reduced intracellular concentrations of ACh have also been reported in primary rat septal cultures following a 12-hr treatment with solubilized $Aβ_{1-42}$ (100 nM) again without evidence of neurotoxicity.[9] In contrast to the aforementioned SN56 model, however, ChAT activity was not affected. Rather, the activity of pyruvate dehydrogenase (PDH), which generates acetyl coenzyme A (acetyl CoA) from pyruvate during mitochondrial glycolysis, was disrupted. Decreased intracellular ACh levels were hypothesized to result from Aβ activation of the $τ$ protein kinase I/glycogen synthase kinase-3β (TPKI/GSK-3β), which is known to phosphorylate and inactivate PDH. Taken together, these data suggest that Aβ peptides may impair ACh synthesis via an additional mechanism, namely by undermining the availability of acetyl CoA (see Fig. 13.2 and Table 13.1). Moreover, it points out a possible similarity between neurotoxic and neuromodulatory actions of Aβ, as prolonged mitochondrial dysfunction would be expected to reduce neuron viability in addition to decreasing acetyl CoA availability. Diminished PDH activity has been reported in AD,[10] stressing the potential significance of these findings.

IMPAIRED METABOLISM AND
CHOLINERGIC TRANSMISSION

It has been reported that rather high concentrations of Aβ (5–50 μM) decrease glucose uptake associated with neurotoxicity in hippocampal and cortical neurons.[11] Hence, Aβ may limit the availability of acetyl CoA by reducing glucose uptake. As the cholinergic neuron is dependent, at least in part, on the same enzyme (PDH) and upstream substrate (glucose) for neurotransmission and cellular metabolism, the impaired metabolism–glucose uptake reported in AD[12] could render affected cholinergic neurons functionally compromised. Indeed, it has been previously pointed out that under pathological circumstances involving metabolic impairment, ACh synthesis is likely to be sensitive to reduced acetyl CoA availability.[13] Accordingly, Aβ possibly influences acetyl CoA availability and ACh synthesis via reduced PDH activity (at nM concentrations) and decreased glucose uptake (at μM concentrations) and may thus contribute to AD-associated cholinergic deficits.

Aβ PEPTIDES MODULATE MUSCARINIC
RECEPTOR SIGNALING

Solubilized Aβ peptides disrupt muscarinic M_1-like receptor signaling pathways (see Fig. 13.2). A 4-hr exposure to $A\beta_{25-35}$ (nM–μM concentrations) reduces carbachol-stimulated guanylyl triphosphate (GTPase) activity in primary rat cortical culture.[14] At higher concentrations, Aβ was shown to reduce M_1 receptor–induced production of IP_1, IP_2, and $IP_{3/4}$ and to prevent carbachol-stimulated increases in intracellular Ca^{2+}. Aβ induced these changes independently of alterations in receptor binding kinetics. Accordingly, muscarinic M_1-like receptor coupled GTPase activity is another level at which relatively low concentrations of solubilized Aβ peptides can inhibit cholinergic neurotransmission (see Table 13.1).

In AD, the levels of M_1-like receptors are equivalent to those of age-matched controls,[15] whereas muscarinic receptor G-protein interaction has been reported to be impaired[16] or unchanged.[17] In support of impaired signaling, severe deficits in IP accumulation have been reported following muscarinic receptor stimulation in AD brain membranes.[18] Moreover, cholinergic agonist-induced IP accumulation is apparently more significantly affected than that associated with other neurotransmitter receptors in AD.[18] The disruption by Aβ peptides of M_1-like muscarinic signaling and the associated IP accumulation suggest that it may contribute to the impaired cholinergic neurotransmission reported in the AD brain.

CAN Aβ PEPTIDES ALTER IN VIVO CHOLINERGIC TRANSMISSION?

Various attempts have been made to model neuropathological, neurochemical, and cognitive aspects of AD in rodents by injecting or infusing Aβ peptides directly into the brain. These studies have reported rather inconsistent data. Nevertheless, cholinergic (and, in one case, dopaminergic) hypofunction has been observed as a consequence of in vitro Aβ treatment.[19–21] Whether these data can be directly compared with the in vitro neuromodulatory actions of low concentrations of soluble Aβ is unclear, especially since some studies used aged Aβ to promote aggregation and neurotoxicity. However, similarities were noted, since Aβ reduced ChAT activities and ACh levels both in vitro and in vivo. It would thus be most relevant to investigate the impact of Aβ on cholinergic neurotransmission in the recently developed mutant amyloid precursor protein (APP) overexpressing mice models, which exhibit both AD-like neuropathologies and behavioral impairments.[22–24]

DOES Aβ SELECTIVELY MODULATE CHOLINERGIC NEURONS?

Most studies concerning the neuromodulatory effects of Aβ have thus far focused on the cholinergic phenotype. This does not rule out the possibility that Aβ may modulate other neurotransmitters. Indeed, an acute treatment with Aβ (3 nM) was shown to reduce stimulated norepinepherine (NE) release[25] and it has been reported that high concentrations of Aβ peptides (μM) can potentiate stimulated glutamate release.[26] The in vivo administration of Aβ peptide was also demonstrated to impair dopaminergic activity.[19] Although the cholinergic system sustains the most marked and consistently observed neurotransmitter deficits in AD, changes have been reported for a variety of other neurotransmitters and neuropeptides.[1] Future research should thus document the selectivity of Aβ neuromodulation for the cholinergic neuron, as well as the potential contribution of both common and phenotype-distinct mechanisms of modulation. For example, investigations of the cholinergic neuron have focused on specific features of ACh synthesis rather than properties common to the release of various neurotransmitters such as vesicle fusion. However, the unique dependence of ACh synthesis on choline levels and acetyl CoA activity imposes upon cholinergic neurons a greater sensitivity to metabolic dysfunction than that of other neurotransmitters, which are not as dependent on such generic precursors for their synthesis.

MECHANISMS INVOLVED IN Aβ NEUROMODULATION

The mechanism(s) underlying the neuromodulatory actions of Aβ is(are) currently unclear but very low concentrations of Aβ have been shown to induce various actions that could influence cellular function. For instance, pM–nM concentrations of Aβ are known to influence signal transduction cascades,[27] whereas higher concentrations (μM) are documented to be neurotoxic, possibly by disrupting intracellular Ca^{2+} levels and by generating reactive oxygen species (ROSs).[28–30] Whether the mechanism of Aβ-induced neurotoxicity and neuromodulation is similar is currently unknown. Nevertheless, both nM and μM concentrations of Aβ are known to influence intracellular ion concentrations.[28]

ROSs are increasingly implicated in Aβ-induced toxicity,[28] but whether these elements are involved in the neuromodulatory action of Aβ is uncertain. Interestingly, a 3-nM concentration of $Aβ_{1-40}$ has been reported to enhance stimulated increases in intracellular Ca^{2+}, this effect being blocked by a spin trapping agent.[25] Furthermore, the ability of Aβ to impair muscarinic M_1-like receptor signaling is nullified by an antioxidant treatment.[14] The production of low quantities of ROSs are not necessarily inconsistent with a physiological function, as free radicals have recently been recognized as atypical messengers with both physiological and neurotoxic properties (i.e., nitric oxide).

It has also recently been reported that Aβ peptides can rather specifically bind to neurons through the receptor for advanced glycosylation end products (RAGE).[31] This interaction has an apparent affinity of ~55 nM, a concentration at which Aβ can generate markers of oxidative stress and is capable of neuromodulatory action. It thus seems possible that Aβ may exert its neuromodulatory actions through a plasma membrane–associated protein. It is of interest here to recall that the inhibition of hippocampal ACh release by Aβ is stereoselective (with the D-isomer being ineffective), this being consistent with a receptor-like mediated mode of action.[2, 3]

SOLUBLE Aβ AND CHOLINERGIC MODULATION: ANY RELATIONSHIP TO Aβ FIBRILS AND NEURITIC PLAQUES?

Studies evaluating the neuromodulatory potential of low concentrations of Aβ have used solubilized Aβ as opposed to fibrillar Aβ, which is better associated with neurotoxic events. Consequently, it is not clear if fibrillary Aβ produces similar neuromodulatory consequences as does soluble Aβ. As previously discussed, toxic μM concentrations of Aβ may reduce ACh synthesis by depressing acetyl CoA production, a mechanism that may not be related to the action of low pM–nM concentrations of Aβ (more likely via decreased choline uptake[5]). Moreover, considering the likely structural difference between low molecular weight soluble Aβ monomers or oligomers and the β-sheet-rich high molecular weight fibrils of aggregated Aβ, they may not bind to an exactly similar selection of potential acceptor proteins.

Most neuritic plaque–associated Aβ peptide in the AD brain is aggregated and highly insoluble. Nevertheless, this Aβ was likely produced during cellular APP metabolism (or mismetabolism) and was probably secreted in a soluble state before the formation of the neuritic plaques. Accordingly, total soluble Aβ peptide level is elevated by 6 fold in the AD brain, whereas soluble $A\beta_{1-42}$ levels are reported to be 12 fold higher.[32] Potentially, this soluble $A\beta$ (in the form of monomers or oligomers) may exert a pathological influence before it is deposited. Cholinergic hypofunction is likely to be one such consequence of an overabundance of soluble $A\beta$ peptides, with enhanced vulnerability to various neuro toxic insults being another.[33]

CHOLINERGIC NEUROMODULATION BY $A\beta$ PEPTIDES: POSSIBLE IMPLICATIONS

The marked potency of $A\beta$ peptides as cholinergic neuromodulators places them among the most, if not the most, potent inhibitors of ACh release currently known. In fact, $A\beta$ peptides have features that warrant further investigation regarding their role as physiologically relevant neuromodulators.

$A\beta$ peptides are synthesized and secreted by brain cells, including neurons, and are present in the cerebrospinal fluid (CSF) of neurologically normal individuals. The precursor, APP, is found in neuron terminals and is processed via either amyloidogenic or nonamyloidogenic pathways.[34] Interestingly, an active domain (APP 319–335; not containing the $A\beta$ sequence) of secreted APP (sAPP) has been shown to increase synaptic density in the rat brain,[35] and nM concentrations of sAPP are neuroprotective in vitro, an effect that involves the stabilization of intracellular Ca^{2+} levels.[36] Higher concentrations (μM) of $A\beta$ peptides are neurotoxic and destabilize intracellular Ca^{2+} mobilization.[28] Thus in response to certain conditions (physiological or pathological), the differential processing of APP could lead to a profile of secreted metabolites (sAPP versus $A\beta$), which perform distinct actions on synaptic activity and transmitter release.

In light of the capacity of $A\beta$ to induce cholinergic hypofunction, it is of interest to add here that APP processing is *modulated* by muscarinic M_1-like receptors. Stimulation of M_1-like receptors or their downstream effectors (i.e., protein kinase C or Ca^{2+}) has been reported to enhance the release of nonamyloidogenic APP fragments[37, 38] and in several models to reduce the secretion of toxic $A\beta$ peptides.[38–40] Hence, acting in a paracrine or autocrine fashion, secreted toxic $A\beta$ peptides could potentially induce conditions more favorable to their own production by inhibiting ACh release and M_1-like receptor signaling leading to further increments in $A\beta$ peptides production (Fig. 13.3). Provided that this relationship exists in the human brain, the feedback loop between $A\beta$ peptides and ACh at the cholinergic synapse could enhance $A\beta$ secretion and exacerbate cholinergic deficits in AD (see Fig. 13.3 as models).

In brief, it is well established that AD is characterized, among other features, by $A\beta$ deposition, marked cholinergic deficits and cognitive impairments. To date, however, it has been rather unclear how these features are related. Since $A\beta$ can directly decrease cholinergic inputs and because ACh is well documented to play a critical role in cognitive processes, the data summarized here may provide some pieces for this puzzle.

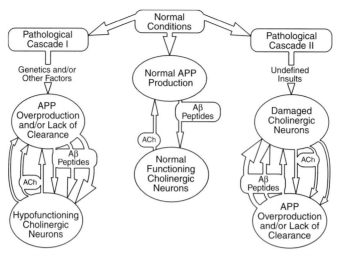

FIGURE 13.3   Possible impact of altered APP production on subsequent cholinergic functions and decreased ACh release leading to further increases in the production in toxic Aβ peptides (cascade I). Alternatively, undefined insults could lead to damaged cholinergic neurons and decreased cholinergic transmission, hence inducing a greater production of aberrant toxic APP metabolites (Aβ peptides) leading to further hypofunctioning of the cholinergic neurons (cascade II). Under normal conditions, these negative feedback loops are operating at levels sufficient to ensure normal cholinergic inputs and the normal maturation of nontoxic APP derivatives. *Aβ,* β-amyloid; *ACh,* acetylcholine; *APP,* amyloid precursor protein.

CHOLINERGIC DEFICITS AND THE ApoE$_4$ GENE

ApoE is a lipophilic plasma protein involved in cholesterol and phospholipid transport. The synthesis of ApoE occurs primarily in the liver and to a lesser extent in other tissues including the brain.[41] Its role in the central nervous system (CNS) is not fully clear but most certainly relates to the mobilization and redistribution of cholesterol and phospholipid during membrane remodeling and plasticity.[41]

Human ApoE is secreted as a 299–amino acid, 34-kDa protein encoded by a 4-exon gene located on the long arm of chromosome 19. Three major isoforms (E$_2$, E$_3$, E$_4$) exist, resulting in six ApoE phenotypes in the population. Functionally, ApoE$_3$ and ApoE$_4$ have higher affinity for the low density lipoprotein (LDL) receptor than does ApoE$_2$.

The identification of the ApoE$_4$ gene allele as a risk factor in AD generated tremendous interest. It was shown that the gene dose (0, 1, or 2) of ApoE$_4$ had a significant impact on the age of onset in both late-onset familial and sporadic forms of AD.[42, 43] Hence, as the copy number increases, the age of onset decreases in AD. Moreover, ApoE$_4$-like immunoreactivity is associated with Aβ in extracellular neuritic plaques, intracellular neurofibrillary tangles, and brain microvasculature in AD.[44] The association of ApoE$_4$ with these key features of the AD brain suggests that ApoE is important to the pathogenesis of this disorder.[41]

BASAL FOREBRAIN CHOLINERGIC NEURONS
AND THE ApoE$_4$ GENE

As discussed above, the major loss of basal forebrain magnocellular cholinergic neurons is one of the key features of AD. These neuronal losses are accompanied by major decreases in cholinergic markers (e.g., ChAT, AChE, ACh) in terminal projection areas including the frontotemporal cortex and the hippocampal formation.

As a first step to establish if the ApoE$_4$ gene dose had an impact on residual cholinergic function in the AD brain, we investigated the existence of possible correlation between the losses of basal forebrain cholinergic neurons and the ApoE$_4$ allele.[45] We observed that ApoE$_4$ AD carriers show a greater reduction in AChE-positive magnocellular neurons in the Ch2 (diagonal band of Broca) and Ch4 (nucleus basalis of Meynert) subareas when compared to ApoE$_4$-negative AD subjects, without an evident alteration in nuclear volume in remaining AChE-positive cells (Fig. 13.4). It suggests that AChE-positive cholinergic neurons of the basal forebrain are particularly at risk in ApoE$_4$ carriers.[41]

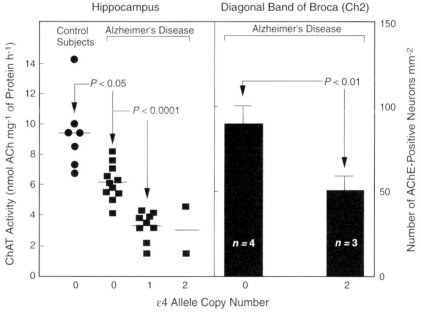

FIGURE 13.4 Apolipoprotein E$_4$ (ApoE$_4$) and cholinergic dysfunction in Alzheimer's disease (AD) subjects. Choline acetyltransferase (ChAT) activity was measured *(left)* in the hippocampus of postmortem control and AD subjects with different ApoE genotypes. Each point refers to one subject: Control subjects are represented by solid circles; AD subjects are represented by solid squares. The acetylcholinesterase (AChE)–positive neuron density in the diagonal band of Broca *(right)* was examined in homozygote E$_{3/3}$ and E$_{4/4}$ AD subjects. Significant differences between groups are indicated by the *P* values on the figure. [Reprinted from Poirier, J. (1994). Apolipoprotein E in animal models of CNA injury and in Alzheimer's disease. *Trends Neurosci., 17,* 525–530. Copyright (1994), with permission from Elsevier Science.]

CHOLINERGIC MARKERS IN PROJECTION AREAS
AND THE ApoE$_4$ GENE

Basal forebrain cholinergic neurons mostly project to the cortex and the hippocampus. We thus investigated various purported markers of the cholinergic synapse versus ApoE$_4$ gene copy numbers. Both ChAT activity and nicotinic receptor binding levels were more significantly decreased as the number of ApoE$_4$ allele increased in these regions of the AD brain (see Fig. 13.4).[45] Since both ChAT and to a lesser extent nicotinic receptors are presynaptically located, it suggests a greater alteration of the long basal forebrain cholinergic projections in the brain of ApoE$_4$ carriers. Levels of muscarinic receptors failed to relate to the ApoE$_4$ gene dosage.[45] Similar results have recently been obtained by others[46] (but see Svenssen et al.[47]) and Gordon et al.[48] recently reported that hippocampal cholinergic markers were significantly altered in ApoE$_4$ knock-out mice; these animals having major cognitive impairments.

IMPACT OF THE ApoE$_4$ GENE ON RESPONSE
TO CHOLINOMIMETIC DRUGS

What is the relevance, if any, of the above findings in the clinics? This critical issue was addressed by comparing responses of ApoE-genotyped AD patients to a cholinergic-based therapy. Accordingly, in our original study, the ApoE genotype was determined in 40 patients enrolled in a 30-week randomized controlled trial of a high dose of tacrine, a potent, centrally active cholinesterase inhibitor, currently on the market for the treatment of AD.[45]

Figure 13.5 depicts individual differences in cognitive measures assessed using the ADAS scale as a function of the ApoE$_4$ genotype. Overall, more than 80% of non-ApoE$_4$ carrying patients showed some improvements during the treatment. In contrast, 60% of the ApoE$_4$ patients were not improved or became worse after the 30 weeks' treatment. Results from this initial study have now been confirmed in a much larger patient sample (more than 400) and with other, more recently marketed esterase inhibitors such as Aricept (unpublished results). Hence, AD patients who do not carry any ApoE$_4$ allele apparently respond better to AChE inhibitors possibly because of greater residual cholinergic functions as established in postmortem studies.[45, 46] Accordingly, it may be preferable to genotype AD patients before using such a drug, ApoE$_4$ bearers being unlikely to be good responders, many even deteriorating under an AChE inhibitor like tacrine (see Fig. 13.5).

It thus appears that a given ApoE genotype has a direct impact on cholinergic functions. Future studies using transgenic animals overexpressing either the human ApoE$_2$, ApoE$_3$, or ApoE$_4$ gene should prove most useful to verify this hypothesis and establish the mechanism involved. Additionally, recent data suggest that the ApoE$_4$ gene allele has an impact on the number of neuritic plaques in cortical areas of the AD brain, with their number being increased with the ApoE$_4$ gene dose.[49, 50]

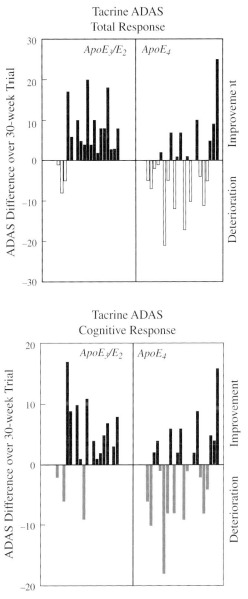

FIGURE 13.5 Effect of ApoE$_4$ allele copy numbers on tacrine drug responsiveness in AD subjects. Forty subjects enrolled in the 30-week randomized controlled trial of high-dose tacrine in AD patients were selected from the original 663-subject cohort. Patients were selected prior to ApoE phenotype determination and were blind to genotype. Phenotypic determination of the ApoE genotype was performed using frozen serum. Graph represents individual variation (ADAS score before and after drug trial) in total and cognitive performances as a function of ApoE$_4$ allele incidence. (From Poirier, J., Delisle, M.-C., Quirion, R., Aubert, I., Farlow, M., Lahiri, D., Hui, S., Bertrand, P., Nalbantoglu, J., Gilfix, B. G., & Gauthier, S. (1995). Apolipoprotein E$_4$ allele as a predictor of cholinergic deficits and treatment outcome in Alzheimer's disease. *Proc. Natl. Acad. Sci. USA, 92,* 12260–12264.)

POSSIBLE LINKS BETWEEN THE
APOE$_4$ GENOTYPE AND
β-AMYLOID PEPTIDES

Recent data suggest the existence of significant interaction between ApoE and Aβ fragments of the amyloid precursor protein. For example, it has been shown that both ApoE$_3$ and ApoE$_4$ form stable SDS-resistant complexes with Aβ peptides[51, 52] and enhance their polymerization in vitro.[53] High concentrations of purified ApoE$_4$ can also reduce Aβ toxicity in neuronal cell cultures.[54] ApoE immunostaining has been localized to extracellular senile plaques and vascular amyloid,[44] and strong correlations were noted between the ApoE$_4$ allele and increased vascular and senile plaques Aβ deposition.[49, 50] We have also reported that ApoE can modulate Aβ internalization in neuronal cell cultures.[49] In a more recent study,[55] we attempted to establish further the impact of Aβ peptides on the uptake of specific ApoE isoforms in both neurons and astrocytes.

Aβ PEPTIDES AND THE
UPTAKE OF ApoE

All three isoforms of human ApoE (in liposomes) are internalized by rat primary hippocampal neuronal culture. Interestingly, however, thebinding and internalization of ApoE$_4$ was significantly increased (2–2.5 fold) in the presence of Aβ $_{1-40}$ (1 μg/ml), whereas the uptake of ApoE$_3$ was less affected.[55] Moreover, the potentialization of the uptake of ApoE isoforms by Aβ was sequence specific with Aβ$_{25-35}$. These effects were not dependent on toxicity as revealed by the lack of alteration of lactate dehydrogenase (LDH) activity.[55]

Confocal microscopy revealed that human ApoE was internalized into hippocampal neurons and more actively in the presence of Aβ$_{1-40}$ (Fig. 13.6). For example, when cultured neurons were incubated in the presence of human ApoE$_4$ liposomes for 1 min, immunoreactive ApoE was evident in punctuate labeling along neuronal processes and diffuse labeling within proximal dendrites. At a later stage (3-min incubation), staining was most prominent in the cell body and soma, although peripheral processes were still labeled. Similar results were obtained after a 24-hr incubation with ApoE$_2$, Aβ$_{1-40}$ apparently increasing further the uptake of the ApoE.

CAN Aβ PEPTIDES MODULATE THE UPTAKE OF
ApoE IN ASTROCYTES?

Type I astrocyte cultures internalized human ApoE liposomes similarly as did neurons with Aβ $_{1-40}$ increasing significantly the uptake of both ApoE$_2$ and ApoE$_4$, ApoE$_3$ being less affected. Aβ$_{25-35}$ was most effective in promoting ApoE$_2$ uptake, whereas Aβ$_{1-28}$ was almost inactive as observed in neuronal cul-

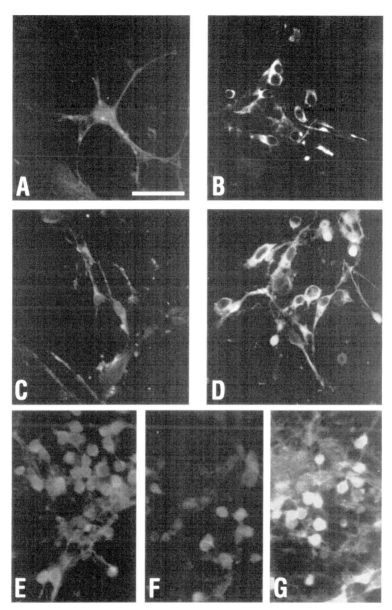

FIGURE 13.6 Confocal microscopy of apolipoprotein E (ApoE) internalization in neurons. Photomicrographs of primary hippocampal neurons treated with combinations of human ApoE, $A\beta_{1-40}$, and anti-low density lipoprotein (anti-LDL) receptor antibodies. (A) Double immunohisto-chemistry for ApoE and GFAP positive astrocyte in control, untreated culture of neurons (scale bar = 25 μm). (B) ApoE negative neurons in the same culture stained red for MAP-2 as a specific neuronal marker. (C) Pulse chase experiment on neuronal cultures treated for 1 min with human $ApoE_4$ liposomes stained for ApoE and GFAP (D) Neuronal cultures treated for 3 min with human $ApoE_4$ liposomes stained for ApoE and for GFAP (E) Neuronal cultures treated for 24 hr with human $ApoE_2$ liposomes. (F) Neuronal cultures with human $ApoE_2$ liposomes and anti-LDL receptor antibodies. (G) Neuronal cultures with human $ApoE_2$ liposomes and $A\beta_{1-40}$. E–G are all stained for ApoE

tures. Interestingly, however, the C7 LDL receptor antibody failed to block the uptake of any ApoE or ApoE + Aβ complexes in astrocytic cultures, suggesting that different ApoE-related receptor binding proteins (e.g., LRP, ApoER2) were involved in the internalization process in neurons versus astrocytes.

Taken together, these findings clearly demonstrate that the uptake profile of the $ApoE_3$ and $ApoE_4$ isoforms can be altered by neurotoxic Aβ peptides such as $Aβ_{1-40}$ and $Aβ_{25-35}$ in both cultured neuronal and astrocytic cells. This adds further to the existence of multiple, complex interactions between the ApoE gene allele and Aβ peptides, those likely having an impact on the etiology and progression of AD.

CONCLUDING REMARKS

In summary, it is clear that direct interactions exist between key phenotypes of the AD brain including Aβ peptides, ApoE isoforms, and cholinergic neurons, each impacting on the other. It would now be most pertinent to investigate possible links with other markers of the AD brain such a neurofibrillary tangles, the presenilins, and immune-derived proteins. Some evidence has already suggested possible interactions between the $ApoE_4$ genes and tau proteins,[56] the presenilins with APP processing,[57] and we reported on the impact of various cytokines on normal cholinergic functions.[58] We firmly believe that only such studies will help to resolve the AD puzzle and lead to truly effective therapies.

ACKNOWLEDGMENTS

Supported by research grants for the Medical Research Council of Canada, the Alzheimer Society of Canada, and the "Reseau Santé Mentale (RSMQ)" of the "Fonds de la Recherche en Santé du Québec (FRSQ)." R. Quirion is a Chercheur-Boursier Mérite Exceptionel, and S. Kar and J. Poirier are Chercheurs-Boursiers of the FRSQ.

REFERENCES

1. Gauthier, S. (1996). *Clinical diagnosis and management of Alzheimer's disease* (p. 372). London: Dunitz.
2. Auld, D. S., Kar, S., & Quirion, R. (1998). β-Amyloid peptides as direct cholinergic neuromodulators: A missing link? *Trends Neurosci., 21,* 43–49.
3. Kar, S., Seto, D., Gaudreau, P., & Quirion, R. (1996). Beta-amyloid-related peptides inhibit potassium-evoked acetylcholine release from rat hippocampal slices. *J. Neurosci., 16,* 1034–1040.
4. Galdzicki, Z., Fukuyama, R., Wadhwani, K. D., Rapoport, S. I., & Ehrenstein, G. (1994). Beta-amyloid increases choline conductance of PC12 cells: Possible mechanism of toxicity in Alzheimer's disease. *Brain Res., 646,* 332–336.
5. Kar, S., Issa, A. M., Seto, D., Auld, D. S., Collier, B., & Quirion, R. (1998). Amyloid β-peptides inhibit acetylcholine release and high-affinity choline uptake in rat hippocampal slices (in press).

6. Wurtman, R. J. (1992). Choline metabolism as a basis for the selective vulnerability of cholinergic neurons. *Trends Neurosci., 15*, 117–122.
7. Nitsch, R. M., Blusztajn, J. K., Pittas, A. G., Slack, B. E., Growdon, J. H., & Wurtman, R. J. (1992). Evidence for a membrane defect in Alzheimer disease brain. *Proc. Natl. Acad. Sci. USA, 89*, 1671–1675.
8. Pedersen, W. A., Kloczewiak, M. A., & Blusztajn, J. K. (1996). Amyloid beta-protein reduces acetylcholine synthesis in a cell line derived from cholinergic neurons of the basal forebrain. *Proc. Natl. Acad. Sci. USA, 93*, 8068–8071.
9. Hoshi, M., Takashima, A., Murayama, M., Yasutake, K., Yoshida, N., Ishiguro, K., Hoshino, T., & Imahori, K. (1997). Nontoxic amyloid beta peptide 1-42 suppresses acetylcholine synthesis. Possible role in cholinergic dysfunction in Alzheimer's disease. *J. Biol. Chem. 272*, 2038–2041.
10. Sorbi, S., Bird, E. E., & Blass, J. P. (1983). Decreased pyruvate dehydrogenase complex activity in Huntington and Alzheimer brains. *Ann. Neurol., 13*, 72–78.
11. Mark, R. J., Pang, Z., Geddes, J. W., Uchida, K., & Mattson, M. P. (1997). Amyloid beta-peptide impairs glucose transport in hippocampal and cortical neurons: Involvement of membrane lipid peroxidation. *J. Neurosci., 17*, 1046–1054.
12. Piert, M., Koeppe, R. A., Giordani, B., Berent, S., & Kuhl, D. (1996). Diminished glucose transport and phosphorylation in Alzheimer's disease determined by dynamic FDG-PET. *J. Nucl. Med., 37*, 201–208.
13. Cooper, J. R. (1994). Unsolved problems in the cholinergic nervous system. *J. Neurochem., 63*, 395–399.
14. Kelly, J. F., Furukawa, K., Barger, S. W., Rengen, M., Mark, R. J., Blanc, E. M., Roth, G. S., & Mattson, M. P. (1996). Amyloid β-peptide disrupts carbachol-induced muscarinic cholinergic signal transduction in cortical neurons. *Proc. Natl. Acad. Sci. USA, 93*, 6753–6758.
15. Araujo, D. M., Lapchak, P. A., Robitaille, Y., Gauthier, S., & Quirion, R. (1988). Differential alteration of various cholinergic markers in cortical and subcortical regions of the human brain in Alzheimer's disease. *J. Neurochem., 50*, 1915–1923.
16. Smith, C. J., Perry, E. K., Perry, R. H., Fairbairn, A. F., & Birdsall, N. J. (1987). Guanine nucleotide modulation of muscarinic cholinergic receptor binding in postmortem human brain— A preliminary study in Alzheimer's disease. *Neurosci. Lett., 82*, 227–232.
17. Pearce, B. D., & Potter, L. T. (1991). Coupling of m1 muscarinic receptors to G protein in Alzheimer disease. *Alz. Dis. Assn. J., 5*, 163–172.
18. Jope, R. S. (1996). Cholinergic muscarinic receptor signaling by the phosphoinositide signal transduction system in Alzheimer's disease. *Alz. Dis. Rev., 1*, 2–14.
19. Itoh, A., Nitta, A., Nadai, M., Nishimura, K., Hirose, M., Hasegawa, T., & Nabeshima, T. (1996). Dysfunction of cholinergic and dopaminergic neuronal systems in beta-amyloid protein-infused rats. *J. Neurochem., 66*, 1113–1117.
20. Abe, E., Casamenti, F., Giovannelli, L., Scali, C., & Pepeu, G. (1994). Administration of amyloid beta-peptides into the medial septum of rats decreases acetylcholine release from hippocampus in vivo. *Brain Res., 636*, 162–164.
21. Giovanelli, L., Casamenti, F., Scali, C., Bartolini, L., & Pepeu, G. (1995). Differential effects of amyloid peptides beta-(1-40) and beta-(25-35) injections into the rat nucleus basalis. *Neuroscience, 66*, 781–792.
22. Games, D., Adams, D., Alessandrini, R., Barbour, R., Berthelette, P., Blackwell, C., Carr, T., Clemens, J., Donaldson, T., Gillespie, F., Guido, T., Hagopian, S., Johnson-Wood, K., Khan, K., Lee, M., Leibowitz, P., Liebergurg, I., Little, S., Masliah, E., McConlogue, L., Montoya-Zavala, M., Mucke, L., Paganini, L., Penniman, E., Power, M., Schenk, D., Seubert, P., Snyder, B., Soriano, F., Tan, H., Vitale, J. Wadsworth, S., Wolozin, B., & Zhao, J. (1995). Alzheimer-type neuropathology in transgenic mice overexpressing V717F β-amyloid precursor protein. *Nature, 373*, 523–527.
23. Hsiao, K., Chapman, P., Nilsen, S., Eckman, C., Harigaya, Y., Younkin, S., Yang, F., & Cole, G. (1996). Correlative memory deficits, Aβ elevation, and amyloid plaques in transgenic mice. *Science, 274*, 99–102.

24. Nalbantoglu, J., Tirado-Santiago, G., Lahsaini, A., Poirier, J., Goncalves, O., Verge, G., Momoli, F., Welner, S. A., Massicotte, G., Julien, J. P., & Shapiro, M. L. (1997). Impaired learning and LTP in mice expressing the carboxy terminus of the Alzheimer amyloid precursor protein. *Nature, 387,* 500–505.

25. Li, M., & Smith, C. P. (1996). Beta-amyloid1-40 inhibits electrically stimulated release of [3H]norepinephrine and enhances the internal calcium response to low potassium in rat cortex: Prevention with a free radical scavenger. *Brain Res. Bull., 39,* 299–303.

26. Arias, C., Arrieta, I., & Tapia, R. (1995). Beta-amyloid peptide fragment 25-35 potentiates the calcium-dependent release of excitatory amino acids from depolarized hippocampal slices. *J. Neurosci. Res., 41,* 561–566.

27. Luo, Y., Sunderland, T., & Wolozin, B. (1996). Physiologic levels of beta-amyloid activate phosphatidylinositol 3-kinase with the involvement of tyrosine phosphorylation. *J. Neurochem., 67,* 978–987.

28. Fraser, S. P., Suh, Y.-H., & Djamgos, M. B. A. (1997). Ionic effects of the Alzheimer's disease beta-amyloid precursor protein and its metabolic fragments. *Trends Neurosci., 20,* 67–72.

29. Mattson, M. P., Barger, S. W., Cheng, B., Lieberburg, I., Smith-Swintosky, V. L., & Rydel, R. E. (1993). Beta-amyloid precursor protein metabolites and loss of neuronal Ca^{2+} homeostasis in Alzheimer's disease. *Trends Neurosci., 16,* 409–414.

30. Hensley, K., Carney, J. M., Mattson, M. P., Aksenova, M., Harris, M., Wu, J. F., Floyd, R. A., & Butterfield, D. A. (1994). A model for beta-amyloid aggregation and neurotoxicity based on free radical generation by the peptide: Relevance to Alzheimer disease. *Proc. Natl. Acad. Sci. USA, 91,* 3270–3274.

31. Yan, S. D., Chen, X., Fu, J., Chen, M., Zhu, H., Roher, A., Slattery, T., Zhao, L., Nagashima, M., Morser, J., Migheli, A., Nawroth, P., Stern, D., & Schmidt, A. M. (1996). RAGE and amyloid-beta peptide neurotoxicity in Alzheimer's disease. *Nature, 382,* 685–691.

32. Kuo, Y.-M., Emmerling, M. R., Vigo-Pelfrey, C., Kasunic, T. C., Kirkpatrick, J. B., Murdoch, G. H., Ball, M. J., & Roher, A. E. (1996). Water-soluble Aβ (1-40, 1-42) oligomers in normal and Alzheimer disease brains. *J. Biol. Chem., 271,* 4077–4081.

33. Paradis, E., Douillard, H., Koutroumanis, M., Goodyer, C., & LeBlanc, A. (1996). Amyloid beta peptide of Alzheimer's disease downregulates Bcl-2 andupregulates bax expression in human neurons. *J. Neurosci., 16,* 7533–7539.

34. Selkoe, D. J. (1993). Physiological production of the beta-amyloid protein and the mechanism of Alzheimer's disease. *Trends Neurosci., 16,* 403–409.

35. Roch, J. M., Masliah, E., Roch-Levecq, A. C., Sundsmo, M. P., Otero, D. A., Veinbergs, I., & Saitoh, T. (1994). Increase of synaptic density and memory retention by a peptide representing the trophic domain of the amyloid beta/Aa4 protein precursor. *Proc. Natl. Acad. Sci. USA, 91,* 7450–7454.

36. Mattson, M. P., Cheng, B., Culwell, A. R., Esch, F. S., Lieburgurg, I., & Rydel, R. E. (1993). Evidence for excitoprotective and intraneuronal calcium-regulating roles for secreted forms of the beta-amyloid precursor protein. *Neuron, 10,* 243–254.

37. Nitsch, R. M., Slack, B. E., Wurtman, R. J., & Growdon, J. H. (1992). Release of Alzheimer amyloid protein precursor derivatives stimulated by activation of muscarinic ACh receptors. *Science, 258,* 304–307.

38. Buxbaum, J. D., Ruefli, A. A., Parker, C. A., Cypess, A. M., & Greengard, P. (1994). Calcium regulates processing of the Alzheimer amyloid protein precursor in a protein kinase C-independent manner. *Proc. Natl. Acad. Sci. USA, 91,* 4489–4493.

39. Hung, A. Y., Haass, C., Nitsch, R. M., Qiu, W. Q., Citron, M., Wurtman, R. J., Growdon, J. H., & Selkoe, D. J. (1993). Activation of protein kinase C inhibits cellular production of the amyloid beta-protein. *J. Biol. Chem., 268,* 22959–22962.

40. Buxbaum, J. D., Koo, E. H., & Greengard, P. (1993). Protein phosphorylation inhibits production of Alzheimer amyloid beta/A4 peptide. *Proc. Natl. Acad. Sci. USA, 90,* 9195–9198.

41. Poirier, J. (1994). Apolipoprotein E in animal models of CNS injury and in Alzheimer's disease. *Trends Neurosci., 17,* 525–530.

42. Corder, E. H., Saunders, A. M., Strittmatter, W. J., Schmechel, D. E., Gaskell, P. C., Small, G. W., Roses, A. D., Haines, J. L., & Pericakvance, M. A. (1993). Gene dose of apolipoprotein E type 4 allele and the risk of Alzheimer's disease in late onset families. *Science, 261,* 921–923.
43. Poirier, J., Davignon, J., Bouthillier, D., Kogan, S., Bertrand, P., & Gauthier, S. (1993). Apolipoprotein E polymorphism and Alzheimer's disease. *Lancet, 342,* 697–699.
44. Namba, Y., Tomonaga, M., Kawasaki, H., Otomo, E., & Ikeda, K. (1991). Apolipoprotein E immunoreactivity in cerebral amyloid deposits and neurofibrillary tangles in Alzheimer's disease and kuru paque amyloid in Creutzfeldt-Jakob disease. *Brain Res., 541,* 163–166.
45. Poirier, J., Delisle, M.-C., Quirion, R., Aubert, I., Farlow, M., Lahiri, D., Hui, S., Bertrand, P., Nalbantoglu, J., Gilfix, B. G., & Gauthier, S. (1995). Apolipoprotein E_4 allele as a predictor of cholinergic deficits and treatment outcome in Alzheimer's disease. *Proc. Natl. Acad. Sci. USA, 92,* 12260–12264.
46. Soininen, H., Kosunen, O., Helisalmi, S., Mannermaa, A., Paljarvi, L., Talasniemi, S., Ryynanen, M., Riekkinen, P., Sr. (1995). A severe loss of choline acetyltransferase in the frontal cortex of Alzheimer patients carrying apolipoprotein E_4 allele. *Neurosci. Lett., 187,* 79–82.
47. Svenssen, A. L., Warpmen, U., Hellstrom-Lindall, E., Bogdanovi, N., Lannfelt, L., & Nordberg, A. (1998). Nicotinic receptors, muscarinic receptors and choline acetyltransferase activity in the temporal cortex of Alzheimer patients with differing apolipoprotein E genotypes (in press). *Neurosci. Lett.*
48. Gordon, I., Grauer, E., Genis, L., Sehayek, E., & Michaelson, D. M. (1995). Memory deficits and cholinergic impairments in apolipoprotein E–deficient mice. *Neurosci. Lett., 199,* 1–4.
49. Beffert, U., & Poirier, J. (1996). Apolipoprotein E, plaques, tangles and cholinergic dysfunction in Alzheimer's disease. *Ann. N. Y. Acad. Sci., 777,* 166–174.
50. Schmechel, D. E., Saunders, A. M., Strittmatter, W. J., Crain, B. J., Hulette, C. M., Joo, S. H., Pericak-Vance, M. A., Goldgaber, D., & Roses, A. D. (1993). Increased amyloid β-peptide deposition in cerebral cortex as a consequence of apolipoprotein E genotype in late-onset Alzheimer disease. *Proc. Natl. Acad. Sci. USA, 90,* 9649–9653.
51. Strittmatter, W. J., Weisgraber, K. H., Huang, D. Y., Dong, L. M., Salvesen, G. S., Pericak-Vance, M. A., Schmechel, D. E., Saunders, A. M., Goldgaber, D., & Roses, A. D. (1993). Binding of human apolipoprotein E to synthetic amyloid beta peptide: Isoform-specific effects and implications for late-onset Alzheimer disease. *Proc. Natl. Acad. Sci. USA, 90,* 8098–8102.
52. LaDu, M. J., Peterson, T. M., Frail, D. E., Reardon, C. A., Getz, G. S., & Falduto, M. T. (1995). Purification of apolipoprotein E attenuates isoform-specific binding to β-amyloid. *J. Biol. Chem., 270,* 9039–9042.
53. Ma, J., Yee, A., Brewer, H. B., Das, S., & Potter, H. (1994). Amyloid-associated proteins alpha(1)-antichymotrypsin and apolipoprotein E promote assembly of Alzheimer beta-protein into filaments. *Nature, 372,* 92–94.
54. Whitson, J. S., Mims, M. P., Strittmatter, W. J., Yamaki, T., Morrisett, J. D., & Appel, S. H. (1994). Attenuation of the neurotoxic effect of Aβ amyloid peptide by apolipoprotein E. *Biochem. Biophys. Res. Commun., 199,* 163–170.
55. Beffert, U., Faure, M. P., Aumont, N., Dea, D., Davignon, J., & Poirier, J. (1998). Apolipoprotein E uptake is increased by β-amyloid peptides in neurons and astrocytes (in press). *J. Neurochem.*
56. Strittmatter, W. J., Roses, A. D. (1996). Apolipoprotein E and Alzheimer disease. *Ann. Rev. Neurosci., 19,* 53–77.
57. Xia, W., Zhang, J., Perez, R., Koo, E. H., & Selkoe, D. J. (1997). Interaction between amyloid presursor protein and presenilins in mammalian cells: Implications for the pathogenesis of Alzheimer's disease. *Proc. Natl. Acad. Sci. USA, 94,* 8208–8213.
58. Hanisch, U., & Quirion, R. (1996). Interleukin-2 as a neuroregulatory cytokine. *Brain Res. Rev., 21,* 246–284.

14

UNRAVELING THE CONTROVERSY OF HUMAN PRION PROTEIN DISEASES

ANDRÉA LEBLANC

Department of Neurology and Neurosurgery, McGill University, and
The Bloomfield Centre for Research in Aging
Lady Davis Institute for Medical Research
Sir Mortimer B. Davis Jewish General Hospital
Montréal, Québec, Canada H3T 1E2

The main claim to fame of prion protein diseases is the transmission of these neurodegenerative diseases through the prion protein in absence of any nucleic acid particle. This hypothesis was deemed heretical in the 1980s, but with growing research and the discovery of similar mechanisms in yeast cells, it is now broadly accepted as a likely possibility for disease transmission. Human forms of the disease can be acquired genetically, iatrogenically, or sporadically, and most are transmissible. Although the iatrogenic forms of the disease can occur in younger individuals, the inherited and sporadic forms usually target individuals between the fourth and sixth decade of life. In this paper, the clinical and pathological manifestation of the classic and newly discovered forms of prion protein diseases will be discussed. In addition, recent molecular genetics discoveries, biochemical and structural features of normal and mutant prion protein, and possible underlying molecular mechanisms of human prion protein diseases obtained from both in vitro and in vivo experimental systems will be presented. The recent studies of prion protein and prion protein diseases are relinquishing some of the mysteries associated with prion protein, but we have yet to understand completely this very fascinating protein.

THREE MANNERS TO CONTRACT PRION PROTEIN DISEASES: INHERITANCE, IATROGENIC TRANSMISSION, AND SPORADIC MANIFESTATION

Human prion protein diseases are a rare form of neurodegenerative diseases that occur at a rate of 0.1 to 1 person per million per year. As in Alzheimer's disease, approximately 10% of these prion protein diseases are inherited. Some mutations can be traced back to the early 1700s, indicating that the disease, although transmissible, is not highly infectious as are some viruses. Iatrogenic transmission has occurred through contamination of brain surgical tools, treatment with human growth and gonadotropin hormones prepared from cadaveric brains, and corneal transplants (reviewed by Lantos[1]). Kuru, discovered in New Guinea, was thought to occur through cannibalism and has almost been eradicated.[2] In absence of inherited or iatrogenic causes for the disease, we group most of the prion protein diseases as sporadic cases of unknown etiology.

Animal prion protein diseases are not inherited and are all transmissible. The oldest form known as sheep scrapie has been around for more than 250 years.[3] Other animals in which prion protein diseases are known to occur include mink, mule deer, elk, cows, and domestic cats.[1, 4–6] The discovery of a new variant of Creutzfeldt-Jakob disease, the most common form of human prion protein diseases, with the recent occurrence of prion protein disease in cows (bovine spongiform encephalopathy or mad cow disease) has raised concerns about the possible transmission to meat-consuming humans.

CLINICAL FEATURES OF HUMAN PRION PROTEIN DISEASES

Prion diseases are characterized clinically by motor and cognitive abnormalities that progressively lead to dementia and provoke death within months to 15 years (Table 14.1). There are three main forms of prion protein diseases that are recognized easily: Creutzfeldt-Jakob disease (CJD), Gerstmann-Sträussler-Scheinker disease (GSS), and fatal familial insomnia (FFI). The most common form, CJD, presents as a rapidly progressive dementia associated with extrapyramidal, pyramidal, and cerebellar signs. The mean duration of the disease is 8 ± 11 months.[7] The age of onset of familial cases varies, but sporadic CJD occurs within a narrow window around 60 years old. Several variants have been identified recently and broaden the spectrum of these human prion protein diseases.[7] The new variant CJD (nvCJD) that is manifested uncharacteristically in young adults is manifested by behavioral changes associated with progressive dementia and ataxia.[8]

GSS has a more prolonged course of cerebellar ataxia or dementia with a mean duration of 5 years.[9–11] These forms always are inherited and usually are manifested between 40 and 60 years of age.[11]

TABLE 14.1 Clinical Symptoms of Inherited Prion Diseases

	Primary symptom	Secondary symptoms	Age of onset	Mean duration
CJD	Dementia	Loss of coordination	60 years	8 months
GSS	Loss of coordination	Dementia	40–60 years	~5 years
FFI	Sleep problems, autonomic nervous system dysfunction	Insomnia and dementia	~48 years	~14 months
FASE	Personality changes	Dementia	~44 years	~4 years

Note: CJD, Creutzfeldt-Jakob disease; *FASE,* familial atypical spongiform encephalopathy; *FFI,* fatal familial insomnia; *GSS,* Gerstmann-Sträussler-Scheinker disease.

FFI, also only known as an inherited form of disease, is characterized by disturbances of sleep, autonomic, endocrine, and motor systems.[12, 13] The age of onset of FFI is around 48 years old (range is 25 to 61 years), and the mean duration of the disease is 13.3 months (ranges from 7 to 33 months).[13]

We have recently identified a familial disease in Brazil that presents as a frontotemporal dementia with early signs of personality change and a rapidly progressing dementia of long duration.[14] Many patients eventually show signs of memory loss, aggressivity, hyperorality, parkisonism, and verbal stereotypies. The age of onset is 44.8 ± 3.8 years, and the mean duration of symptoms of 4.2 ± 2.4 years. In this manuscript, I designate this disease familial atypical spongiform encephalopathy (FASE).

PATHOLOGICAL FEATURES OF HUMAN PRION PROTEIN DISEASES

The pathological features of prion protein diseases include variable degrees and region-specific cerebral and cerebellar spongiform change, neuronal loss, gliosis, and prion amyloid plaques.[10] Tubulovesicular structures are observed by electron microscopy.[15] Different forms of prion protein disease do not necessarily display each of these pathological features (Table 14.2).

Most CJD cases show spongiform change, neuronal loss, and gliosis of the cerebral cortex.[7] Prion plaques are present in 15% of CJD cases, and a "florid" plaque reminiscent of kuru plaques is a pathological highlight of nvCJD.[8, 16, 17] Many variants of inherited and sporadic CJD display specific clinicopathological traits,[7] and there is growing evidence that the clinical and pathological presentation of the disease is associated with the molecular makeup of the prion protein.[18]

Prion amyloid plaques define the neuropathological diagnosis of GSS disease.[11, 19] The prion amyloid in the Indiana Kindred family is made of an 11-kDa fragment of the prion protein.[19] It remains to be seen if all prion amyloid are

TABLE 14.2 Pathological Manifestations of Prion Diseases

	CJD	GSS	FFI	FASE	nvCJD	Kuru
Spongiform change	+	+	+/-	+	+	+
Gliosis	+	+	+	—	+	+
Neuronal loss	+	+	+	+	+	+
Amyloid plaques	(15% of cases)	+	-	+/-	+	+

Note: CJD, Creutzfeldt-Jakob disease; *FASE,* familial atypical spongiform encephalopathy; *FFI,* fatal familial insomnia; *GSS,* Gerstmann-Sträussler-Scheinker disease; *nvCJD,* new variant Creutzfeldt-Jakob disease.

identical. Gliosis and neuronal loss (but not always spongiform change) accompany the plaques. Some plaques display both prion and β-amyloid–containing peptides and the GSS–Indiana Kindred family also shows cerebral neurofibrillary tangles; features usually associated with Alzheimer's disease.[20–22] A variant also displays amyloid deposition in the cerebrovascular tissues.[23]

Spongiform change is most often sparse or absent in FFI, and plaques are not detected.[13] However, in cases of long duration, it is possible to detect the characteristic spongiform change of prion diseases.[13] Neuronal loss and gliosis are localized specifically in the mediodorsal and anterior ventral nuclei of the thalamus.[24]

FASE displays the characteristic spongiform change and neuronal loss in the deep layers of the cortex.[14] The frontal and temporal lobes are most severely affected, whereas the occipital and parietal lobes, globus pallidus, and entorhinal cortex are affected mildly. However, even in the most affected areas, there is no evidence of the expected gliosis. This feature is quite unusual as gliosis usually is accepted as a common marker of neurodegenerative diseases.

Prion protein immunoreactivity that resists formic acid and hydrolytic autoclaving treatments is accepted for the identification of prion protein diseases in addition to or in the absence of the typical pathological features of prion diseases.[10, 25, 26] The immunoreactivity for prion protein can be detected in plaques and synaptic and perivacuolar areas.[10]

Neuropathological confirmation of prion diseases strongly supports the clinical and molecular evaluations, especially in inherited forms of the disease where rare cases prevent statistical confirmation of the association of the disease with the mutation in the prion protein gene.

MOLECULAR BIOLOGY OF PRION PROTEIN DISEASE

The human prion protein gene is contained within 2 exons separated by a 10–13-kb intron on chromosome 20.[27–29] The second exon of the prion gene contains the entire coding sequence of the protein.[30] Expression of a 2.1–2.5-kb prion protein mRNA is highest in brain, followed by lung, and heart.[31–33] Prion protein mRNA is found at lower levels in pancreas, spleen, testes, kidney, and liver.[32] In brain, neurons express high levels of prion protein mRNA,[30] bur reports also indicate high levels in nonneuronal cells such as ependymal, meninges, astrocytes, and microglia.[34, 35] Prion protein mRNA increases with brain development, [36, 37] indicating a function for prion protein in brain maturation.

The prion protein contains three hydrophobic sequences: the N-terminal signal peptide, which is deleted during translation; a stop transfer effector domain at amino acids 112–136; and a C-terminal glycophosphatidylinositol anchor protein (GPI) signal at amino acids 231–253 that is removed during the transit of the prion protein in the endoplasmic reticulum upon addition of the GPI-anchor (Fig.14.1).[38–48] The N-terminal region of the prion protein also contains an

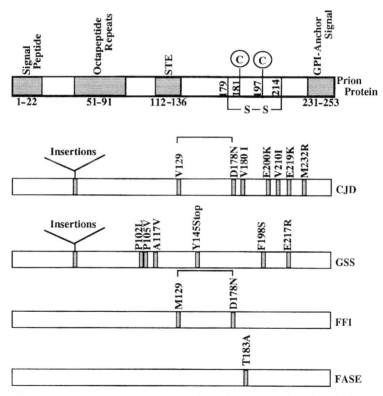

FIGURE 14.1   Prion protein characteristics and mutations associated with inherited prion protein diseases. The schematic diagram indicates the various properties of the prion protein. In the first panel, the localization of each characteristic is indicated by the amino acid number under or in the boxed diagram. The last four boxed diagrams indicate the location and type of mutation in each of the inherited prion protein diseases. *C*, N-linked glycosylation; *CJD*, Creutzfeldt-Jakob disease; *FASE*, familial atypical spongiform encephalopathy; *FFI*, fatal familial insomnia; *GSS*, Gerstmann-Sträussler-Scheinker disease; *S*, disulfide bond, *STE*, stop transfer effector domain.

octapeptide repeat between amino acids 51–91 whose function is unknown presently. Posttranslational modifications include N-linked glycosylation at amino acids 181 and 197 and a disulfide bond formation between cysteine 179 and 214.

Detection of mutations within the prion protein coding region is somewhat easy, since exon 2 comprises only 800 nucleotides. So far, six different single point mutations have been associated with CJD, [16, 17, 49–56] six with GSS, [16, 17, 54–62] one with FFI,[63–65] and one in FASE[14] (see Fig. 14.1). Many CJD and GSS families also contain insertional or deletion mutations of the octapeptide repeat. Insertional mutations of the octapeptide repeats are associated with CJD and GSS, whereas deletions are considered as a polymorphism.[66–70]

A polymorphic amino acid at codon 129 results in either a methionine or valine at this amino acid with codon. Homozygosity at codon 129 is common in sporadic CJD.[71] In FFI and CJD associated with the codon D178N mutation, the polymorphic codon 129 amino acid determines the phenotypic manifestation of the disease.[69, 70] CJD occurs when the D178N allele contains V129, whereas FFI occurs when the D178N allele contains M129. In many GSS cases the mutant allele contains V129 (reviewed by Tateishi[72]).

BIOCHEMICAL HALLMARKS OF HUMAN PRION PROTEIN IN NORMAL AND AFFECTED BRAINS

Since transmission of prion diseases is believed to occur through the prion protein, the biochemical properties of the prion protein in disease have been highly investigated. Prion protein is a 25-kDa protein, which after posttranslational modifications migrates as a broad band of 30–45 kDa.[73–75] The GPI-anchored protein is normally releasable from the cell surface membrane with phosphatidylinositol phospholipase C, but this feature is altered in abnormal forms of prion protein.[76–78] The normal protein is sensitive to proteinase K digestion and is detergent soluble; however, partial resistance to proteinase K digestion and detergent insolubility are acquired in prion protein isolated from affected brains or cell culture.[79, 80] The nuclear magnetic resonnance (NMR) spectra of amino acids 121–231 of normal prion protein shows three α-helices and two β-sheets, and it is predicted that abnormal prion protein has increased β-sheet content.[81–83] It is fascinating that these characteristics of the abnormal prion protein are the same in iatrogenically acquired, sporadic, or inherited prion protein diseases. Although it is possible that the abnormal form of the prion protein triggers the disease, the fact that this abnormal protein arises independent of the mode of transmission could also indicate that the abnormal prion protein is a consequence rather than the cause of the disease. The strongest evidence that the prion protein is the transmissible agent and is the cause of the disease is based on the fact that purified prion protein is capable of transmitting the disease.[84] In addition, there is considerable evidence that the abnormal prion protein conformation can be acquired in a cell-free system supporting a conformational change through protein–protein interaction.[85–86]

On the other hand, it has been impossible to show transmissibility of units of the prion protein, since prion protein loses its infectivity when disaggregated.[87] It is possible that disaggregation changes the conformation of the protein toward a noninfectious form or that proteinase K resistance is the result of strong protein–protein interactions that protect regions of the prion protein against protease digestion. One alternative explanation is that aggregated prion protein causes cell death, resulting in the abnormal prion protein. A jeopardized neuron is likely to be dysfunctional at many levels, including translational and posttranslational modifications of proteins. For example, the prion protein may not be properly folded in a sick cell, resulting in proteinase K resistance and changes in the tertiary structure of the protein. Many proteins are rendered detergent insoluble in dying cells, and proteinase K resistance has also been observed in the β-amyloid protein of Alzheimer's disease.[88] A prion protein fragment comprised of amino acids 106–126 induces apoptosis of postmitotic cells.[89] The physiological significance of this neurotoxic function is not clear at the present time. The concentration used to induce neurotoxicity was between 25 and 80 μM, a concentration that is not physiological in free-floating protein. However, it can be argued that the prion amyloid deposited in the plaques is concentrated enough to induce neuronal degeneration or apoptosis of neighboring neurons.

Prion protein is normally localized on the cell surface and attached through the GPI-anchor. This protein, present in caveolae, clathrin-coated pits and/or rafts,[90–92] is reinternalized and undergoes proteolytic processing. It is believed that the abnormal form of the protein occurs through reinternalization of the cell surface prion protein. The abnormal prion protein accumulates in lysosomes and other cytoplasmic structures.[93, 94]

MOLECULAR SIGNATURE OF PRION PROTEIN DISEASES

One interesting characteristic of the abnormal prion protein in either sporadic, iatrogenic, or inherited prion protein diseases is its partial resistance to proteinase K. However, even more unexpected was the finding that each disease gives a characteristic pattern of proteinase K-resistant peptides after deglycosylation, which can be resolved simply on polyacrylamide gels. This pattern is not only characteristic of the disease within the human family affected but is also transmissible to animal species, thereby constituting a "molecular signature" of the disease. This feature was first observed between the D178N/129V CJD and D178N/129M FFI variants[95] and later observed in cell-free systems and in a variety of other prion protein diseases such as sporadic CJD, GSS, and nvCJD.[18, 86, 96, 97, 99] Proteinase K resistance is explained on the basis of abnormal conformation of the prion protein. Therefore, these results indicate that the conformation of the protein is transmitted with the disease and implies that the

abnormal protein induces an identical conformational change on normal endogenous prion protein, resulting in the onset of disease.

FUNCTIONS OF THE NORMAL
PRION PROTEIN

Despite the extensive characterization of the abnormal prion protein, there is
still no elucidation of how this abnormal form of prion protein causes neuronal
dysfunction and demise. Is the abnormal prion protein or is the lack of normal
prion protein causing cell death? The seeming "normality" of prion protein
null mice would indicate that the abnormal prion protein is at the root of the
neuronal cell death. However, in null mice, there is always the possibility that
the surviving mice have developmentally adapted to overcome the protein
deficiency. If a lack of normal prion protein is the cause of the disease, it is essential to understand the role of prion protein. The function of the prion protein
is still unclear but must be important, since this protein is highly expressed in
the brain.

SYNAPTIC TRANSMISSION

The localization of the prion protein on the cell surface through the GPI-
anchored protein hints at a possible receptor function. Whatley et al.[100] show
increased internal calcium ions in synaptosomes treated with recombinant prion
protein, indicating a role for prion protein in signal transduction resulting in the
regulation of intracellular free calcium. In fact, when probed further, prion protein null mice show a weak γ-aminobutyric acid (GABA) receptor–mediated fast
inhibition and impaired long-term potentiation in hippocampal neurons.[98]

MITOGEN

Cashman et al.[101] showed a role for prion protein in lymphocyte activation.
Peptides of the prion protein, $PrP_{106-126}$ can induce astrocyte proliferation and
hypertrophy, indicating that prion protein can act as a mitogen.[102] It is not clear
whether the protein has to be free or if the mitogenic role is present in cell surface proteins. However, there is some release of prion protein in cell culture
media from human primary neuron cultures (our laboratory—unpublished observations) and in human cerebrospinal fluid, indicating a possible in vivo role for
soluble prion protein in cell proliferation.[103]

APOPTOSIS AND PRION PROTEIN

We have observed strong homology between the octapeptide repeat region of
prion protein and the bcl-2 antiapoptotic functional BH2 domain of bcl-2 protein

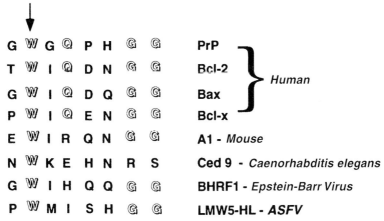

G	W	G	Q	P	H	G	G	PrP	
T	W	I	Q	D	N	G	G	Bcl-2	
G	W	I	Q	D	Q	G	G	Bax	
P	W	I	Q	E	N	G	G	Bcl-x	
E	W	I	R	Q	N	G	G	A1 - *Mouse*	
N	W	K	E	H	N	R	S	Ced 9 - *Caenorhabditis elegans*	
G	W	I	H	Q	Q	G	G	BHRF1 - *Epstein-Barr Virus*	
P	W	M	I	S	H	G	G	**LMW5-HL - *ASFV***	

FIGURE 14.2 Homology between prion protein octapeptide repeat and bcl-2–related proteins. The arrow indicates an essential amino acid for bcl-2 and bax interaction, as well as for the antiapoptotic function of bcl-2.[104] *ASFV,* African swine fever virus.

(Fig. 14.2). Since the BH2 domain is responsible for the antiapoptotic function of bcl-2, as well as for its interaction with bax, protein–protein interactions are a logical mechanism for the role of bcl-2–related proteins in mediating cell survival and cell death.[104] Because of the homology between prion protein and bcl-2, it is possible that prion protein also mediates cell survival and cell death. Since prion protein is highly expressed in neurons, it may provide a backup protective system to neurons, especially in aging, since bcl-2 protein decreases in the aging central nervous system.[105] Kurshner and Morgan[106] showed that the C-terminal region of bcl-2 can bait the N-terminal of prion protein in the yeast matchmaker 2-hybrid system. Since the BH2 domain of bcl-2 is located at the C-terminus, these results support a possible interaction of the octapeptide repeat of prion protein with the BH2 domain of bcl-2. We have confirmed the interaction of full-length human prion and bcl-2 proteins in our laboratory using the yeast 2-hybrid matchmaker system. In addition, we have co-immunoprecipitated prion protein and bcl-2 from human fetal and adult brain and human primary neuron cultures and achieved in vivo cross-linking of bcl-2 to prion protein in the primary neuron cultures.[107] We also see, in contrast to the results of Kurshner and Morgan[106, 108] that prion protein interacts with bax protein. These interactions could indicate a role for prion protein in neuronal survival and cell death. In fact, the strong homology with the BH2 domain of bcl-2–related proteins may even indicate that prion proteins are part of this family of proteins. We have yet to confirm that the interaction confers an antiapoptotic response. This will not be an easy task, since it is likely that the interaction of prion protein with the bcl-2 proteins (just like the interaction of all the other known members of the bcl-2 related proteins) is quite complex.

TRANSMISSIBILITY MODELS FOR
PRION PROTEIN DISEASES

The most intriguing feature of prion diseases is its transmission in the absence of nucleic acids. The finding that a protein extract devoid of nucleic acid could transmit the neurodegenerative scrapie disease to hamster challenged a fundamental belief of molecular biology: that transmission of disease occurs through bacterial or viral infections.[84] Despite major experimental attempts in the last 10 years, no nucleic acids have been found in the transmissible scrapie prion isolate.[34, 35, 109] However, no one has proven beyond the shadow of a doubt that transmissible prion isolates do not contain a small amount of highly protected nucleic acids. Various models of prion protein disease transmission have been proposed; although evidence supports many, none have been conclusively proven. We add to these models one that is also based on protein–protein interaction and addresses a possible fundamental role of the normal prion protein.

CONVERSION OF PrP^C TO PrP* TO PrP^{SC}

First, it is proposed that prion diseases occur through the transformation of the normal cellular form of the prion protein (cellular PrP; PrP^C) into a proteinase K-resistant protein that can form amyloid fibrils and often becomes a transmissible agent for the disease (scrapie PrP;Prp^{SC} or PrP^{RES}) (Fig. 14.3, A). The initial formation of PrP^{SC} could result from prion protein mutations or occur spontaneously at low rates. Since the transmissible inoculate comprises mainly PrP^{SC} protein

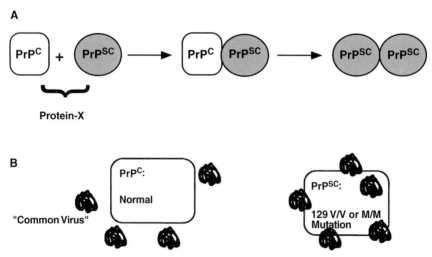

FIGURE 14.3   Models of prion protein disease transmissibility. (A) Protein only hypothesis. (B) Viral theory. (C) Virion theory. (D) PrP^{SC} seed theory. (E) Prion protein–bcl-2 and related protein interaction. PrP^C indicates normal cellular prion protein, PrP^{SC} represents the abnormal proteinase K resistant form of the protein. Description of each model is found in the text.   *(Continues)*

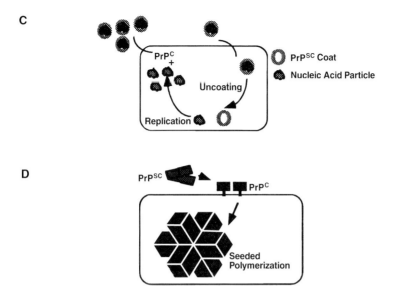

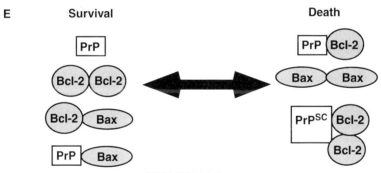

FIGURE 14.3 *(Continued)*

without detectable levels of nucleic acids, the transformation of PrP^C into PrP^{SC} presumably occurs by a conformational change of PrP^C through interaction with PrP^{SC}.[5, 6, 110–114] It is proposed that an intermediate partially unfolded structure of the prion (PrP*) occurs before the transformation to PrP^{SC}.[115] The dimer hypothesis is strongly supported by the resistance of prion protein knock-out transgenic mice to PrP^{SC} infection,[116] metabolic labeling experiments showing that synthesis of PrP^{SC} arises, albeit very slowly, from labeled PrP^C,[85] and conformational transmission of human disease to mice.[97, 99]

The transformation of PrP^{SC} from PrP^C requires localization of the PrP^C to the plasma membrane[39,43–45, 117, 118] and an intact secretory pathway. Glycosylation is not essential.[43–45, 85, 119, 120] In fact, inhibition of N-glycosylation by tunicamycin accelerates the appearance of PrP^{SC}.[45] The production of PrP^{SC} occurs between the plasma membrane and the secondary lysosome, where it eventually accumulates undegraded.[39, 73, 74, 93, 118, 121] The fact that a second gene controls the time of incubation of the disease[122-126] indicates that additional factors, although unknown, are essential to the evolution of prion diseases.

Interestingly, two yeast systems have been discovered where a conformational change of a protein results in a change of phenotype similar to that observed in prions[127] (reviewed by Tuite[128]). (1) The ure2p protein, involved in the metabolism of nitrogenous compounds, spontaneously converts to the inactive form, ure2p*. (2) A nonsense suppressor [psi+] converts to an inactive form [psi-]. The presence of [psi+] depends on a third gene product, sup35, which contains an N-terminal nonapeptide tandem repeat that is highly homologous to the prion octapeptide repeat. The two yeast systems also share similar N-terminal secondary structures with the prion protein. Overexpression of ure2p and [psi+] results in an increased conversion to ure2p* and psi*, respectively. Therefore the yeast systems and prion protein, share the spontaneous conversion of a protein, which alters the phenotype of a cell.

A HIDDEN NUCLEIC ACID GENOME CO-PURIFIES WITH THE INFECTIOUS PrP^{SC}: THE VIRION THEORY

A second model supposes the existence of a nucleic acid genome in the infectious prion preparation (see Fig. 14.3, *B*). This hypothesis is based on the fact that the existing strains of scrapie are different thereby suggesting the presence of a nucleic acid component. However, as discussed above, this nucleic acid component, although plausible, has never been detected in the transmissible prion isolates, and nongenetic propagation of strain-specific properties has been shown.[86, 109]

VIRAL THEORY

Another group proposes that cells housing abnormal prion protein whether arising through a mutation, spontaneously or by transmission, become susceptible to a yet unknown but common small virus[129] (see Fig. 14.3, *C*). The viral hypothesis is supported by the lack of penetrance of certain mutations of the prion protein, indicating that factors other than the prion protein mutation are required for the development of disease. Evidence for viral particles is weak, but reports indicate the presence of a small viral-like particle in hamster-infected brains and sporadic and familial CJD, as well as human infectious preparations with viral properties.[130-132]

A "UNIFIED" THEORY: HOLOPRION REQUIRES AN APOPRION (PrP^{SC}) AND CO-PRION (A NUCLEIC ACID COMPONENT)

A third theory suggests that conversion of PrP^C into PrP^{SC} is mediated by a holoprion made of an apoprion (PrP^{SC}) and a co-prion (a nucleic acid), which is either associated with the apoprion upon infection or recruited by the cell. The co-prion would have the ability to replicate within the cell, and replication would be regulated by the apoprion.[133] This model combines the virion theory model and the viral theory model.

SEED THEORY

Finally, a number of investigators believe that PrP^{SC} serves as a seed upon which PrP^{C} would be transformed into PrP^{SC} in a "crystal-like" fashion (see Fig. 14.3, *D*). This hypothesis is supported by the cell-free generation of proteinase K–resistant prion protein.[85] This model supposes that disaggregation of the "crystal" will result in the loss of proteinase K resistance of prion protein.

Each of these proposed models are logical and possible, but despite accumulating evidence, it has been impossible to confirm any of these conclusively. Either prion protein diseases have different etiologies and all models are correct, or there is still a missing link to the prion protein mystery.

ALTERNATE MODEL EXPLAINING NORMAL AND ALTERED PRION FUNCTION

I propose an alternate model. This model is based solely on protein–protein interactions without any nucleic acid component. The proteins involved include the prion protein and bcl-2–related proteins. I propose that prion proteins have a similar function than bcl-2 and bcl-2–related proteins in the maintenance of cell survival. The model is based on the fact that (1) the four prion octapeptide tandem repeats are highly homologous to the sequence responsible for heterodimerization of antiapoptotic protein bcl-2 with proapoptotic protein bax (see Fig. 14.2), which is required for the viable state of the cell,[104] (2) increased numbers of octapeptide repeats are associated with CJD, (3) prion protein interacts with bcl-2, and (4) prion protein interacts with bax. I propose that failure to maintain proper interactions between prion protein and bcl-2–related protein will result in a cell death situation similar to that observed with bcl-2 and bax.

The prion protein with an abnormal conformation may not interact normally with the bcl-2–related proteins and therefore promote cell death. The fact that bcl-2 expression is down regulated in aging may explain why inherited prion diseases are generally not manifested before the fourth decade of life, since the expression of bcl-2 earlier may compensate for the prion protein mutations.[105] The abnormal prion protein may also sequester the antiapoptotic proteins by interacting too strongly with these proteins and thus promote cell death. Alternatively, normal prion protein may bind the proapoptotic protein bax and prevent its role in cell death. The prion protein–bax interaction may be abolished in abnormal forms of the prion protein. Transgenic mice overexpressing prion protein develop a myopathy and neuropathy.[134] It is possible that high levels of prion protein in a cell drives the "death" mode. Again, the expression of bcl-2 would protect the mice in early development, since only older mice are affected.[134]

CONCLUSIONS

Many studies on prion protein and prion diseases have initiated ideas and revealed mechanisms that are not only relevant to the understanding of prion diseases but further advance our general understanding of the molecular mechanisms of disease. Molecular genetics has shown the influence of an inoffensive polymorphic amino acid on the clinical and pathological manifestation of a disease. The initially proposed mechanism of prion disease transmission through protein–protein interactions seems less bizarre today with the recent understanding of a number of proteins related to bcl-2 that also function through protein–protein interaction to either promote cell survival or cell death. Whereas discoveries are usually made in simple organisms and then extended to humans, the prion hypothesis has reversed this order with the finding of yeast prions long after the discovery of the mammalian prion protein. Until resolved, the uniqueness of the prion protein disease transmission will certainly fuel other imaginative discoveries in the future.

REFERENCES

1. Lantos, P. L. (1992). From slow virus to prion: A review of transmissible spongiform encephalopathies [see comments]. *Histopathology, 20,* 1–11.
2. Anders, K. H. (1993). Human prion diseases. *West. J. Med. 158,* 295.
3. Hadlow, W. J. (1995). Neuropathology and the Scrapie-Kuru connection. *Brain Pathol., 5,* 27–31.
4. Bessen, R. A., & Marsh, R. F. (1992). Biochemical and physical properties of the prion protein from two strains of the transmissible mink encephalopathy agent. *J. Virol., 66,* 2096–2101.
5. Prusiner, S. B. (1993). Biology of prion diseases. *J. Acquir. Immune Defic. Syndr., 6,* 663–665.
6. Prusiner, S. B. (1993). Genetic and infectious prion diseases. *Arch. Neurol., 50,* 1129–1153.
7. Richardson, E. P., & Masters, C. L. (1995). The nosology of Creutzfeldt-Jakob disease and conditions related to the accumulation of PRP^CJD in the nervous system. *Brain Pathol., 5,* 33–41.
8. Will, R., Ironside, J., Zeidler, M., Cousens, S., Estibeiro, K., Alperovitch, A., Poser, S., Pocchiari, M., Hofman, A., & Smith, P. (1996). A new variant of Creutzfeldt-Jakob disease in the UK. *Lancet, 347,* 921–925.
9. Hsiao, K., & Prusiner, S. B. (1990). Inherited human prion diseases. *Neurology, 40,* 1820–1827.
10. Budka, H., Aguzzi, A., Brown, P., Brucher, J. M., Bugiani, O., Gullota, F., Haltia, M., Hauw, J.-J., Ironside, J. W., Jellinger, K., Kretzschmar, H. A., Lantos, P. L., Masullo, C., Schlote, W., Tateishi, J., & Weller, R. O. (1995). Neuropathological diagnosis criteria for Creutzfeldt-Jakob disease (CJD) and other human spongiform encephalopathies (prion diseases). *Brain Pathol., 5,* 459–466.
11. Ghetti, B., Dlouhy, S. R., Giaconne, G., Bugiani, O., Frangione, B., Farlow, M. R., & Tagiavini, F. (1995). Gerstmann-Sträussler-Shceinker disease and the Indiana kindred. *Brain Pathol., 5,* 61–75.
12. Lugaresi, E., Medori, R., Montagna, P., Baruzzi, A., Cortelli, P., Lugaresi, A., Tinuper, P., Zucconi, M., & Gambetti, P. (1986). Fatal familial insomnia and dysautonomia with selective degeneration of thalamic nuclei. *N. Engl. J. Med., 315,* 997–1003.
13. Gambetti,, P., Parchi, P., Petersen, R., Chen, S., & Lugaresi, E. (1995). Fatal familial insomnia and familial Creutzfeldt-Jakob disease: Clinical pathological and molecular features. *Brain Pathol., 5,* 43–51.

14. Nitrini, R., Rosemberg, S., Passos-Bueno, M. R., Lughetti, P., Papadopoulos, M., Carrilho, P. E., Caramelli, P., Albrecht, S., Zatz, M., & LeBlanc, A. C. (1997). Familial spongiform encephalopathy with distinct clinico-pathological features associated with a novel prion gene mutation at codon 183. *Ann. Neurol., 42,* 138–146.
15. Narang, H. K. (1992). Scrapie-associated tubulofilamentous particles in human Creutzfeldt-Jakob disease. *Res. Virol., 143,* 387–395.
16. Hsiao, K. K., Cass, C., Schellenberg, G. D., Bird, T., Devine, G. E., Wisniewski, H., & Prusiner, S. B. (1991). A prion protein variant in a family with the telencephalic form of Gerstmann-Sträussler-Scheinker syndrome. *Neurology, 41,* 681–684.
17. Hsiao, K., Meiner, Z., Kahana, E., Cass, C., Kahana, I., Avrahami, D., Scarlato, G., Abramsky, O., Prusiner, S. B., & Gabison, R. (1991). Mutation of the prion protein in Libyan Jews with Creutzfeldt-Jakob disease. *N. Engl. J. Med., 324,* 1091–1097.
18. Parchi, P., Castellani, R., Capellari, S., Ghetti, B., Young, K., Chen, S., Farlow, M., Dickson, D., Sima, A., Trojanowski, J., Petersen, R., & Gambetti, P. (1996). Molecular basis of phenotypic variability in sporadic Creutzfeldt-Jakob disease. *Ann. Neurol., 39,* 767–778.
19. Tagliavini, F., Prelli, F., Ghiso, J., Bugiani, O., Serban, D., Prusiner, S. B., Farlow, M. R., Ghetti, B., & Frangione, B. (1991). Amyloid protein of Gerstmann-Sträussler-Scheinker disease (Indiana kindred) is an 11 kd fragment of prion protein with an N-terminal glycine at codon 58. *Embo. J., 10,* 513–519.
20. Ghetti, B., Tagliavini, F., Masters, C. L., Beyreuther, K., Giaconne, G., Verga, L., Farlow, M. R., Conneally, P. M., Dlouhy, S. R., Azzarelli, B., & Bugiani, O. (1989). Gerstmann-Sträussler-Scheinker disease. II. Neurofibrillary tangles and plaques with PrP-amyloid coexist in an affected family. *Neurology, 39,* 1453–1461.
21. Giaccone, G., Tagliavini, F., Verga, L., Frangione, B., Farlow, M. R., Bugiani, O., & Ghetti, B. (1990). Neurofibrillary tangles of the Indiana kindred of Gerstmann-Sträussler-Scheinker disease share antigenic determinants with those of Alzheimer's disease. *Br. Res., 530,* 325–329.
22. Miyazono, M., Kitamoto, T., Iwaki, T., & Tateishi, J. (1992). Colocalization of prion protein and beta protein in the same amyloid plaques in patients with Gerstmann-Sträussler syndrome. *Acta Neuropathol. (Berl.), 83,* 333–339.
23. Ghetti, B., Piccardo, P., Spillantini, M., Ichimiya, Y., Porro, M., Perini, F., Kitamoto, T., Tateishi, J., Seiler, C., Frangione, B., Bugiani, O., Giaccone, G., Prelli, F., Goedert, M., Dloughy, S. R., & Tagliavini, F. (1996). Vascular variant of prion protein cerebral amyloidosis with tau-positive neurofibrillary tangles: The phenotype of the stop codon 145 mutation in PRNP. *Proc. Natl. Acad. Sci. USA, 93,* 744–748.
24. Manetto, V., Medori, R., Cortelli, P., Montagna, P., Baruzzi, A., Hauw, J., Rancruel, G., Vanderhaeghen, J., Mailleux, P., Bugiani, O., Tagliavini, F., Bouras, C., Rizzuto, N., Lugaresi, E., & Gambetti, P. (1992). Fatal familial insomnia: Clinical and pathological study of five new cases. *Neurology, 42,* 312–319.
25. Kitamoto, T., Shin, R. W., Doh, U. K., Tomokane, N., Miyazono, M., Muramoto, T., & Tateishi, J. (1992). Abnormal isoform of prion proteins accumulates in the synaptic structures of the central nervous system in patients with Creutzfeldt-Jakob disease. *Am. J. Pathol., 140,* 1285–1294.
26. Lantos, P. L., McGill, I. S., Janota, I., Doey, L. J., Collinge, J., Bruce, M. T., Whatley, S. A., Anderton, B. H., Clinton, J., Roberts, G. W., & Rossor, M. N. (1992). Prion protein immunocytochemistry helps to establish the true incidence of prion diseases. *Neurosci. Lett., 147,* 67–71.
27. Basler, K., Oesch, B., Scott, M., Westaway, D., Walchli, M., Groth, D., McKinley, M., Prusiner, S., & Weissmann, C. (1986). Scrapie and cellular PrP isoforms are encoded by the same chromosomal gene. *Cell, 46,* 417–428.
28. Puckett, C., Concannon, P., Casey, C., & Hood, L. (1991). Genomic structure of the human prion protein gene [see comments]. *Am. J. Hum. Genet., 49,* 320–329.
29. Liao, Y. C., Lebo, R. V., Clawson, G. A., & Smuckler, E. A. (1986). Human prion protein cDNA: Molecular cloning, chromosomal mapping, and biological implications. *Science, 233,* 364–367.

30. Kretzschmar, H. A., Stowring, L. E., Westaway, D., Stubblebine, W. H., Prusiner, S. B., & Dearmond, S. J. (1986). Molecular cloning of a human prion protein cDNA, *DNA, 5,* 315–324.

31. Robakis, N. K., Sawh, P. R., Wolfe, G. C., Rubenstein, R., Carp, R. I., & Innis, M. A. (1986). Isolation of a cDNA clone encoding the leader peptide of prion protein and expression of the homologous gene in various tissues. *Proc. Natl. Acad. Sci. USA, 83,* 6377–6381.

32. Oesch, B., Westaway, D., Walchli, M., McKinley, M., Kent, S., Aebersold, R., Barry, R., Tempst, P., Teplow, D., Hood, L., Prusiner, S., & Weissmann, C. (1985). A cellular gene encodes scrapie PrP 27-30 protein. *Cell, 40,* 735–746.

33. Chesebro, B., Race, R., Wehrly, K., Nishio, J., Bloom, M., Lechner, D., Bergstrom, S., Robbins, K., Mayer, L., Keith, J. M., et al. (1985). Identification of scrapie prion protein-specific mRNA in scrapie-infected and uninfected brain. *Nature, 315,* 331–333.

34. Brown, P., Liberski, P., Wolff, A., & Gadjusek, C. (1990). Conservation of infectivity in purified fibrillary extracts of scrapie-infected hamster brain after sequential enzymatic digestion or polyacrylamide gel electrophoresis. *Proc. Natl. Acad. Sci. USA, 87,* 7240–7244.

35. Brown, H., Goller, N., Rudelli, R., Merz, G., Wolf, G., Wisniewski, H., & Robakis, N. (1990). The mRNA encoding the scrapie agent protein is present in a variety of non-neuronal cells, *Acta Neuropathol., 80,* 1–6.

36. Lazarini, F., Deslys, J. P., & Dormont, D. (1991). Regulation of the glial fibrillary acidic protein, beta actin and prion protein mRNAs during brain development in mouse. *Brain Res. Mol. Brain Res., 10,* 343–346.

37. Manson, J., West, J. D., Thomson, V., McBride, P., Kaufman, M. H., & Hope, J. (1992). The prion protein gene: A role in mouse embryogenesis? *Dev., 115,* 117–122.

38. Caughey, B., Race, R. E., Ernst, D., Buchmeier, M. J., & Chesebro, B. (1989). Prion protein biosynthesis in scrapie-infected and uninfected neuroblastoma cells. *J. Virol., 63,* 175–181.

39. Caughey, B., & Raymond, G. (1991). The scrapie-associated form of PrP is made from a cell surface precursor that is both protease- and phospholipase-sensitive. *J. Biol. Chem., 266,* 18217–18223.

40. Haraguchi, T., Fisher, S., Olofsson, S., Endo, T., Groth, D., Tarentino, a., Borchelt, D. R., Teplow, D., Hood, L., Burlingame, A., & Prusiner, S. (1989). Asparagine-linked glycosylation of the scrapie and cellular prion proteins. *Arch. Biochem. Biophys., 274,* 1–13.

41. Pan, K. M., Stahl, N., & Prusiner, S. B. (1992). Purification and properties of the cellular prion protein from Syrian hamster brain. *Protein Sci., 1,* 1343–1352.

42. Rogers, M., Taraboulos, A., Scott, M., Groth, D., & Prusiner, S. B. (1990). Intracellular accumulation of the cellular prion protein after mutagenesis of its Asn-linked glycosylation sites. *Glycobiology, 1,* 101–109.

43. Taraboulos, A., Rogers, M., Borchelt, D. R., McKinley, M. P., Scott, M., Serban, D., & Prusiner, S. B. (1990). Acquisition of protease resistance by prion proteins in scrapie-infected cells does not require asparagine-linked glycosylation. *Proc. Natl. Acad. Sci. USA, 87,* 8262–8266.

44. Taraboulos, A., Serban, D., & Prusiner, S. B. (1990). Scrapie prion proteins accumulate in the cytoplasm of persistently infected cultured cells. *J. Cell Biol., 110,* 2117–2132.

45. Taraboulos, A., Raeber, A. J., Borchelt, D. R., Serban, D., & Prusiner, S. B. (1992). Synthesis and trafficking of prion proteins in cultured cells. *Mol. Biol. Cell, 3,* 851–863.

46. Baldwin, M. A., Stahl, N., Reinders, L. G., Gibson, B. W., Prusiner, S. B., & Burlingame, A. L. (1990). Permethylation and tandem mass spectrometry of oligosaccharides having free hexosamine: Analysis of the glycoinositol phospholipid anchor glycan from the scrapie prion protein. *Anal. Biochem., 191,* 174–181.

47. Rogers, M., Yehiely, F., Scott, M., & Prusiner, S. B. (1993). Conversion of truncated and elongated prion proteins into the scrapie isoform in clutured cells. *Proc. Natl. Acad. Sci. USA, 90,* 3182–3186.

48. Stahl, N., Baldwin, M. A., Hecker, R., Pan, K. M., Burlingame, A. L., & Prusiner, S. B. (1992). Glycosylinositol phospholipid anchors of the scrapie and cellular prion proteins contain sialic acid. *Biochem., 31,* 5043–5053.

49. Goldfarb, L., Mitrova, E., Brown, P., Toh, B., & Gadjusek, D. (1990). Mutation in codon 200

of scrapie amyloid protein gene in two clusters of Creutzfeldt Jakob disease in Slovakia. *Lancet, 336,* 514–515.

50. Goldfarb, L., Haltia, M., Brown, P., Nieto, A., Kovanen, J., McCombie, W., Trapp, S., & Gadjusek, C. (1991). New mutation in scrapie amyloid precursor gene (at codon 178) in Finnish Creutzfeldt-Jakob kindred. *Lancet, 337,* 425.

51. Brown, P. Goldfarb, L. G., Kovanen, J., Haltia, M., Cathala, E., Sulima, M., Gibbs, C., & Gadjusek, C. (1992). Phenotypic characteristics of familial Creutzfeldt-Jakob disease associated with the codon 178asn PRNP mutation. *Ann. Neurol. 31,* 282–285.

52. Korczyn, A. D., Chapman, J., Goldfarb, L. G., Brown, P., & Gajdusek, D. C. (1991). A mutation in the prion protein gene in Creutzfeldt-Jakob disease in Jewish patients of Libyan, Greek, and Tunisian origin. *Ann. N. Y. Acad. Sci., 640,* 171–176.

53. Pocchiari, M., Salvatore, M., Cutruzolla, F., Genuardi, M., Travaglini, A. C., Masullo, C., Macchi, G., Alema, G., Galgani, S., Xi, Y. G., Petraroli, R., Silvestrini, M. C., & Brunori, M. (1993). A new mutation of the prion protein gene in Creutzfeldt-Jakob disease. *Ann. Neurol., 34,* 802–807.

54. Kitamoto, T., Amano, N., Terao, Y., Nakazato, Y., Isshiki, T., Mizutani, T., & Tateishi, J. (1993). A new inherited prion disease (PrP-P105L mutation) showing spastic parapesis. *Ann. Neurol., 34,* 808–813.

55. Kitamoto, T., Iizuka, R., & Tateishi, J. (1993). An amber mutation of prion protein in Gerstmann-Sträussler syndrome with mutant PrP plaques. *Biochem. Biophys. Res. Commun., 192,* 525–531.

56. Kitamoto, T., Ohta, M., Doh, U. K., Hitoshi, S., Terao, Y., & Tateishi, J. (1993). Novel missense variants of prion protein in Creutzfeldt-Jakob disease or Gerstmann-Sträussler syndrome. *Biochem. Biophys. Res. Commun., 191,* 709–714.

57. Hsiao, K., Baker, H. F., Crow, T. J., Poulter, M., Owen, F., Terwilliger, J. D., Westaway, D., Ott, J., & Prusiner, S. B. (1989). Linkage of a prion protein missense variant to Gerstmann-Sträussler syndrome. *Nature, 338,* 342–345.

58. Mastrianni, J., Curtis, M., Oberholtzer, J., Da Costa, M., DeArmond, S., Prusiner, S., & Garbern, J. (1995). Prion disease (PrP-A117V) presenting with ataxia instead of dementia. *Neurology, 45,* 2042–2050.

59. Doh, U. K., Tateishi, J., Sasaki, H., Kitamoto, T., & Sakaki, Y. (1989). Pro----leu change at position 102 of prion protein is the most common but not the sole mutation related to Gerstmann-Sträussler syndrome. *Biochem. Biophys. Res. Commun., 163,* 974–979.

60. Dlouhy, S. R., Hsiao, K., Farlow, M. R., Foroud, T., Conneally, P. M., Johnson, P., Prusiner, S. B., Hodes, M. E., & Ghetti, B. (1992). Linkage of the Indiana kindred of Gerstmann-Sträussler-Scheinker disease to the prion protein gene. *Nat. Genet., 1,* 64–67.

61. Kretzschmar, H. A., Kufer, P., Riethmuller, G., DeArmond, S., Prusiner, S. B., & Schiffer, D. (1992). Prion protein mutation at codon 102 in an Italian family with Gerstmann-Sträussler-Scheinker syndrome. *Neurology, 42,* 809–810.

62. Tranchant, C., Doh, U. K., Steinmetz, G., Chevalier, Y., Kitamoto, T., Tateishi, J., & Warter, J. M. (1991). [Mutation of codon 117 of the prion gene in Gerstmann-Sträussler-Scheinker disease]. *Rev. Neurol. (Paris), 147,* 274–278.

63. Medori, R., Tritschler, H. J., LeBlanc, A., Villare, F., Manetto, V., Chen, H. Y., Xue, R., Leal, S., Montagna, P., Cortelli, P., et al. (1992). Fatal familial insomnia, a prion disease with a mutation at codon 178 of the prion protein gene [see comments]. *N. Engl. J. Med., 326,* 444–449.

64. Medori, R., Montagna, P., Tritscheler, H., LeBlanc, A., Cortelli, P., Lugaresi, E., & Gambetti, P. (1992). Fatal familial insomnia: A second kindred with mutation of prion protein gene at codon 178. *Neurology, 42,* 669–670.

65. Petersen, R. B., Tabaton, M., Berg, L., Schrank, B., Torack, R. M., Leal, S., Julien, J., Vital, C., Deleplanque, B., Pendlebury, W. W., Drachman, D., Smith, T. W., Martin, J. J., Ocla, M., Montagna, P., Ott, J., Autilio-Gambetti, L., Lugaresi, E., & Gambetti, P. (1992). Analysis of the prion protein gene in the thalamic dementia. *Neurology, 42,* 1859–1863.

66. Palmer, M. S., Mahal, S. P., Campbell, T. A., Hill, A. F., Sidle, K. C., Laplanche, J. L., &

Collinge, J. (1993). Deletions in the prion protein gene are not associated with CJD. *Hum. Mol. Genet. 2*, 541–544.

67. Owen, F., Poulter, M., Collinge, J., Leach, M., Shah, T., Lofthouse, R., Chen, Y. F., Crow, T. J., Harding, A. E., Hardy, J., & Rossor, M. N. (1991). Insertions in the prion protein gene in atypical dementias. *Exp. Neurol., 112*, 240–242.

68. Owen, F., Poulter, M., Shah, T., Collinge, J., Lofthouse, R., Baker, H., Ridley, R., McVey, J. & Crow, T. J. (1990). An in-frame insertion in the prion protein gene in familial Creutzfeldt-Jakob disease. *Brain Res. Mol. Brain Res., 7*, 273–276.

69. Goldfarb, L., Brown, P., Vrbovska, A., Baron, H., McCombie, R., Cathala, F., Gibbs, C., & Gadjusek, C. (1992). An insert mutation in the chromosome 20 amyloid precursor protein in a Gerstmann-Sträussler-Scheinker family. *J. Neurol. Sci., 111*, 189–194.

70. Goldfarb, L. G., Petersen, R. B., Tabaton, M., Brown, P., LeBlanc, A. C., Montagna, P., Cortelli, P., Julien, J., Vital, C., Pendelbury, W. W., Haltia, M., Willis, P. R., Hauw, J. J., McKeever, P. E., Monari, L., Schrank, B., Swergold, G. D., Autilio-Gambetti, L., Gajdusek, C., Lugaresi, E., & Gambetti, P. (1992). Fatal familial insomnia and familial Creutzfeldt Jakob disease: Disease phenotype determined by a DNA polymorphism. *Science, 258*, 806–808.

71. Palmer, M. S., Dryden, A. J., Hughes, J. T., & Collinge, J. (1991). Homozygous prion protein genotype predisposes to sporadic Creutzfeldt-Jakob disease [published erratum appears in 1991. *Nature 8*, 352(6335), 547] [see comments]. *Nature, 352*, 340–342.

72. Tateishi, J. (1995). Prion diseases. *Microbiol. Immunol., 39*, 923–928.

73. Caughey, B., Neary, K., Buller, R., Ernst, D., Perry, L., Chesebro, B., & Race, R. (1990). Normal and scrapie-associated forms of prion protein differ in their sensitivities to phospholipase and proteases in intact neuroblastoma cells. *J. Virol., 64*, 1093–1101.

74. Caughey, B., Neary, K., Buller, R., Ernst, D., Perry, L. L., Chesebro, B., & Race, R. E. (1990). Normal and scrapie-associated forms of prion protein differ in their sensitivities to phospholipase and proteases in intact neuroblastoma cells. *J. Virol., 64*, 1093–1101.

75. Jendroska, K., Keinzel, F. P., Torchia, M., Stowring, L., Kretzschmar, H. A., Kon, A., Stern, A., Prusiner, S. B., & DeArmond, S. J. (1991). Proteinase-resistant prion protein accumulation in Syrian hamster brain correlates with regional pathology and scrapie infectivity. *Neurology, 41*, 1482–1490.

76. Stahl, N., Borchelt, D. R., Hsiao, K., & Prusiner, S. B. (1987). Scrapie prion protein contains a phosphatidylinositol glycolipid. *Cell, 51*, 229–240.

77. Stahl, N., Borchelt, D. R., & Prusiner, S. B. (1990). Differential release of cellular and scrapie prion proteins from cellular membranes by phosphatidylinositol-specific phospholipase C. *Biochemistry, 29*, 5405–5412.

78. Lehmann, S., & Harris, D. A. (1995). A mutant prion protein displays an aberrant membrane association when expressed in cultured cells. *J. Biol. Chem., 270*, 24589–24597.

79. McKinley, M. P., Bolton, D. C., & Prusiner, S. B. (1983). A protease-resistant protein is a structural component of the scrapie prion. *Cell, 35*, 57–62.

80. Lehmann, S., & Harris, D. A. (1996). Mutant and infectious prion proteins display common biochemical properties in cultured cells. *J. Biol. Chem., 271*, 1633–1637.

81. Pan, K., Baldwin, M., Nguyen, J., Gasset, M., Serban, A., Groth, D., Mehlhorn, I., Huang, Z., Fletterick, R., Cohen, F., & Prusiner, S. (1993). Conversion of alpha-helices into ß-sheets features in the formation of the scrapie prion proteins. *Proc. Natl. Acad. Sci. USA, 90*, 10962–10966.

82. Huang, Z., Gabriel, J. M., Baldwin, M. A., Fletterick, R. J., Prusiner, S. B., & Cohen, F. E. (1994). Proposed three-dimensional structure for the cellular prion protein. *Proc. Natl. Acad. Sci. USA, 91*, 7139–7143.

83. Riek, R., Hornemann, S., Wider, G., Billeter, M., Glockshuber, R., & Wuthrich, K. (1996). NMR structure of the mouse prion protein domain PrP (121–231). *Nature, 282*, 180–182.

84. Bolton, D. C., McKinley, M. P., & Prusiner, S. B. (1982). Identification of a protein that purifies with the scrapie prion. *Science, 218*, 1309–1311.

85. Kocisco, D. A., Come, J. H., Priola, S. A., Chesebro, B., Raymond, G. J., Lansbury, P. T., &

Caughey, B. (1994). Cell-free formation of protease-resistant prion protein. *Nature, 370,* 471–474.

86. Bessen, R. A., Kocisko, D. A., Raymond, G. J., Nandan, S., Lansbury, P. T., & Caughey, B. (1995). Nongenetic propagation of strain-specific properties of scrapie prion protein. *Nature, 375,* 698–700.

87. Riesner, D., Kellings, K., Post, K., Wille, H., Serban, H., Groth, D., Baldwin, M., & Prusiner, S. (1996). Disruption of prion rods generates 10-nm spherical particles having high a-helical content and lacking scrapie infectivity. *J. Virol., 70,* 1714–1722.

88. Knauer, M., Soreghan, B., Burdick, D., Kosmoski, J., & Glabe, C. (1992). Intracellular accumulation and resistance to degradation of the Alzheimer amyloid A4/ß protein. *Proc. Natl. Acad. Sci. USA, 89,* 7437–7441.

89. Forloni, G., Angeretti, N., Chiesa, R., Monzani, E., Salmona, M., Bugiani, O., & Tagliavini, F. (1993). Neurotoxicity of a prion protein fragment. *Nature, 362,* 543–546.

90. Shyng, S. L., Heuser, J. E., & Harris, D. A. (1994). A glycolipid-anchored prion protein is endocytosed via clathrin coated pits. *J. Cell. Biol., 125,* 1239–1250.

91. Vey, M., Pilkuhn, S., Wille, H., Nixon, R., DeArmond, S., Smart, E., Anderson, R., Taraboulos, A., & Prusiner, S. (1996). Subcellular colocalization of the cellular and scrapie prion proteins in caveolae-like membranous domains. *Proc. Natl. Acad. Sci. USA, 93,* 14945–14949.

92. Naslavsky, N., Stein, R., Yanai, A., Friedlander, G., & Taraboulos, A. (1997). Characterization of detergent-insoluble complexes containing the cellular prion protein and its scrapie isoform. *J. Biol. Chem., 272,* 6324–6331.

93. McKinley, M. P., Taraboulos, A., Kenaga, L., Serban, D., Stieber, A., DeArmond, S. J., Prusiner, S. B., & Gonatas, N. (1991). Ultrastructural localization of scrapie prion proteins in cytoplasmic vesicles of infected cultured cells. *Lab. Invest., 65,* 622–630.

94. Laszlo, L., Lowe, J., Self, T., Kenward, N., Landon, M., McBride, T., Farquhar, C., McConnell, I., Brown, J., Hope, J., & Mayer, J. R. (1992). Lysosomes as key organelles in the pathogenesis of prion encephalopathies. *J. Pathol., 166,* 333–341.

95. Monari, L., Chen, S. G., Brown, P., Parchi, P., Petersen, R. B., Mikol, J., Gray, F., Cortelli, P., Montagna, P., Ghetti, B., Goldfarb, L. G., Gajdusek, D. C., Lugaresi, E., Gambetti, P., & Autilio-Gambetti, L. (1994). Fatal familial insomnia and familial Creutzfeldt-Jakob disease: Different prion proteins determined by a DNA polymorphism. *Proc. Natl. Acad. Sci. USA, 91,* 2839–2942.

96. Piccardo, P., Seiler, C., Dloughy, S. R., Young, K., Farlow, M., Prelli, F., Frangione, B., Bugiani, O., Tagliavini, F., & Ghetti, B. (1996). Proteinase-K-resistance prion protein isoforms in Gerstmann-Sträussler-Scheinker disease (Indiana Kindred). *J. Neuropathol. Exp. Neurol., 55,* 1157–1163.

97. Collinge, J., Sidle, K., Heads, J., Ironside, J., & Hill, A. (1996). Molecular analysis of prion strain variation and the aetiology of "new variant." *CJD, 383,* 685–690.

98. Collinge, J., Whittington, M. A., Sidle, K. C. L., Smith, C. J., Palmer, M. S., Clarke, A. R., & Jefferys, J. G. R. (1994). Prion protein is necessary for normal synaptic function. *Nature, 370,* 295–297.

99. Telling, G., Parchi, P., DeArmond, S., Cortelli, P., Montagna, P., Gabizon, R., Mastriani, J., Lugaresi, E., Gambetti, P., & Prusiner, S. (1996). Evidence for the conformation of the pathologic isoform of the prion protein enciphering and propagating prion diversity. *Science, 274,* 2079–2082.

100. Whatley, S. A., Powell, J. G., Politopoulos, G., Campbell, I., Brammer, M., & Percy, N. (1995). Regulation of intracellular free calcium levels by the cellular prion protein. *NeuroReport, 6,* 2333–2337.

101. Cashman, N., Loertscher, R., Nalbantoglu, J., Shaw, I., Kascsak, R., Bolton, D., & Bendheim, P. (1990). Cellular isoform of the scrapie agent protein participates in lymphocyte activation. *Cell, 61,* 185–192.

102. Forloni, G., Del Bo, R., Angeretti, N., Chiesa, R., Smiroldo, S., Doni, R., Ghibaudi, E., Salmona, M., Porro, M., Verga, L., Giaccone, G., Bugiani, O., & Tagliavini, F. (1994). A neu-

rotoxic prion protein fragment induces rat astroglial proliferation and hypertrophy. *Euro. J. Neurosci., 6,* 1415–1422.

103. Tagliavini, F., Prelli, F., Porro, M., Salmona, M., Bugiani, O., & Frangione, B. (1992). A soluble form of prion protein in human cerebrospinal fluid: Implications for prion-related encephalopathies. *Biochem. Biophys. Res. Commun., 184,* 1398–1404.

104. Yin, X.-M., Oltval, Z. N., & Korsmeyer, S. J. (1994). BH1 and BH2 domains of bcl-2 are required for inhibition and heterodimerization with bax. *Nature, 369,* 321–323.

105. Merry, D. E., Veis, D. J., Hickey, W. F., & Korsmeyer, S. J. (1994). Bcl-2 protein expression is widespread in the developing nervous system and retained in the adult PNS. *Dev., 120,* 301–311.

106. Kurshner, C., & Morgan, J. (1995). The cellular prion protein (PrP) selectively binds to Bcl-2 in the yeast two-hybrid system. *Mol. Br. Res., 30,* 165–168.

107. Papadopoulos, M, & LeBlanc, A. Prion protein interaction with bcl-2 and bax (ms in prep).

108. Kurshner, C., & Morgan, J. (1996). Analysis of interaction sites in homo- and heteromeric complexes containing Bcl-2 family members and the cellular prion protein. *Mol. Br. Res., 37,* 249–258.

109. Bellinger-Kawahara, C., Cleaver, J., Diener, T., & Prusiner, S. (1987). Purified scrapie prions resist inactivation by UV irradiation. *J. Virol., 61,* 159–166.

110. Prusiner, S. (1987). Prions and neurodegenerative diseases. *N. Engl. J. Med., 317,* 1571–1581.

111. Prusiner, S. B. (1987). Prion diseases and central nervous system degeneration. *Clin. Res., 35,* 177–191.

112. Prusiner, S. B., Stahl, N., & DeArmond, S. J. (1988). Novel mechanisms of degeneration of the central nervous system—Prion structure and biology. *Ciba Found. Symp., 135,* 239–260.

113. Prusiner, S. B., & DeArmond, S. J. (1990). Prion diseases of the central nervous system. *Monogr. Pathol., 1990,* 86–122.

114. Prusiner, S. B., & Westaway, D. (1991). Infectious and genetic manifestations of prion diseases. *Mol. Plant. Microbe Interact., 4,* 226–233.

115. Cohen, F. E., Pan, K. M., Huang, Z., Baldwin, M. Fletterick, R. J., & Prusiner, S. B. (1994). Structural clues to prion replication. *Science, 264,* 530–531.

116. Bueler, H., Aguzzi, A., Sailer, A., Greiner, R., Autenried, P., Aguet, M., & Weissmann, C. (1993). Mice devoid of PrP are resistant to scrapie. *Cell, 73,* 1339–1347.

117. Caughey, B. (1991). *In vitro* expression and biosynthesis of prion protein. *Curr. Top. Microbiol. Immunol., 172,* 93–107.

118. Borchelt, D. R., Scott, M., Taraboulos, A., Stahl, N., & Prusiner, S. B. (1990). Scrapie and cellular prion proteins differ in their kinetics of synthesis and topology in cultured cells. *J. Cell. Biol., 110,* 743–752.

119. Bolton, D., Meyer, R., & Prusiner, S. (1985). Scrapie PrP 27-30 is a sialoglycoprotein. *J. Virol., 53,* 596–606.

120. Endo, T., Groth, D., Prusiner, S. B., & Kobata, A. (1989). Diversity of oligosaccharide structures linked to asparagines of the scrapie prion protein. *Biochemistry, 28,* 8380–8388.

121. Borchelt, D. R., Taraboulos, A., & Prusiner, S. B. (1992). Evidence for synthesis of scrapie prion proteins in the endocytic pathway. *J. Biol. Chem., 267,* 16188–16199.

122. Carlson, G. A., Ebeling, C., Torchia, M., Westaway, D., & Prusiner, S. B. (1993). Delimiting the location of the scrapie prion incubation time gene on chromosome 2 of the mouse. *Genetics, 133,* 979–988.

123. Carlson, G. A., Goodman, P. A., Lovett, M., Taylor, B. A., Marshall, S. T., Peterson, T. M., Westaway, D., & Prusiner, S. B. (1988). Genetics and polymorphism of the mouse prion gene complex: Control of scrapie incubation time. *Mol. Cell. Biol., 8,* 5528–5540.

124. Carlson, G. A., Kingsbury, D. T., Goodman, P. A., Coleman, S., Marshall, S. T., DeArmond, S., Westaway, D., & Prusiner, S. B. (1986). Linkage of prion protein and scrapie incubation time genes. *Cell, 46,* 503–511.

125. Westway, D., Goodman, P. A., Mirenda, C. A., McKinley, M. P., Carlson, G. A., & Prusiner, S.

B. (1987). Distinct prion proteins in short and long scrapie incubation period mice. *Cell, 51*, 651–662.

126. Hunter, N., Dann, J., Bennett, A., Sommerville, R., McConnell, I., & Hope, J. (1992). Are Sinc ans the PrP gene congruent? Evidence from PrP gene analysis in Sinc congenic mice. *J. Gene. Virol., 73*, 2751–2755.

127. Wickner, R. B. (1994). [URE3] as an altered URE2 protein: Evidence for a prion analog in *Saccharomyces cerevisiae. Science, 264*, 566–569.

128. Tuite, M. F. (1994). Psi no more for yeast prions. *Nature, 370*, 327–328.

129. Chesebro, B. (1997). Human TSE disease—Viral or protein only? *Nature Med., 3*, 491–492.

130. Ozel, M., Xi, Y. G., Baldauf, E., Diringer, H., & Pocchiari, M. (1994). Small virus-like structures in brains from cases of sporadic and familial Creutzfeldt-Jakob disease. *Lancet, 344*, 923–924.

131. Ozel, M., & Diringer, H. (1994). Small virus-like structures in fractions from scrapie hamster brain. *Lancet, 343*, 894–895.

132. Manueldis, L., Sklaviadis, T., Akowitz, A., & Fritch, W. (1995). Viral particles are required for infection in neurodegenerative Creutzfeldt-Jakob disease. *Proc. Natl. Acad. Sci. USA, 92*, 5124–5128.

133. Weissmann, C. (1991). A 'unified theory' of prion propagation [see comments]. *Nature, 352*, 679–683.

134. Westway, D., DeArmond, S., Cayetano-Canlas, J., Groth, D., Foster, D., Yang, S., Torchia, M., Carlson, G., & Prusiner, S. (1994). Degeneration of skeletal muscle, peripheral nerves, and the central nervous system in transgenic mice overexpressing wild-type prion proteins. *Cell, 76*, 117–129.

15

TRANSLATIONAL CONTROL, APOPTOSIS, AND THE AGING BRAIN

EUGENIA WANG

The Bloomfield Centre for Research in Aging
Lady Davis Institute for Medical Research
Sir Mortimer B. Davis Jewish General Hospital
Department of Medicine, McGill University
Montréal, Québec, Canada H4H 1R3

Neurons function in their host tissue, the brain, for the entire life span of the animal. How these cells maintain their long-lived (sometimes more than 100 years, in the case of humans) terminally-differentiated state, and avoid both replicating and dying, is the greatest mystery of molecular programming. We suggest that surveillance by molecular mechanism(s) may function to maintain gene expression at the levels needed to contribute to this long-lived state. S1, a homolog of the translational factor, EF-1α, may be a candidate maintenance gene, possibly by acting as a "dimmer switch" to suppress its sister's protein presence, allowing itself to replace EF-1α's function in protein translation without functioning in other EF-1α roles such as microtubule severing and actin bundling, which exert detrimental effects on cytoskeletal organization, and produce a cellular milieu unfavorable to neuronal function and survival.

CLONING, SEQUENCING, AND ANALYZING FULL-LENGTH HUMAN, RAT, AND MOUSE EF-1α AND S1 cDNA CLONES AND RAT S1 GENOMIC CLONES

In 1985 we cloned and sequenced a full-length rat cDNA clone, which we called S1, from a rat brain γ-gt11 library.[1] Sequence analysis showed that this clone shares high homology with human EF-1α (the only available sequence of mammalian EF-1α at the time). We further cloned the full-length rat EF-1α and found that rat S1 and EF-1α are two different genes, with S1 transcripts measuring 2.0 kB, whereas EF-1α is 1.8 kB.[2] We found that EF-1α is a 20-member supergene family, of which only EF-1α and S1 are expressed, while the rest are retropseudogenes.[1] Cloning the corresponding genomic clone shows that S1 cDNA is encoded by a single-copy gene, whose transcription unit is 12 kilobase pairs, containing seven introns.[1] The organization of the eight exons is virtually identical between S1 and the human EF-1α gene; in contrast, however, neither a TATA box nor a CAAT box is found in the proximal 5′-flanking regions from positions −1 to −1359 of the S1 gene. There is no evident sequence homology in the regions of the seven introns, and the two genes, S1 and EF-1α, are indeed different. Further cloning of the cosmic library allowed us to extend the 5′-flanking region to the −9.3 kB position; future experiments will allow us to map the regions responsible for the transcriptional regulation of S1 gene expression.

So far, beside human and mouse EF-1α cDNA clones (available to us as gifts),[3] we have cloned human, rat, and mouse S1, as well as rat EF-1α. All these clones are full length, and amino acid sequence analysis shows that EF-1α and S1 share high homology (~92%) in their coding region, with major differences residing in the C-terminal half. At the nucleotide level, the 5′ and 3′ untranslated regions (UTRs) of EF-1α and S1 diverge significantly. However, sequence comparison among human, rat, and mouse S1 or EF-1α shows that individually they are highly conserved, and among different mammalian species only a few base changes are found.

CHARACTERIZATION OF TISSUE SPECIFICITY AND AGE-DEPENDENT CHANGES OF S1 EXPRESSION

RNase protection assays with specific riboprobes show that S1 message is limited to brain, muscle, and heart, whereas EF-1α message is found in all tissues; this is verified by in situ hybridization. We further find that the tissue-specific expression of S1 is due to the presence of its message in neurons and myotubes, the major cell components of their respective tissues.[4–6] S1 message presence is developmentally dependent, activated at embryonic Day 20 and reaching a stable level at postnatal Day 14 in brain, muscle, and heart. During this period, the message level for EF-1α varies among these three tissues; in brain it declines slightly, whereas in heart and muscle the decline is significant. The ratio of EF-1α to S1 messages never varies beyond 2:1 or 1:1, and remains constant until 26 months of age in rats.

S1 IS FOUND IN NEURONS ONLY, NOT IN ASTROCYTES OR MICROGLIA

We have shown that the expression of S1 in brain is mainly due to the presence of its transcripts exclusively in neurons, not in microglia or astrocytes.[6] This determination was performed by identifying transcripts of S1 and β-amyloid precursor protein (APP) found in isolated primary cultures of human neurons, not in astrocytes or microglia.[6] A further proof that S1 appears only in neurons is that S1 is not found in astrocyte cultures with glial filament acidic (GFA)-positivity. Our immunoblotting work with these isolated primary human cultures further shows that S1 protein is found only in neuronal but not in astrocyte protein extracts.

In our earlier publications, S1 was termed "statin-like," but this was misleading. The two genes, statin and S1, are distinct and, as we now know, share little similarity in their molecular and biochemical properties.[1, 2, 4–19] For example, statin is a 57 kDa protein with two isoforms having pI values 6.5 and 7.0, whereas S1 protein is 50 kDa with a pI value of 9.8.[20, 21] S1 expression is restricted to neurons, myotubes, and cardiomyocytes, whereas statin is found in all nonproliferating cells, including S1-negative tissues such as kidney, intestine, liver, skin, and thyroid.[10–21] The statin-specific monoclonal antibodies, S30 and S44, do not cross-react with glutathione-S-transferase (GST)-S1 protein; vice versa, S1-specific antibody CB5 does not recognize statin. The amino acid compositions of statin and S1 show little similarity between the two proteins. Finally, our recent success at cloning and sequencing putative statin clones shows that similarity is not observed at the nucleotide level either, and none of the statin cDNA isolates can hybridize with S1. Thus, statin and S1 are distinct, and our earlier reference to S1 as statin-like was due to our fortuitous selection of an S1 cDNA clone with the statin monoclonal antibody. Therefore we have dropped the "statin-like" term from all our publications since 1992.[22] However, sequence analysis shows that S1 is identical to the human cDNA clone termed EF-1α2, reported by Knudsen et al.[20] Thus, S1 is distinct from statin but identical to EF-1α2.

EF-1α/S1 SUPERGENE FAMILY

EF-1α is part of the elongation factor-1 complex, which also includes EF-1β and EF-1γ, and promotes guanosine triphosphate (GTP)-dependent binding of aminoacyl-tRNAs to ribosomes during peptide elongation.[23] Mammalian species share virtually 100% amino acid similarity, whereas nonmammalian species share between 75% and 95%; this high degree of homology resides in the N-terminal half, containing functional domains for translation.[24, 25] In many species, including *Drosophila* and *Xenopus*, beside the constitutive isoform, various genes encoding EF-1α's sister forms are characterized as being development- or differentiation-dependent in their expression, and their precise functions are not necessarily implicated in translation.[26–34] Our work so far shows that EF-1α is a supergene family of 20 members, 18 of which are retropseudo-genes, leaving EF-1α and S1 as the only two expressed genes.[1] EF-1α gene expression is found at various levels in all tissues, whereas S1 gene expression is tissue specific for brain, heart, and muscle.[4–9] Conventional EF-1α's of all species are interchangeable in their function; this suggests that S1 may also function in the traditional role of EF-1α.

MULTIPLE FUNCTIONAL
INVOLVEMENT OF EF-1α

Increasingly, results show that beside its traditional role in peptide chain elongation, EF-1α is associated with actin,[26–28] mitotic spindle,[29] hn-ribonucleoproteins (RNPs),[30] the endoplasmic reticulum,[31] calcium-calmodulin kinases, and phosphatidylinositol 4 kinase.[32, 33] Overexpression of EF-1α potentiates cell transformation.[34] Surprisingly, purified EF-1α is found to literally cut microtubules into pieces.[35] This multiple involvement may be the reason why, in growing cells, the abundance of EF-1α is as high as 2% of total protein, freeing cells from rate limitation in protein translation, and allowing constant cytoskeletal remodeling and active signal transduction in a replicating environment. Here, S1 may not only function as a molecular "dimmer switch," suppressing EF-1α abundance, but also replace its protein translational function without the damaging effects on cytoskeletal organization. The postdevelopmental switch from EF-1α to S1 expression is the precise mechanistic design needed for the scenario in adult neurons, operating to maintain these cells in their terminally differentiated state, instead of replicating.

GENERAL PRINCIPLES
FOR TRANSLATION REGULATION

Although we do not rule out control related to the stability and turnover of messages and proteins, the core of this level of regulation is certainly how the translational machinery is carried out.[36–38] Knowledge so far shows that the rate-limiting steps for translation of mRNA are: (1) assembly of initiation complex at the 5′ cap site; (2) ribosome binding to proximal 5′ cap sites, for entry to small ribosomal subunits; (3) polyadenylation (polyA) at the 3′ end; and (4) 5′ and 3′ end-to-end communication. Regulation of these four steps is: (1) for the cap-binding complex eIF-4F, composed of eIF-4E, the RNA helicase eIF-4A, and eIF-4G, the rate-limiting factor is the first one[39]; (2) total translational silencing can be accomplished if the specific protein binding to the proximal-5′ cap site sequesters ribosomal entry[40, 41]; (3) the length of polyA tail may be important in binding to the polysomes and successful end-to-end communication, and furthermore iron-binding and therefore adenylation properties may also be intimately associated with translational efficiency[42–44]; and (4) the UTRs at both 5′ and 3′ ends have been recognized as crucial elements for determining regulation at the translational level.[45–48]

Obviously, the essence of 5′ or 3′ UTR translational regulation presents us with the scenario that designated sequences proximal to the 5′ cap binding or in the 3′ UTR may be candidate target *cis*-elements to bind to specific proteins or RNAs serving as *trans*-factors, thus providing a physical or biochemical hindrance to successful ribosomal entry or end-to-end communication, and causing a repressing or silencing effect. Furthermore, the realization that strategic cellular sites are needed for translational efficiency suggests that some *cis*-elements of the 3′ UTR may be regarded as the address code, whereas the corresponding RNA binding proteins serve as the "mailman" for destined cytoplasmic locations for translation.[49–55] Future research will elucidate how these many regulatory elements and factors of the 5′ and 3′ UTRs operate and orchestrate the exquisite regulation of gene expression.

Many factors identified as responsible for translational repression are proteins, transcripts of themselves, or close relatives with highly homologous sequences. For example, in *Caenorhabditis elegans*, the *lin*4 transcript binds to its sister gene *lin*14 to repress the translation of the latter,[56–58] and in *Drosophila* the *bcd* protein serves as the translational repressor to bind the 3′ UTR of its sister, *cad*, to prevent translational initiation.[59, 60] Also, it has been suggested that p53 may serve as its own translational repressor, possibly binding at the proximal region of the 5′ cap sites to prevent ribosomal entry.[61–66]

APOPTOSIS, CELL CYCLE TRAVERSE, AND CYTOSKELETAL ORGANIZATION

Growth factor deprivation-induced apoptosis (used here as a synonym for programmed cell death) requires the transcriptional activation of immediate early gene expression such as c-*fos*, c-*myc*, and c-*jun*, which typify the G_1 phase of cell cycle traverse.[67] It seems that the initiating phases of both apoptosis and cell cycle traverse share the same G_1 biochemical features,[68–70] and the triage point determining life or death may be at the G_1/S boundary. Emerging results show that cytoskeletal disorganization may be one of the cellular events accompanying apoptosis activation. Thus, initiating apoptosis may require translational regulation of elongation factor translational efficiency, new protein translation of immediate early genes, active transduction of signals, and dismantling of cytoskeletal structures; EF-1α's roles, in the elongation step of protein synthesis, signal transduction, and association with cytoskeletal elements such as actin and tubulin, make this protein an obviously essential functional player in apoptosis.

APOPTOSIS, CELL CYCLE TRAVERSE, PROTEIN
TRANSLATION, AND SIGNAL TRANSDUCTION

Due to its abundance, EF-1α's role in protein synthesis is not appreciated because it has not been recognized as a rate-limiting factor in protein translation. However, if one links the facts that EF-1α is one of the few genes whose transcripts are repressed translationally during protein synthesis inhibition-associated G_1 phase growth arrest, and that G_1 phase progression is needed for both cell cycle traverse and apoptosis, it is obvious that reduction of EF-1α expression can elicit arrest at the G_1/S boundary, preventing cells from either entering the death pathway or proceeding to DNA replication in S phase. G_1/S arrest associated with repression of EF-1α translation may then serve as a double-edged sword, preventing both death and growth, an ideal molecular strategy for terminally differentiated cells such as neurons. What better way to provide regulatory control of EF-1α translation efficiency than by requiring the expression of a sister gene such as S1, using its own UTR elements to compete with that of EF-1α? Future experiments will reveal whether this is indeed the case, and if so, how it may be part of the regulation controlling cellular EF-1α protein levels.

REENTRY TO CELL CYCLE TRAVERSE,
APOPTOSIS SUSCEPTIBILITY, AND RESISTANCE

In general, cells in mammalian systems can be classified into three types: (1) those that are constantly involved in replication, (2) those that are temporarily quiescent and will resume cell division when proper mitogenic stimulation is provided, and (3) those that are permanently withdrawn from cell cycle traverse. Clearly, the decision for any cell to remain in one of these three cellular states is dependent upon the balance of signals to enter or not to enter the G_1 phase of cell cycle traverse. Once a cell has entered this G_1 phase, it needs to orchestrate the activation of a repertoire of gene expression to ensure the completion of G_1 phase. However, as described above, the experience of entering into G_1 phase is also seen in cells receiving apoptotic signaling, and as suggested previously, commitment to life or death seems to occur at the G_1/S border (Fig. 15.1). The molecular governing principle for entering S phase or detouring to apoptotic death may be dependent upon the presence of survival factors such as *bcl2*, and the absence of killing factors such as the cysteine proteases. Integrating into this complex scheme for life or death, there is yet a third choice for every cell, which is arrest at the G_1/S border, proceeding neither to S phase for replication, nor to apoptotic death. One noted example of the third choice is senescent human fibroblasts. Upon being presented either mitogen stimulation via serum addition or apoptosis signaling via serum deprivation, these cells cycle through the entire G_1 phase and even express a number of G_1-specific genes; however, due to the lack of key molecular events such as c-*fos* expression,[71] as well as the down regula-

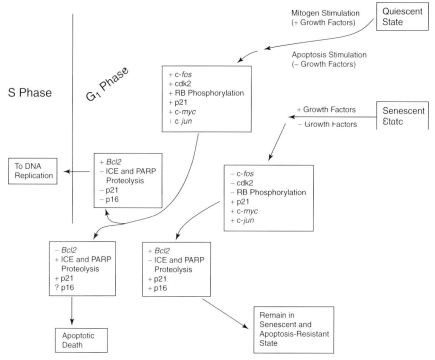

FIGURE 15.1   Diagrammatic illustration of the gene expressions involved in the G_1/S border as the triage point for replicative life, death, or neither event. *ICE,* interleukin converting enzyme; *PARP,* poly-ADP-ribose polymerase; *RB,* retinoblastoma gene product.

tion of *bcl2*[72] and the activation of cysteine protease activity, these cells then linger in this apoptosis-resistant and replication-repressed state, which can last for as long as 2 to 3 months. Therefore, the very fact that senescent human fibroblasts can maintain this long-lived state attests to the resilient nature required for the signal transduction network to operate toward the long duration of the nonproliferating state, a deliberate cellular operation to keep cells in this state for their destined functions.

CONTROL OF LONG-LIVED PHENOTYPES AND NEURONAL SURVIVAL

Once neurons have completed their program of terminal differentiation and made successful contact with targeted neighboring cells, there must be a molecular operation to keep them contact-stable and permanent, with no degrees of freedom. It is by millions of these interactions that our neuronal networks are built. Given these structural constraints, how is neuroplasticity or adaptability possible? Here, we might borrow the transportation highway as a metaphoric interpretation; the structure of the highway layout has no flexibility once it is constructed, but vehicles traveling over the highway have many options and possible actions, dependent upon the skills or whims of the individuals guiding the vehicles. A similar analogy can be made for the information highway: The highway involved in computer networking is rigid in its hardware infrastructure, with zero degrees of freedom, but the information traveling on it may be highly flexible. Returning to our neuronal network, the network as an infrastructural entity composed of millions of neurons has few degrees of freedom, but the transmission of impulses is highly fluid, giving maximal latitude for the information to be relayed.

If we accept the suggestion that the hardwiring of neuronal networks allows little room for freedom, how does nature allow such a rigid architecture? Here, the salvation may lie in the redundancy of structural formation allowing the integration of many parallel or multiple pathways to perform the same function. To ensure that the entire functioning system does not collapse if one single route fails to operate, redundancy, multiple lanes if you will, is the best means. Here, one might suggest that multiple pathways might even be the most constructionally wise way for neuronal network communication. If this is true, nature needs to provide a means such as apoptosis to get rid of neurons if and when they become damaged. With such an arrangement, one small subroutine in the communication route may be dismantled with no harm to the whole or may allow sister cells to take over the functional capacity to transmit information. Taken together, there emerges a hypothetical model image (i.e., the neuronal network composed of individual neurons interconnected by redundant transmission pathways). The success of the entire neuronal operation may depend upon the stability of backbone structures, which are defined by individual neurons. Then, it becomes a task at the molecular level to keep these connecting neurons long-lived and functionally active. The strategy may be to provide a cellular mechanism for the maintenance of neurons in a state that is antiapoptotic, and one that keeps the signal transduction for this state balanced for neuronal transmission without venturing onto the route leading to death. Neuronal apoptosis may only be activated when cellular damage is beyond repair.

SIGNAL TRANSDUCTION AND CHAOS THEORY

Subconsciously, when we discuss the complexity of neuronal networks, we realize there is yet another layer of complexity (i.e., signal transduction pathways, ever-important pathways essential for the faithful transmission of signals both intracellularly and intercellularly). We are advised by an explosion of literature reports that no cellular action operates by a single modality (i.e., no single gene expression can do it all). Each cellular action can only be implemented by a cascade of gene expression that functions in a chain reaction of events in a single pathway or parallel pathways, working toward the final execution of the action. The end result of these multiple activations in multiple layers may be the activation of transcriptional events, the operation of translational activity, post-translational modification for phosphorylation or dephosphorylation, etc. Obviously, individual gene expression players may then be incorporated together to form the final programs directing the processes of cell replication, apoptotic death, terminal differentiation, and even intercellular communication. Therefore for the cellular theatrical production to make cell division occur or terminal differentiation proceed may require a cast of hundreds of gene activations. Among this mass, where is the order? Which is the controlling step? And how is the operation of the controlling steps implemented? Here, we may want to borrow a much-discussed mathematical model based on the theory of chaos and fractals. Suppose one hypothetically uses an initial signal on the cell plasma membrane as the starting point, each subsequent step brings bifurcated routes, and each in turn splits into other bifurcated routes (Fig. 15.2). The end result is that all pathways of signal transduction are activated. However, signals may converge to selected key stations, whose functions are to coordinate the "in" and "out" signals, and the final orchestration of these key stations may then be the ultimate implementation of the molecular actions for the outcome determined by the very initial signals at the plasma membrane. If we accept this model of inter-

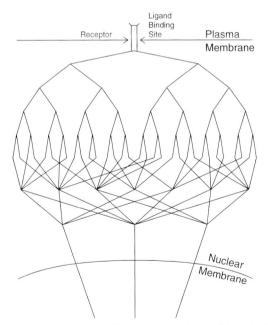

FIGURE 15.2   Cartoon sketch of the bifurcation plan as the possible model for signal transduction network organization.

pretation for biological action, we are then forced to acknowledge that no single cellular stage can be accomplished without the mobilization and expression of hundreds of genes in exquisite order, like a production of Beethoven's *Ninth Symphony* or Tchaikovsky's *1812 Overture*. The publication[73, 74] of the molecular anatomy of normal and cancerous cells, where unique neoplasm-associated gene expressions are identified by the hundreds, attests to the needs of hundreds of genes for a single cellular action.

SIGNAL TRANSDUCTION AND NEURONAL NETWORK FUNCTION

Even after a neuron completes its terminal differentiation program and withdraws from cell replication events completely, the molecular tasks for signal transduction may be even more demanding. These tasks entail the signal transduction framework being set up for networking, resulting in fulfilling the demands of stable cytoskeletal organization, coordinated intercellular communication at synapses, fluid retrograde axonal transport, and proper transcriptional and translational operation of those gene expressions vital to the long-term survival and operation of working neurons. Thus, the signal transduction network within functioning neurons may be more demanding than in fibroblasts, whose "activities" are focused on the single-mindedness of self-replication. Presently, we recognize four distinct pathways of signal transduction dictating individual events; three of them start with insulin-receptor signaling, and diverge to (1) the Ras/Rac/Rho pathway for cytoskeletal organization; (2) the Ras/Raf/MAP kinase and Src/Jnk kinase pathway for transcriptional activation of proto-oncogenes such as Fos, Jun, and Myc; and (3) the PI3K/AKT pathway directing S6 kinase for protein translational activation. This last pathway is also directed to the glycogen syntheses kinase 3 (GSK3), which serves as a bridge to the fourth pathway, working from the signaling event governed by Notch and Frizzled gene expressions, whose main signals derive from intercellular or cell-to-substratum contact. This last pathway, which I call the determining route for "cellular social biology," may be one of the key events dominating the synaptic junction where intercellular communication operates. The complexity of signal relays in neuronal networking is then clearly involved in not only the above four pathways, devoted to the processing of neuronal transmission from one neuron to the next, but also others from one cytoplasmic component to another intracellularly. In order to allow neurons to function, there must be the infrastructure to support the success of transport and communication intercellularly and intracellularly. The cast of hundreds of gene expressions is then no less than those involved in the maintenance of structures such as cytoskeletal architecture, the endoplasmic reticulum system, mitochondrial machinery, and finally the entire nucleus. If one accepts the premise as I have developed it, the consequent detrimental effect of one of the above cellular structures being less than perfect is unfathomable. Therefore, I submit that even though neurofibrillary tangles or amyloid plaques are notorious signs of neuronal degeneration, they may be only the footprint of damage beyond repair. We may be more desperately handicapped by neurons that have less-than-perfect cytoskeletal organization, and these neurons, which I term as "half-dead or functionally decrepit," may be asymptomic warnings of abnormality of the neuronal network, leading to neuronal death.

SIGNAL TRANSDUCTION, APOPTOSIS, AND FUNCTIONAL IMPAIRMENT IN NEURONS

As suggested earlier, some of the same signal transductions needed for cell cycle traverse are also needed for the signaling event of apoptosis. The shared nature of these signaling steps clearly demonstrates that there must be a "decision-making process" that has the effect of determining whether a cell lives or dies. We also suggest that senescent human fibroblasts represent a third choice for cellular decision (i.e., surviving without either replicating or dying). Superficially, one might equate the senescent human fibroblast to the condition of the neuron, in that both are permanently withdrawn from cell cycle traverse and both are apoptosis refractile. However, the similarity between the two cellular states stops here. Senescent human fibroblasts arrive at their state by serial passaging and are "socially isolated" (i.e., seldom forming contacts with neighboring cells). Neurons, on the other hand, arrive at their particular physiological state by a program of terminal differentiation, and their social contacts are the very means of their surviving and functioning. The shut-down of replicative machinery is a deliberate cellular action in neurons, instead of the consequence of serial passaging in senescent human fibroblasts. For an adult neuron, the activation of signaling for the cell replication program may be suicidal, because by design it has previously removed all the supporting machinery required for the mechanics of DNA synthesis. The end result is then an abortive cell cycle traverse, with apoptotic death as the only route of final action (Fig. 15.3). Is this the actual molecular regulation for neuronal apoptosis? Does the accumulation of damage beyond repair occur in apoptotic neurons in the aging brain? Does neurodegeneration such as transpires in Alzheimer's disease result solely from ill-afforded neuronal cell death? All of these questions require us to examine the entire regulation of signal transduction for neuronal repair versus neuronal death with precision, to evaluate the expression of many genes rather than examining alterations in single gene expression. What we may find is that subsets of neurons differ in their molecular anatomic profile, and it is this gene profiling that contributes to the different patterns programmable for the different vital functions of healthy neurons.

Dangers of G₁ Phase Reentry

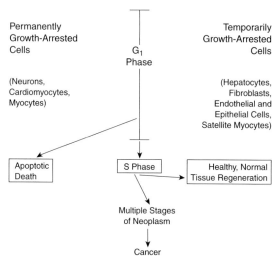

FIGURE 15.3   Diagrammatic illustration of possible modes of abortive cell cycle traverse leading to apoptotic death for neurons, while successful traverse leads to departure from the quiescent state to cell replication for cells such as hepatocytes.

SIGNAL TRANSDUCTION AND TRANSLATIONAL CONTROL

The above discussion emphasizes the coordination and orchestration of many genes into a single neuronal action. Here, I shall discuss the opposing scenario, in that among the many gene expressions, there are obviously dominant ones (i.e., the global regulators whose functions cause many downstream targeted changes). One might term these regulators as "upstream" or "master" controllers, which may act at the transcriptional, translational, or posttranslational levels. Here, I return to focus my discussion on EF-1α and how the regulation of its translation, controlled by a signal transduction pathway, may affect multiple downstream events. As described previously, the very first phase of response to mitogen stimulation is the translational up-regulation of selected genes, including EF-1α; this event precedes all other G₁ phase activities, including the activation of c-*fos*, c-*myc*, retinoblastoma gene product (RB) phosphorylation, and activation of various cyclin-dependent kinases. It is interesting that genes targeted for translational up regulation at this first step are those whose cellular functions mostly direct the translational machinery. This fact may not be coincidental, but rather deliberate, preparing the cells to be able to cope with the next step, which is the transcriptional activation of the many genes whose newly synthesized messages need to be processed for translation, eventually producing func-

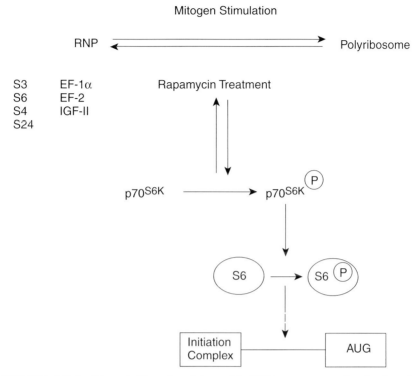

FIGURE 15.4   Schematic illustration of the model of EF-1α message translocation from the untranslatable ribonucleoprotein (RNP) pool to the translatable polyribosome pool and how S6 kinase (S6k) may affect this shift. *AUG*, translation initiation codon; *IGF*, insulin-like growth factor; *p70*, protein of 70 kDa; *S*, ribosomal protein.

tional proteins. Therefore before any massive transcriptional activation, there may be the need of a stage of translational action of factors involved in the translational machinery; the most efficient way of operation is to shift the messages for these translational factors from the storage ribonucleoprotein (RNP) pool to the translatable polyribosome pool. The key to regulating this shift is then the signal transduction pathway required for S6 kinase phosphorylation, involved not only in the initiation factor activity for the selected translational factors, but also in the binding of key UTRs such as the polypyrimidine tract of EF-1α to the polyribosomes. Clearly, the key dominant or global factor is p70$^{\text{S6kinase}}$, which functions to move EF-1α messages from the untranslatable to the translatable pool (Fig. 15.4). The end result is the production of more EF-1α protein molecules, which may result in increased activity of protein elongation, severing microtubules, bundling actin filaments, and association of phospho-inositol-4 (PI4) kinase, all of which are essential for cell replication to succeed.

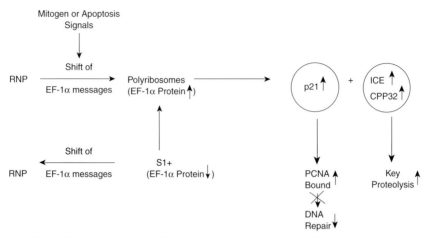

FIGURE 15.5 Diagrammatic illustration of the possible modes for S1 protein to regulate the EF-1α message shift from the ribonucleoprotein (RNP) to the polyribosome pool. *CPP*, apopain; *ICE*, interleukin 1β converting enzyme; *p21*, protein of 21 kDa (Sdi1, WAF1, CIP1, or Pic1); *PCNA*, proliferating cell nuclear antigen; *CPP32*, a capsase.

TRANSLATIONAL CONTROL AND NEURONAL REPAIR

Clearly, the foremost task for the neuronal system is not to be concerned with neurons that are already dead, but to focus on the task of rescuing neurons that are en route to death. Before embarking on this rescue attempt, one must examine how healthy neurons might venture out on this dangerous route. A simple explanation may again be based on the complexity of signal transduction pathways. A misfired or misdirected signal can easily cause the shift of EF-1α messages from the RNP pool to the polyribosome fractions, resulting in the up regulation of messages needed for apoptosis, at least the traverse of some part if not all of G_1 phase; once this happens, the cell is en route to death unless an interceding event can stop it. Here we suggest, as shown in Fig. 15.5, that the interceding event may be the involvement of the S1 protein, whose main function is not only substituting for its sister, EF-1α, in translation, but also competing with its polyribosome binding site. This competition in securing the same resource for translation may be the very mode of translational repression for EF-1α protein production and may act as a countermeasure to deter the signal transduction activation causing the shift of EF-1α messages from the storage pool to the translatable polysome fraction. Thus, S1 may function to preserve the beneficial translational regulation and to remove the detrimental cytoskeletal disorganization by repressing the translatability of its sister's messages, a molecular check-and-balance to ensure that signals for apoptosis are never implemented.

NEURONAL FUNCTIONALITY, TRANSLATIONAL CONTROL, AND THE AGING BRAIN'S REVISITING OF EMBRYONIC GENE EXPRESSION

Many reports have shown that gene expressions in the brain of extremely old animals exhibit some of the same features experienced in their earlier life, most notably the embryonic stages. This phenomenon I term here as the aging brain's revisiting of embryonic life, a sort of molecular evolution of the very old mirroring the very young. Not clear are the reasons why in the very old, molecular mechanisms regress from adult to embryonic regulatory modes. The old Chinese saying that "the very old return to childlike nature" seems to imply that one can relive life, starting all over again from childhood. Ludicrous as this idea may seem, it occurs in many popular ideas centering around the folklore of the "fountain of youth." Now, if we reject this "return-to-childhood" notion, how then may one explain the revisiting of the aging brain, in expressing some early development-specific genes? Here, the explanation of S1's function in the check-and-balance control of EF-1α expression may be the simplest interpretation. As we have observed, in extreme old age, there is a decline in S1 gene expression, and this decline seems to be transcriptionally regulated. A simple hypothesis is that with age, there is an age-dependent transcriptional repression of S1 gene expression, which in turn results in decreased repression of EF-1α translation. The end result is an increase of EF-1α protein product, and consequentially increased translational elongation, signal transduction, and cytoskeletal disorganization, all of which are characteristic phenotypes typical of embryonic development. Nevertheless, this change is not programmed and lacks the entire repertoire of gene expression required for development. We cannot assume that such a return to some but not all aspects of embryonic gene expression is designed to recapture the youthful program; it is rather more likely to lead to the misdirected paths of neuronal dysfunction, impairment, and degeneration.

SUMMARY

Our working hypothesis is that for cell replication to succeed, EF-1α protein needs to be present in abundance to provide a cellular environment with active protein translation, engaged signal transduction, and flexible cytoskeletal organization. In contrast, for adult neurons to maintain their permanently non-replicating state, the strategy must be the opposite, provided by a milieu fit for steady-state protein translation, specialized signal transduction, and fixed cytoskeletal organization. In order to maintain this specialized neuronal phenotype, we hypothesize that EF-1α's protein abundance must be kept at a controlled level, optimized to the need of a consistent cellular milieu. Thus, we suggest that S1 may function to keep EF-1α's protein presence at a consistently reduced level, providing the low EF-1α/S1 ratio necessary to the neuronal long-lived phenotype. Finally, we suggest that failure to maintain this low EF-1α/ S1 ratio may occur during neuronal apoptosis, and the elevated EF-1α due to translational derepression may be the prerequisite upstream effector facilitating the gene expressions committing neurons to die. Future experiments will allow us to test our hypothesis that a low EF-1α/S1 ratio is crucial to neuronal survival and that aging neurons may suffer the dysregulation of this ratio, tilting toward death. Therapeutic treatment to reverse the high EF-1α/S1 ratio to the low level, as well as using the EF-1α/S1 ratio as a prognostic and diagnostic index, may be ultimate goals in keeping the aging brain healthy and functional as long as possible.

ACKNOWLEDGMENTS

The author wishes to thank Mr. Alan N. Bloch for not only proofreading the manuscript, but also numerous times of encouragement and discussion. This work is supported by a grant (R01 AG09278) to E. W. from the National Institute on Aging of the National Institutes of Health, U.S.A.

REFERENCES

1. Ann, D. K., Moutsatsos, I. K., Nakamura, T., Mao, P.-L., Lee, M. J., Chin, S., Liem, R., & Wang, E. (1991). Isolation and characterization of a rat chromosomal gene for a polypeptide antigenically related to statin. *J. Biol. Chem., 266,* 10429–10437.
2. Ann, D. K., Lin, H. H., Lee, S., Tu, Z.-J., & Wang, E. (1992). Characterization of the statin-like S1 and rat elongation factor 1 as two distinctly expressed messages in rat. *J. Biol. Chem., 267*(2), 699–702.
3. Lu, X. A., & Werner, D. (1989). The complete sequence of mouse elongation factor-1 alpha (EF-1 alpha) mRNA. *Nucleic Acids Res., 17,* 442–446.
4. Lee, S., Francoeur, A. M., Liu, S., & Wang, E. (1992). Tissue-specific expression in mammalian brain, heart, and muscle of S1, a member of the EF-1 gene family. *J. Biol. Chem., 267,* 24064–24068.
5. Lee, S., Wolfraim, L. A., & Wang, E. (1993). Characterization of the differential expression of S1 and elongation factor-1α during rat development. *J. Biol. Chem., 268*(32), 24453–24459.
6. Lee, S., LeBlanc, A., Duttaroy, A., & Wang, E. (1995). Terminal differentiation-dependent alteration in the expression of translation elongation factor-1α and its sister gene, S1, in neurons. *Exp. Cell Res., 219,* 589–597.
7. Lee, S., Stollar, E., & Wang, E. (1993). Localization of S1 and elongation factor-1α mRNA in rat brain and liver by non-radioactive *in situ* hybridization. *J. Histochem. Cytochem., 41*(7), 1093–1098.
8. Lee, S., Ann, D., & Wang, E. (1994). Cloning of human and mouse cDNAs for S1, the second member of the mammalian elongation factor-1α gene family: Analysis of a possible evolution pathway. *Biochem. Biophys. Res. Commun., 203,* 1371–1377.
9. Kawashima, T., Berthet-Colominas, C., Wulff, M., Cusack, S., & Leberman, R. (1996). The structure of the *Escherichia coli* EF-Tu.EF-Ts complex at 2.5 Å resolution. *Nature, 379,* 511–518.
10. Wang, E., & Krueger, J.G. (1985). Application of an unique monoclonal antibody as a marker for nonproliferating subpopulations of cells of some tissues. *J. Histochem. Cytochem., 33,* 587–594.

11. Lee, M.-J., Sandig, M., & Wang, E. (1992). Statin, a protein marker specific for nonproliferating cells, is a phosphoprotein and forms an *in vivo* complex with a 45 kilodalton serine/threonine kinase. *J. Biol. Chem., 267,* 21773–21781.

12. Tsanaclis, A. M. C., Brem, S., Gately, S., Schipper, H., & Wang, E. (1991). Statin immunolocalization in human brain tumors: Detection of noncycling cells using a novel marker of cell quiescence. *Cancer, 68,* 786–792.

13. Schipper, H. M., Mauricette, R., Liang, J.-J., Lee, M.-J., & Wang, E. (1992). Expression of the non-proliferation-specific protein, statin, in grey matter neuroglia of the aging rat brain. *Brain Res., 591,* 129–139.

14. Kyzer, S., Mitmaker, B., Gordon, P. H., Schipper, H. M., & Wang, E. (1992). Proliferation activity of colonic mucosa at different distances from primary adenocarcinoma as determined by the presence of statin: A nonproliferation specific nuclear protein. *Dis. Colon Rectum, 35,* 879–883.

15. Mitmaker, B., Kyzer, S., Gordon, P. H., & Wang, E. (1992). Histochemical localization of Statin—a non-proliferation-specific nuclear protein—in nuclei of normal and abnormal human thyroid tissue. *Euro. J. Histochem., 36,* 123–136.

16. Yang, G., & Wang, E. (1993). Expression of terminin in the rat brain during neuronal differentiation. *Brain Res., 615,* 71–79.

17. Trudel, M., Oligny, L., Caplan, S., Caplan, C., Schipper, H. M., & Wang, E. (1994). Statin: A novel marker of nonproliferation, expression in nonneoplastic lymphoid tissues and follicular lymphomas. *Am. J. Clin. Pathol., 101,* 421–425.

18. Palanca-Wessels, M. C. A., Gown, A. M., Wang, E., & Coltrera, M. D. (1994). Immunocytochemical detection of statin, a nuclear protein of G_0 phase, in deparaffinized, methacarn-fixed human breast cancer tissues: Correlation with expression of PCNA and markers of clinical outcome. *Appl. Immunohistochem., 2,* 248–253.

19. Danova, M., Pellicciari, C., Bottone, M. G., Mangiarotti, R., Gibelli, N., Riccardi, A., Zibera, C., Mazzini, G., & Wang, E. (1994). Multiparametric assessment of the cell cycle effects of tamoxifen on MCF-7 human breast cancer cell line. *Oncol. Rep., 1,* 739–745.

20. Knudsen, S. M., Frydenberg, J., Clark, B. F., & Leffers, H. (1993). Tissue-dependent variation in the expression of elongation factor-1α isoforms: Isolation and characterisation of a cDNA encoding a novel variant of human elongation-factor 1α. *Eur. J. Biochem., 215,* 549–554.

21. Celis, J. E., Gesser, B., Rasmussen, H. H., Madsen, P., Leffers. H., Dejgaard, K., Honore, B., Olsen, E., Ratz, G., Lauridsen, J. B., Basse, B., Mouritzen, S., Hellerup, M., Andersen, A., Walbum, E., Celis, A., Bauw, G., Puype, M., Van Damme, J., & Vandekerckhove, J. (1990). Comprehensive two-dimensional gel protein databases offer a global approach to the analysis of human cells: The transformed amnion cells (AMA) master database and its link to genome DNA sequence data. *Electrophoresis, 12,* 989–1071.

22. Lee, S., Duttaroy, A., & Wang, E. (1995). EF-1α/S1 and regulation of protein synthesis during aging. In Holbrook, Martin & Lockshin (Eds.), *Cellular aging and cell death* (pp. 139–151). New York: Wiley.

23. Hershey, J. W. B. (1991). Translational control in mammalian cells. *Ann. Rev. Biochem., 60,* 717–755.

24. Kawashima, T., Berthet-Colominas, C., Wulff, M., Cusack, S., & Leberman, R. (1996). The structure of the *Escherichia coli* EF-Tu.EF-Ts complex at 2.5A resolution. *Nature, 379,* 511–518.

25. Rattan, S. I. S. (1991). Protein synthesis and the components of protein synthetic machinery during cellular aging. *Mutat. Res., 256,* 115–125.

26. Edmonds, B. T., Murray, J., & Condeelis, J. (1995). pH regulation of the F-actin binding properties of *Dictyostelium* elongation factor-1α. *J. Biol. Chem., 270,* 15222–15230.

27. Yang, F., Demma, M., Warren, V., Dharmawardhane, S., & Condeelis, J. (1990). Identification of an actin-binding protein from *Dictyostelium* as elongation factor 1α. *Nature, 347,* 494–496.

28. Condeelis, J. (1995). Elongation factor 1α, translation and cytoskeleton. *Trends Biochem. Sci., 20,* 169–170.

29. Ohta, K., Toriyama, M., Miyazaki, M., Murofushi, H., Hosoda, S., Endo, S., & Sakai, H. (1990).

The mitotic apparatus-associated 51-kDa protein from Sea Urchin eggs is a GTPase. *J. Biol. Chem., 265,* 3240–3247.

30. Herrera, F., Correia, H., Triana, L., & Fraile, G. (1991). Association of ribosomal subunits. *Eur. J. Biochem., 200,* 321–327.

31. Hayashi, Y., Urade, R., Utsumi, S., & Kito, M. (1989). Anchoring of peptide elongation factor EF-1α by phosphatidyl-inositol at the endoplasmic reticulum membrane. *J. Biochem. (Tokyo), 106,* 560–563.

32. Yang, W., Burkhart, W., Cavallius, J., Merrick, W. C., & Boss, W. F. (1993). Purification and characterization of a phosphatidyl-inositol 4-kinase activator in carrot cells. *J. Biol. Chem., 268,* 392–398.

33. Yang, W., & Boss, W. F. (1994). Regulation of phosphatidyl-inositol 4-kinase by the protein activator PIK-A49: Activation requires phosphorylation of PIK-A49. *J. Biol. Chem., 269,* 3852–3857.

34. Tatsuka, M., Mitsui, H., Wada, M., Nagata, A., Nojima, H., & Okayama, H. (1992). Elongation factor-1 gene determines susceptibility to transformation. *Nature, 359,* 333–336.

35. Shiina, N., Gotoh, Y., Kubomura, N., Iwamatsu, A., & Nishida, E. (1994). Microtubule severing by elongation factor 1α. *Science, 266,* 282–285.

36. Curtis, D., Lehmann, R., & Zamore, P. D. (1995). Translational regulation in development. *Cell, 81,* 171–178.

37. Tarun, S. Z., Jr., & Sachs, A. B. (1995). A common function for RNA 5′ and 3′ ends in translation initiation in yeast. *Genes Develop., 9,* 2997–3007.

38. Dreyfuss, G., Hentze, M., & Lamond, A. I. (1996). From transcript to protein. *Cell, 85,* 963–972.

39. Sonenberg, N. (1996). mRNA 5′cap-binding protein eIF-4E and control of cell growth. In J. W. B. Hershey, M. B. Mathews, & N. Sonenberg (Eds.), *Translational control* (pp. 246–269). Cold Spring Harbor, NY: Cold Spring Harbor Laboratory Press.

40. Stripecke, R., Oliveira, C. C., McCarthy, J. E. G., & Hentza, M. M. (1994). Proteins binding to 5′-untranslated region sites: A general mechanism for translational regulation of mRNAs in human and yeast cells. *Mol. Cell. Biol., 14,* 5898–5909.

41. Gray, N. K., & Hentze, M. W. (1994). Iron regulatory protein prevents binding of the 43S translation pre-initiation complex to ferritin and eALAS mRNAs. *EMBO J., 13,* 3882–3891.

42. Casey, J. L., Hentze, M. W., Koeller, D. M., Caughman, S. W., Rouasult, T. A., Kausner, R. D., & Harford, J. B. (1988). Iron-responsive elements: Regulatory RNA sequences that control mRNA levels and translation. *Science, 240,* 924–928.

43. Mullner, K. L., & Kuhn, I. C. (1988). A stem loop in the 3′ untranslated region mediates iron-dependent regulation of transferin receptor mRNA stability in the cytoplasm. *Cell, 53,* 815–825.

44. Munroe, D., & Jacobson, A. (1991). Tales of poly(A): A review. *Gene, 91,* 151–158.

45. Munroe, D., & Jacobson, A. (1990). mRNA poly(A) tail, a 3′ enhancer of translational initiation. *Mol. Cell. Biol., 10,* 3441–3455.

46. Kwon, Y. K., & Hecht, N. B. (1993). Proteins homologous to the *Xenopus* germ cell-specific RNA-binding proteins p54/p56 are temporally expressed in mouse male germ cells. *Dev. Biol., 158,* 90–100.

47. Yang, J., Porter, L., & Rawls, J. (1995). Expression of the dihydroörotate dehydrogenase gene, *dhod,* during spermatogenesis in *Drosophila melanogaster. Mol. Gen. Genet., 246,* 334–341.

48. Brown, E. J., & Shreiber, S. L. (1996). A signaling pathway to translational control. *Cell, 86,* 517–520.

49. Gavis, E. R., & Lehmann, R. (1992). Localization of nanos RNA controls embryonic polarity. *Cell, 71,* 301–313.

50. Gavis, E. R., & Lehmann, R. (1994). Translational regulation of nanos by RNA localization. *Nature (London), 369,* 315–318.

51. Kislauskis, E. H., Li, Z., Singer, R. H., & Taneja, K. L. (1993). Isoform-specfic 3′-untranslated sequences sort α-cardiac and β-cytoplasmic actin messenger RNAs to different cytoplasmic compartments. *J. Cell Biol., 123,* 165–172.

52. Mowry, K. I., & Melton, D. A. (1992). Vegetal messenger RNA localization directed by a 340 nt RNA sequence element in *Xenopus* oöcytes. *Science, 255,* 991–994.

53. Tanguay, R. L., & Gallie, D. R. (1996). Translational efficiency is regulated by the length of the 3′ untranslated region. *Mol. Cell. Biol., 16,* 146–156.

54. Bassell, G. J., Taneja, K. L., Kislauskis, E. H., Sundell, C. L., & Singer, R. H. (1994). Actin filaments and the spatial positioning of mRNAS. *Adv. Exp. Med. Biol., 358,* 183–189.

55. Kislauskis, E. H., Zhu, X., & Singer, R. H. (1994). Sequences responsible for intracellular localization of beta-actin messenger RNA also affect cell phenotype. *J. Cell Biol., 127,* 441–451.

56. Ambros, V. (1989). A hierarchy of regulatory genes controls a larva to adult developmental switch in *C. elegans. Cell, 57,* 49–57.

57. Ambros, V., & Moss, E. G. (1994). Heterochromic genes and the temporal control of *C. elegans* development. *Trends Genet., 10,* 123–127.

58. Lee, R. C., Feinbaum, R. L., & Ambros, V. (1993). The *C. elegans* heterochronic gene lin-4 encodes small RNAs with antisense complementary to lin-14. *Cell, 75,* 843–854.

59. Struhl, G., & Dubnau, J. (1996). RNA recognition and translational regulation by a homeodomain protein. *Nature, 379,* 694–699.

60. Rivera-Pomar, R., Niessling, D., Schmidt-Orr, U., Gehring, W., & Jackie, H. (1996). RNA binding and translational suppression by bicoid. *Nature, 379,* 746–749.

61. Ewen, M., & Miller, S. J. (1996). p53 and translational control. *Biochim. Biophys. Acta, 1242,* 181–184.

62. Levy, S., Aveni, D., Harharan, N., Perry, R. P., & Meyuhas, O. (1991). Oligopyrimidine tract at the 5′ end of mammalian ribosomal protein mRNAs is required for their translational control. *Proc. Natl. Acad. Sci. USA, 15,* 3319–3323.

63. Jefferies, H. B., Reinhard, C., & Kozma, S. C. (1994). Rapamycin selectively represses translation of the polypyrimidine tract mRNA family. *Proc. Natl. Acad. Sci. USA, 91,* 4441–4445.

64. Terada, N., Takase, K., Papst, P. Nairn, A. C., & Gelfand, E. W. (1995). Rapamycin inhibits ribosomal protein synthesis and induces G_1 prolongation in mitogen-activated T lymphocytes. *J. Immunol., 155,* 3418–3426.

65. Terada, N., Patel, H. R., Takase, K., Kohno, K., Nairu, A. C., & Gelfand, E. W. (1994). Rapamycin selectively inhibits translation of mRNA encoding elongation factors and ribosomal proteins. *Proc. Natl. Acad. Sci. USA, 91,* 11477–11481.

66. Hariharan, N., & Perry, R. P. (1990). Functional dissection of a mouse ribosomal protein promoter: Significance of the polypyrimidine initiator and an element in the TATA-box region. *Proc. Natl. Acad. Sci. USA, 87,* 1526–1530.

67. Pandey, S., & Wang, E. (1995). Cells *en route* to apoptosis are characterized by the upregulation of c-*fos,* c-*myc,* c-*jun, cdc2,* and RB phosphorylation, resembling events of early cell cycle traverse. *J. Cell. Biochem., 58,* 135–150.

68. Zbigniew, D. (1995). Apoptosis in antitumor strategies: Modulation of cell cycle or differentiation. *J. Cell. Biochem., 58,* 151–159.

69. Meikrantz, W., & Schlegel, R. (1995). Apoptosis and the cell cycle. *J. Cell. Biochem., 58,* 160–174.

70. King, K. L., & Cidiowski, J. A. (1995). Cell cycle and apoptosis: Common pathways to life and death. *J. Cell. Biochem., 58,* 175–180.

71. Seshadri, T., & Campisi, J. (1990). Repression of c-*fos* transcription and altered genetic program in senescent human fibroblasts. *Science, 247,* 205–209.

72. Wang, E. (1995). Failure to undergo programmed cell death in senescent human fibroblasts is related to inability to down-regulate *bcl*2 presence. *Cancer Res., 55,* 2284–2292.

73. Zhang, L., Zhou, W., Velculescu, V. E., Kern, S. E., Hruban, R. H., Hamilton, S. R., Vogelstein, B., & Kinzler, K. W. (1997). Gene expression profiles in normal and cancer cells. *Science, 276,* 1268–1272.

74. Pennisi, E. (1997). A catalog of cancer genes at the click of a mouse. *Science, 276,* 1023–1024.

16

ASTROCYTE SENESCENCE AND THE PATHOGENESIS OF PARKINSON'S DISEASE

HYMAN M. SCHIPPER

The Bloomfield Centre for Research in Aging
Lady Davis Institute for Medical Research
Sir Mortimer B. Davis Jewish General Hospital
Departments of Neurology and Neurosurgery and Medicine (Division of Geriatrics)
McGill University, Montréal, Québec, Canada H4H 1R3

PARKINSON'S DISEASE

Parkinson's disease (PD) is a movement disorder of uncertain etiology characterized by the accelerated loss of dopamine-containing neurons in the pars compacta of the substantia nigra. As in the case of the other major human neurodegenerative disorders (Alzheimer's disease and amyotrophic lateral sclerosis), aging is perhaps the single most important risk factor for the development of PD. In North America, it has been estimated that approximately 1% of the population older than the age of 60 and as many as 3% of all people older than age 85 are afflicted with PD. The cardinal clinical features of the disease include bradykinesia (paucity of movement), rigidity, and a resting tremor of the head and limbs. In addition, PD victims may be variably affected with disturbances of speech and writing, depression, dementia, autonomic nervous system dysfunction, and respiratory complications. Invariably, patients with advanced PD experience severe gait disturbances and postural instability progressing to complete immobility and institutionalization. For a limited number of years in the course of their illness, significant symptomatic relief can be afforded to these patients by implementing levodopa replacement therapy, treatment with anticholinergic drugs and dopamine agonists, or surgical ablation of relevant brain (basal ganglia) loci. With the possible exception of the drug, l-deprenyl (see the following), there currently exists no treatment that unequivocally slows the progression of neuronal cell loss in this degenerative disorder.

The development of appropriate therapies for arresting the progression of PD would be facilitated by a more in-depth understanding of the basic mechanisms responsible for the cascade of biological and chemical abnormalities that characterize this disorder. A number of critical issues in this regard remain unresolved (Table 16.1). For example, although aging is a significant risk factor for the development of PD, the cellular and molecular changes that render the senescent nervous system prone to this degenerative process remain enigmatical. Although there is abundant evidence of enhanced oxidative stress (free radical damage) in the substantia nigra of PD subjects, it remains unclear whether the latter plays a causative role in the neuronal degeneration or whether it represents an effect or epiphenomenon of a more fundamental disease process. Elevated levels of tissue iron have been reported in brain regions affected by PD; however, much remains to be learned concerning the molecular nature of these abnormal iron deposits and their cellular and subcellular substrates. In addition, membrane receptors for transferrin, the major iron transport protein in mammalian tissues, are mysteriously deficient in PD-affected brain tissues exhibiting iron overload (see the following). It is now well established that mitochondrial function is impaired in the substantia nigra of PD patients, although it remains to be resolved whether this mitochondrial insufficiency is primary (causative) or secondary (e.g., due to oxidative injury) in nature. Finally, as in the case of the other major human neurodegenerative disorders, astrocyte hypertrophy (gliosis) is a characteristic pathological feature of PD-affected neural tissues. It is uncertain, however, to what extent these neuroglial changes represent a mere passive response to neuronal degeneration and to what degree, if any, these nonneuronal cells contribute to the degenerative process. These and other issues germane to the pathogenesis

TABLE 16.1   Parkinson's Disease

Facts	Enigmas
Aging related	Mechanism?
Dopaminergic cell loss	Etiopathogenesis?
	Genetic versus environmental
Mitochondrial insufficiency	Primary or secondary?
Oxidative stress	Source?
	Cause or effect?
Iron sequestration	Cellular substrate(s)?
	Subcellular compartments?
	Transferrin-receptor discordance
	Mechanism of iron deposition?
Astrogliosis	Passive or active role?

of PD are discussed in the remainder of this chapter. In addition, an experimental model of astroglial senescence is presented, which may bring to light important cause-and-effect relationships among the various pathological features of PD and other aging-related neurodegenerations that are difficult to discern on examination of postmortem human materials.

OXIDATIVE STRESS HYPOTHESIS OF PARKINSON'S DISEASE

Free radicals are unstable, highly reactive molecules characterized by the presence of unpaired electrons in their outermost shells. The reduction of molecular oxygen is a major source of biologically-active radical species including superoxide (O_2^-), singlet oxygen, and the hydroxyl radical ($\cdot OH$). In addition to these oxyradicals, organisms are frequently exposed to organic free radicals, which derive from the environment (e.g., the herbicide, paraquat, and other xenobiotics) or are generated within tissues in the course of normal cellular metabolism (e.g., cytotoxic semiquinones derived from the oxidation of catechols). A host of intracellular and extracellular free radical scavengers and antioxidants, including α-tocopherol and the various dismutases, catalases, and peroxidases, have evolved to counteract the potentially destructive effects of redox reactions in biological systems. Oxidative stress resulting from excessive free radical generation or a deficiency in the cellular defenses may promote lipid peroxidation and dissolution of plasma and mitochondrial membranes, DNA cross-links with resultant mutagenesis, and inactivation of other essential macromolecules. Free radicals have been implicated in the pathogenesis of a broad spectrum of human diseases involving virtually every organ system. A relatively large compartment of unsaturated fat present in mammalian brain renders this organ particularly vulnerable to oxidative stress. In the normal central nervous system (CNS), lipid peroxidation is involved in the process of neuromelanization and in the accumulation of the aging pigment, lipofuscin. Oxygen-derived and organic free radicals may also play significant roles in various neurodegenerative disorders, in the development of epileptic foci, in cerebral vasospasm secondary to subarachnoid hemorrhage, and in reperfusion brain injury following cerebral ischemia.

In the early 1980s, André Barbeau at the Institut de Recherches Cliniques de Montréal, Gerald Cohen at the Mt. Sinai School of Medicine in New York City, and others have outlined an hypothesis for the pathogenesis of PD based on the premise that accelerated breakdown of dopamine in surviving nigrostriatal neurons (the primary neural pathway that undergoes degeneration in PD) generates excessive amounts of neurotoxic free radicals. This oxidative stress, in turn, conceivably exhausts local cellular defenses and thereby promotes lipid peroxidation and rancidification of neuronal membranes, cytoskeletal damage, and a cascade of nigrostriatal cell degeneration. There is currently a broad consensus implicating oxidative stress as a major factor in the pathogenesis of PD. The free radical hypothesis of PD draws support from the following observations: (1) Accelerated breakdown of dopamine in surviving nigrostriatal neurons has been demonstrated in rodents following partial lesioning of this neuronal population with the neurotoxin, 6-hydroxydopamine. (2) The natural breakdown of dopamine by the glial enzyme, monoamine oxidase B (MAO-B), yields dihydroxyphenylacetaldehyde (DOPAc), ammonia (NH_3), and the pro-oxidant species, H_2O_2 (Fig. 16.1, Pathway 1). H_2O_2 may be directly neurotoxic or it may undergo further metabolism in the presence of redox-active transition metals (such as iron and copper) to more reactive oxyradicals such as ·OH. Furthermore, the catechol ring of dopamine may undergo spontaneous auto-oxidation or peroxidase/H_2O_2–catalyzed oxidation to neurotoxic quinones and orthosemiquinone radicals (see Fig. 16.1, Pathway 2). Indeed, the toxicity of specific catecholamines to cultured neuroblastoma cells appears to be directly related to their rates of auto-oxidation to quinones. In the course of catechol auto-oxidation, O_2^- and H_2O_2 are generated from molecular oxygen exerting an additional oxidative stress. (3) The neurotoxins, 6-hydroxydopamine and manganese, induce parkinsonism in animals via the generation of free radicals. Free radicals may also play a role in human and experimental parkinsonism resulting from intoxication with the heroin-like substance, 1-methyl-4-phenyl-1,2,3,6–tetrahydropyridine (MPTP). MPTP is a protoxin that is readily oxidized by MAO-B in astrocytes to the neurotoxin, 1-methyl-4-phenyl-pyridium (MPP+). Upon its extrusion from the astroglial compartment, MPP^+ is selectively taken up by dopaminergic neurons wherein it inhibits mitochondrial respiration. In addition to poisoning the formation of intracellular energy substrates (adenosine triphosphate [ATP]), MPP^+ promotes the accumulation of cytotoxic free radicals by interfering with normal electron transport within the inner mitochondrial membrane. (4) Basal lipid peroxidation in the substantia nigra of PD subjects is significantly elevated relative to non-PD controls matched for age and postmortem interval. Free radical scavenger enzymes (such as catalase) and intracellular reducing substances (such as reduced glutathione) are reportedly deficient in the basal ganglia of patients with

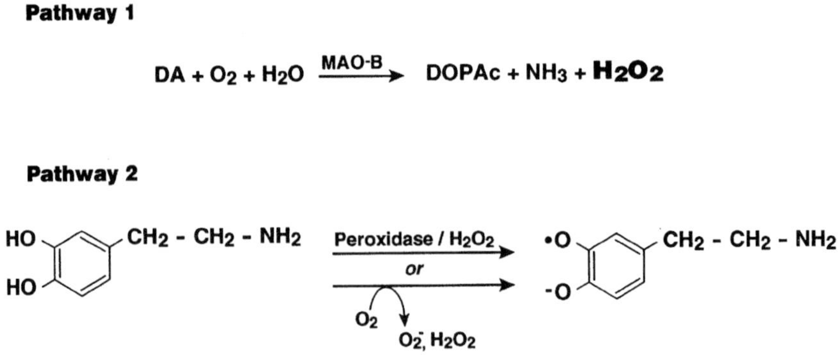

Pathway 1

$$DA + O_2 + H_2O \xrightarrow{\text{MAO-B}} DOPAc + NH_3 + H_2O_2$$

Pathway 2

FIGURE 16.1   Dopamine oxidation.

PD. Whether these deficiencies constitute primary biochemical lesions in PD that enhance neuronal vulnerability to free radical assault or are themselves the result of a more fundamental disease process has not yet been determined. Interestingly, and in contradistinction to other antioxidant defence mechanisms, the mitochondrial enzyme, manganese superoxide dismutase (MnSOD), appears to be augmented in the basal ganglia of PD subjects; this has been interpreted as an adaptive response aimed at protecting the mitochondrial compartment from oxidative injury. (5) Although still controversial, results of several multicenter clinical trials suggest that treatment of early PD with the MAO-B inhibitor, l-deprenyl, may slow the progression of this neurodegenerative disorder by curtailing the production of dopamine-derived H_2O_2 (see Fig. 16.1). Thus, a variety of factors, both genetic and environmental, may disrupt the balance between rates of endogenous free radical production and the activity of local antioxidant defense mechanisms and thereby influence the progression of nigrostriatal degeneration in PD. As discussed in the following section, iron stores are pathologically elevated in PD-affected brain regions and may serve as a major generator of reactive oxygen species in this condition.

IRON DEPOSITION IN PARKINSON'S DISEASE

Abnormally high levels of tissue iron have been consistently reported in the substantia nigra and basal ganglia of PD subjects (see Table 16.1). Using conventional histochemical stains, the abnormal nigral iron appears to be predominantly deposited within astrocytes, microglia, macrophages, and microvessels and correlates with depletion of dopaminergic neurons in this brain region. In general, increased expression of tissue ferritin, the major intracellular iron storage protein, parallels the distribution of the excess iron and largely implicates non-neuronal (glial) cellular compartments. The extracellular transport of ferric iron and its delivery to virtually all mammalian tissues is mediated by a second iron binding protein, transferrin. To maintain normal tissue iron homeostasis, plasma membrane transferrin receptor densities and intracellular ferritin concentrations are tightly regulated at transcriptional and posttranscriptional levels by iron bioavailability and intracellular iron stores. In normal rat and human brain tissues, there appears to be an overt mismatch between local brain iron concentrations and the densities of cell surface transferrin binding sites. Moreover, in contrast to the ferritin data, the density of transferrin binding sites remains unchanged or varies inversely with increased iron stores in the substantia nigra and striatum of PD subjects. Thus, in contrast to most peripheral tissues, transferrin and its receptor may play a limited role, if any, in the trapping of iron within the aging and degenerating CNS. Indeed, attention is shifting toward alternative iron transport mechanisms such as that mediated by lactoferrin and the lactoferrin receptor, which are reportedly augmented in neurons, astrocytes, and blood vessels of PD-affected neural tissues. Unlike transferrin, lactoferrin binding to its receptor is not affected by degrees of tissue iron saturation and could, theoretically, permit toxic levels of this metal to accumulate within the degenerating basal ganglia of PD subjects. By participating in what chemists refer to as Fenton reactions, the aberrantly sequestered brain iron could promote oxidative stress and lipid peroxidation and thereby directly contribute to the neurodegenerative process. Specifically, the excess iron may facilitate the reduction of H_2O_2, derived from the oxidation of dopamine and from the mitochondrial electron transport chain, to the highly cytotoxic hydroxyl radical. Furthermore, in the presence of H_2O_2, ferrous iron may behave as a nonenzymatic peroxidase activity capable of converting relatively innocuous catecholamines, such as dopamine, to potentially neurotoxic quinones and semiquinone radicals (see Fig. 16.1). Efforts to ameliorate iron-mediated neuronal injury in PD presupposes some understanding of the regulatory mechanisms subserving iron metabolism and sequestration in the aging and degenerating CNS. Resolution of the following important, but as yet unanswered, questions should provide novel insight into the pathophysiology of PD and possibly expedite the implementation of effective neuroprotection for the management of this debilitating disease:

1. What is the role of heme versus nonheme iron in PD and other aging-related neurodegenerative disorders?
2. Which cell type(s) and subcellular compartments are responsible for the abnormal trapping of brain iron in these degenerative disorders?
3. Does induction of a cellular stress (heat shock) response facilitate trapping of redox-active iron in neural tissues?
4. What is the role of transferrin and other iron binding proteins in this process?

As reviewed in the following, profound biochemical and structural changes occur in aging subcortical astrocytes and in astroglial cultures subjected to oxidative stress, which may explain the abnormal pattern of iron deposition and possibly unify many of the seemingly disparate pathological features of PD and other aging-related neurodegenerative conditions.

ASTROCYTES IN HEALTH AND DISEASE

Astrocytes represent a major class of nonneuronal brain cells (neuroglia) that can be readily identified in fetal and mature CNS using antibodies directed against cytoskeletal components such as vimentin and glial fibrillary acid protein (GFAP). Classically, astrocytes have been subdivided into protoplasmic and fibrous types on the basis of their morphological characteristics and localization to gray or white matter, respectively. With the application of immunolabeling techniques, a classification scheme has evolved based on the differential expression of surface protein markers by a host of progenitor and mature cells of the astroglial lineage. Astrocytes are the most abundant of all cell types in the mammalian nervous system and perform a wide range of adaptive functions. These cells play important roles in the elaboration of a scaffolding for neuronal migration during embryogenesis and are crucial for the maintenance of the blood-brain barrier and normal ion homeostasis They contain much of the enzyme machinery necessary for the metabolism of a variety of neurotransmitters and neuropeptides, and they both produce and respond to a host of immunomodulatory molecules (cytokines). On the other hand, astrocytes may, under certain circumstances, mediate deleterious effects within the CNS and thereby contribute to a decline in neurologic function. Examples of the latter include formation of epileptogenic scar tissue in response to CNS injury, neoplastic transformation and malignant behavior, and the accumulation and release of excitotoxic amino acids following tissue hypoxia, oxidative stress, and metal exposure. As mentioned previously, astrocytes may also convert relatively harmless protoxins (such as MPTP) to potent neurotoxins (MPP+).

Astrocyte hypertrophy, accumulation of GFAP-positive intermediate filaments and possibly proliferation (reactive gliosis) are characteristic pathological features of all the major aging-related neurodegenerative disorders including Alzheimer's disease, PD, Huntington's disease, and amyotrophic lateral sclerosis. Gliosis also occurs, albeit to a lesser extent, in the course of normal brain aging. In the aging mammalian brain, in Alzheimer's disease, and in experimental models of PD, reactive gliosis is accompanied by significant increases in the activity of MAO-B. Under these circumstances, excessive H_2O_2 derived from the accelerated breakdown of dopamine and other monoamines may foster further neuronal injury. The astrocytic compartment may also contribute to the development of Alzheimer-like pathology by serving as an important site for lysosomal enzyme-mediated protein degradation, β-amyloid metabolism, abnormal phosphate group addition to cytoskeletal proteins, and the synthesis of molecules believed to influence neuronal survivability such as clusterin and apolipoprotein E. Occasionally, astrocytes may also exhibit specific pathological changes, suggesting that they may be the primary neural targets of the disease process as in the case of liver failure and certain rare neurodegenerative conditions of childhood. As discussed in the following section, astrocytes in many subcortical brain regions exhibit progressive, aging-related changes that may directly implicate these cells in the development of PD and other free radical–based neurodegenerative processes.

PEROXIDASE-POSITIVE SUBCORTICAL GLIAL SYSTEM

In rats, humans, and other vertebrates, granule-laden astrocytes exhibiting an affinity for chrome-alum hematoxylin and aldehyde fuchsin (Gomori stains) reside in limbic and periventricular brain regions, including areas affected by PD pathology, such as the striatum, globus pallidus, and substantia nigra, and sites implicated in Alzheimer's disease, such as the CA1 region of the hippocampus. (Please see the author's review article in *Neurobiology of Aging* [1996] cited in Further Readings for a comprehensive reference list and figures illustrating the morphology, histochemistry, and molecular biology of these cells.) The astrocytic inclusions are rich in sulfhydryl groups, emit orange-red autofluorescence, and stain intensely with diaminobenzidine (DAB), a marker of endogenous peroxidase activity. Histochemical analyses of this peroxidase activity indicate that the latter is nonenzymatic in nature and is likely mediated by ferrous iron and/or other redox-active transition metals. In humans and rodents, numbers of peroxidase-positive astrocytes and their granular content progressively increase as a function of advancing age. In female rodents, early castration attenuates, and exogenous estrogen accelerates, the senescence-dependent accumulation of these glial inclusions in the hypothalamic arcuate nucleus, a neuroendocrine locus concerned with the regulation of pituitary gonadotropin secretion. In the rat substantia nigra, the DAB-positive glial inclusions increase in abundance by a factor of 4 between 3 and 15 months of age. Despite their consistent increase with aging, the results of various histochemical and morphological studies indicate that the peroxidase-positive granules are a unique form of glial inclusion constitutively different from the aging pigment, lipofuscin. Increased numbers of peroxidase-positive glia have also been documented in rodent periventricular brain regions after cranial X-irradiation and in rat spinal cord following contusion injury. As discussed in the following section, our ability to generate primary brain cell cultures highly enriched for peroxidase-positive astroglia has significantly advanced our knowledge concerning the origin of the peroxidase-positive inclusions, the mechanisms responsible for their biogenesis, and the role(s) these cells may play in brain aging and neurodegenerative disease.

PEROXIDASE-POSITIVE ASTROCYTES IN PRIMARY CULTURE

Using dissociated fetal or neonatal rat brain cell cultures, our laboratory has shown that exposure to the sulfhydryl agent, 2-mercaptoethylamine or cysteamine (CSH), induces a massive proliferation of astrocytic inclusions, which are structurally and histochemically identical to those which naturally accumulate in subcortical astroglia of the intact aging brain. Both in situ and in the CSH-treated glial cultures, the inclusions appear membrane-bound, variable in size and shape, and exhibit an intensely electron-dense granular matrix. Elemental iron is detected in the inclusions by electron microprobe analysis, and the presence and concentration of the metal correlates closely with the presence and intensity of DAB (peroxidase) staining. These astrocyte granules exhibit little or no affinity for Prussian blue, a marker of ferric and hemosiderin iron, arguing that *ferrous* iron is responsible for the nonenzymatic peroxidase activity in the cells.

SUBCELLULAR PRECURSORS
OF PEROXIDASE-POSITIVE
ASTROGLIAL INCLUSIONS

In CSH-treated astroglia, the earliest morphological changes appear restricted to the *mitochondrial* compartment. Within 24–72 hr of CSH exposure, many astroglial mitochondria exhibit progressive swelling, rearrangement or dissolution of their cristae, subcompartmental sequestration of redox-active iron, and in some cases, fusion with lysosomes or strands of the endoplasmic reticulum. In young adult rats, subcutaneous CSH injections induce 2–3-fold increases in numbers of peroxidase-positive astrocyte granules in the basal ganglia, hippocampus, and other brain regions. As in the case of the CSH-treated cultures, peroxidase-positive glial granules in the intact rat and human brain invariably exhibit mitochondrial proteins in immunohistochemical preparations. Taken together, these observations indicate that (1) the iron-laden astrocyte granules are derived from abnormal mitochondria engaged in a complex autophagic process and (2) CSH accelerates the appearance of a senescent phenotype in these cells.

ASTROCYTE GRANULE BIOGENESIS AND THE
CELLULAR STRESS RESPONSE

Evidence from our laboratory indicates that activation of the cellular heat shock (stress) response plays an important role in the biogenesis of peroxidase-positive astrocytic inclusions both in situ and in primary glial cultures. Within several hours of CSH exposure, cultured rat astroglia exhibit robust increases in the expression of heat shock protein (HSP)27, HSP90, heme oxygenase-1 (HO-1), and ubiquitin as determined by Western blotting and immunofluorescence microscopy. In spite of the mitochondrial injury incurred by CSH exposure (see previous discussion), CSH-pretreated astroglia (paradoxically) exhibit enhanced resistance to H_2O_2 toxicity and trypsinization-related cell death relative to controls, providing physiological evidence of an antecedent cellular stress response in the former. We also determined that systemic administration of CSH to young adult rats elicits significant increases in concentrations of peroxidase-positive astrocyte granules and numbers of GFAP-positive astrocytes coexpressing HSP27, HSP72, and HSP90 in various subcortical brain regions. Along similar lines, estrogen treatment induces both a heat shock response and subsequent granulation in astrocytes residing in estrogen receptor-rich brain regions such as the hypothalamic arcuate nucleus and the third ventricular subependymal zone. Both in the CSH-treated glial cultures and in the intact periventricular brain, confocal microscopy revealed consistent colocalization of HSP27 and ubiquitin (but not HSP90 or the small stress protein, αB-crystallin) to the peroxidase-positive astrocytic inclusions. The latter observations further extend previous histochemical studies underscoring the identical origin of this CSH-induced astroglial inclusions and those that spontaneously accumulate in the aging subcortical brain.

A considerable body of evidence suggests that intracellular oxidative stress may be the "final common pathway" responsible for the transformation of normal astrocyte mitochondria to peroxidase-positive inclusions in vitro and in the intact aging brain: (1) prior to eliciting astrocyte granulation, both CSH and estrogen up regulate stress proteins that typically respond to oxidative stress (e.g. HSP27, HSP90, and HO-1) but have little or no effect on redox-insensitive proteins such as glucose-regulated protein (GRP)94. (2) CSH undergoes redox cycling in the presence of transition metals resulting in the generation of thiyl radicals, H_2O_2, and other pro-oxidant species. H_2O_2 induces HSP and HO-1 expression in rat astrocytes and stimulates the accumulation of peroxidase-positive astroglial granules in primary culture akin to the effects of CSH. (3) MnSOD, a mitochondrial antioxidant enzyme inducible in various cell types by oxidative stress, is up-regulated in cultured astroglia and in the intact rat brain following CSH treatment. (4) Numbers of peroxidase-positive granules exhibit dose-dependent increases in the rat hypothalamus in response to ionizing radiation, a known generator of intracellular pro-oxidant intermediates. Taken together, these observations suggest that the topography and intensity of endogenous glial peroxidase activity may provide a marker for CNS regions that are particularly susceptible to chronic oxidative stress during normal aging and under pathological conditions.

IRON SEQUESTRATION IN AGING ASTROGLIA

As described previously, CSH induces the accumulation of iron-rich astroglial inclusions in vitro and in situ which are indistinguishable from those that normally accrue in subcortical astrocytes with advancing age. Thus, careful analysis of the mechanisms by which CSH perturbs glial iron homeostasis and promotes sequestration of the metal provides a unique opportunity to delineate fundamental processes subserving iron deposition in the aging and degenerating nervous system. Several laboratories, including our own, have previously concluded on the basis of histochemical and spectrofluorometric data that porphyrins and heme ferrous iron are responsible, respectively, for the orange-red autofluorescence and nonenzymatic peroxidase activity in these glial inclusions. However, we subsequently determined that CSH suppresses the incorporation of the heme precursors, δ-amino [^{14}C]-levulinic acid and [^{14}C]glycine into astroglial porphyrin and heme in primary culture prior to and during the time when increased iron content is detectable in swollen astrocyte mitochondria by microprobe analysis. Thus, contrary to earlier hypothesis, de novo biosynthesis of porphyrins and heme is not responsible for the increased mitochondrial iron content, autofluorescence, and peroxidase activity observed in these cells after CSH exposure. Following inhibition of porphyrin-heme biosynthesis, CSH significantly augments the incorporation of ^{59}Fe (or ^{55}Fe) into astroglial mitochondria without affecting transfer of the metal into whole-cell and lysosomal compartments. Interestingly, this CSH effect was only demonstrable when inorganic ^{59}FeCl$_3$ but not ^{59}Fe-diferric transferrin, served as the metal donor. These observations are consistent with previous reports that intracellular transport of low molecular weight, inorganic iron may be many times more efficient than that of transferrin-bound iron in certain tissues, including melanoma cells, Chinese hamster ovary cells, and K562 cells. Our findings support the conclusion of others that inhibition of heme biosynthesis stimulates the selective transport of low molecular weight iron from the cytoplasm to the mitochondrial compartment.

Recent observations in our laboratory suggest that dopamine may be an important endogenous stressor mediating nigrostriatal glial iron trapping in PD and, to a lesser extent, in the course of normal aging. As in the case of CSH, physiologically relevant concentrations of dopamine stimulate the sequestration of nontransferrin bound ^{55}Fe in the mitochondrial compartment of cultured astroglia without affecting the disposition of transferrin-derived ^{55}Fe. The effects of dopamine on glial iron trapping can be blocked by co-administration of ascorbate, but not by dopamine receptor antagonists, indicating that, akin to the action of CSH, dopamine-derived free radicals promote the sequestration of nontransferrin derived iron within astroglial mitochondria. That such dopamine–astrocyte interactions may be operational in vivo is supported by (1) immunocytochemical evidence of direct contact between tyrosine hydroxylase–positive (dopaminergic) neuronal processes and iron-laden astrocytes in the rat basal ganglia and hypothalamus and (2) studies demonstrating increased elemental iron in the substantia nigra of 6-hydroxydopamine-lesioned rats.

ROLE OF HO-1 IN BRAIN IRON DEPOSITION

HO-1 is a 32-kDa member of the stress protein superfamily that catalyses the rapid conversion of heme to biliverdin in brain and other tissues. In response to oxidative stress, induction of HO-1 may protect cells by breaking down pro-oxidant molecules such as heme to bile pigments (biliverdin, bilirubin) with free radical scavenging capabilities. But this reaction is a double-edged sword; HO-1–mediated heme degradation liberates free iron and carbon monoxide (CO), which under certain circumstances may amplify intracellular oxidative stress by stimulating free radical generation within the *mitochondrial* compartment! We and George Perry's group at Case Western Reserve University in Cleveland have shown that HO-1 is markedly up regulated in neurons and astrocytes of Alzheimer-diseased human cerebral cortex and hippocampus (but not in unaffected substantia nigra) relative to age-matched, nondemented controls. Conversely, we noted that the percentage of GFAP-positive astrocytes expressing HO-1 in substantia nigra (but not in other brain regions) of PD subjects is significantly increased in comparison with age-matched controls. Although HO-1 synthesis in these conditions may confer some degree of cellular protection by degrading pro-oxidant heme to antioxidant bile pigments, it is conceivable that heme-derived free iron and CO may contribute, at least partly, to the development of mitochondrial electron transport chain deficiencies and excess mitochondrial DNA mutations reported in the brains of AD and PD subjects.

The up regulation of HO-1 may have important implications for the biogenesis of mitochondria-derived astrocytic inclusions in senescent and oxidatively challenged astroglia. Within several hours of CSH exposure, cultured astroglia exhibit 3–10-fold increases in HO-1 mRNA and protein levels and enzymatic activity. As in the case of CSH, dopamine, H_2O_2, and menadione (which generates both O_2- and H_2O_2) consistently up-regulate HO-1 in cultured astroglia prior to promoting the sequestration of nontransferrin bound ^{55}Fe by the mitochondrial compartment. We hypothesize that in dopamine-exposed glial cultures and in senescent subcortical astroglia in situ, the liberation of free iron and CO resulting from HO-1–catalyzed heme degradation may promote early oxidative injury to mitochondrial membranes and thereby facilitate the transformation of normal astrocyte mitochondria to iron-rich cytoplasmic inclusions. In support of this notion, we observed that dopamine-induced sequestration of mitochondrial iron in cultured rat astroglia is preventable by co-administration of the competitive heme oxygenase inhibitor, tin-mesoporphyrin, or the HO-1 transcriptional suppressor, dexamethasone.

PROTOXIN BIOACTIVATION BY AGING ASTROGLIA

Using electron spin resonance spectroscopy, we demonstrated that, in the presence of H_2O_2, nonenzymatic (iron-mediated) peroxidase activity induced in cultured astroglia by CSH pretreatment promotes the oxidation of dopamine and other catechol ring–containing compounds to potentially neurotoxic orthosemiquinone radicals. These observations are consistent with previous reports that dopamine and other monoamines are readily converted to semiquinones with proven neurotoxic activity via peroxidase-mediated reactions. Because aging subcortical astrocytes may exhibit both enhanced MAO-B activity (see the previous discussion) *and* abundant mitochondrial iron, it is entirely conceivable that H_2O_2 produced by MAO-B oxidation of dopamine serves as a co-factor for further dopamine degradation to neurotoxic semiquinone radicals by peroxidase-mediated reactions (see Fig. 16.1). In 1995 our laboratory in collaboration with Donato DiMonte and William Langston at the Parkinson's Institute (Sunnyvale, California), published evidence suggesting that redox-active glial iron may also facilitate the bioactivation of the protoxin, MPTP, to the dopaminergic neurotoxin, MPP+, in the presence of MAO blockade. This observation may have important clinical implications for neuroprotective therapy in PD: Many PD subjects are currently treated with l-deprenyl or other MAO inhibitors designed to arrest the conversion of endogenous dopamine or putative MPTP-like environmental pro-toxins to dopaminergic neurotoxins. Our findings indicate, however, that patients so treated may still experience chronic, low-level exposure to dopaminergic neu-rotoxins generated nonenzymatically in glia via iron-mediated reactions.

The high stress protein content of peroxidase-positive astrocytes and com-pensatory increases in MnSOD observed in these cells (see the previous discus-sion) could serve to limit the extent of oxidative injury within the glia them-selves. However, the damaging effects of reactive oxygen species need not be confined to the cellular compartment in which they are generated. For example, superoxide can be extruded from cells via anion channels, and H_2O_2, being lipid soluble, can easily traverse plasma membranes to reach the intercellular space. Indeed, we recently observed that catecholamine-secreting PC12 cells grown atop monolayers of CSH-pretreated (iron-rich) astrocytes are far more vulnera-ble to dopamine/H_2O_2–related killing than PC12 cells co-cultured with control (iron-poor) astroglia. Presumably for reasons mentioned above, astroglial death was not significantly augmented by dopamine-H_2O_2 exposure in both co-culture paradigms. Extrapolating from these in vitro findings to the intact basal ganglia, we hypothesize that leakage of free radicals from peroxidase-positive astrocytes into the surrounding neuropil may promote oxidative injury and degeneration of nearby dopaminergic terminals and other vulnerable neuronal constituents. In this regard, the progressive increase in numbers of peroxidase-positive astrocytes that have been documented in the basal ganglia and other subcortical regions of the aging rodent and human brain may enhance the susceptibility of the latter to parkinsonism and other free radical–related neurodegenerative disorders.

PATHOLOGICAL NEURONAL–GLIAL INTERACTION IN PARKINSON'S DISEASE

The sequence of biochemical and structural changes observed in CSH-stressed astroglia suggest a model for inclusion formation, iron trapping, and the perpetuation of oxidative injury in the aging and PD-afflicted nervous system (see Fig. 16.2): (1) In the aging basal ganglia, dopamine and/or other unidenti-fied oxidative stressors (simulated by CSH exposure) induce a cellular stress response in subpopulations of astroglia characterized by induction of various HSPs and HO-1. In these cells, free iron and CO derived from HO-1–mediated heme degradation may initiate or exacerbate injury to the mitochondrial com-partment. (2) Stress-related suppression of porphyrin–heme biosynthesis and direct oxidative injury to mitochondrial membranes foster the selective transport of nontransferrin derived, nonheme iron into the mitochondrial matrix. Compensatory up-regulation of MnSOD may provide some degree of protection to the glial mitochondria by limiting the accumulation of superoxide. (3) By pro-moting further oxidative stress, the redox-active mitochondrial iron participates

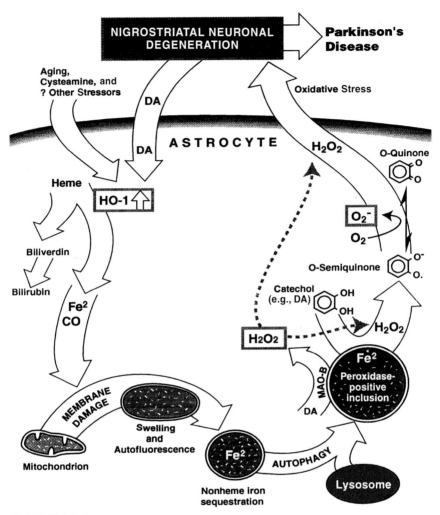

FIGURE 16.2   A model for glial inclusion formation, iron sequestration, and oxidative injury in the aging and degenerating nervous system.

in a vicious circle of pathological events whereby damage to glial mitochondria and to the surrounding neuropil is perpetuated. This model of astrocyte senescence is consistent with the prevailing mitochondrial hypothesis of aging, which states that oxidative injury to mitochondria results in bioenergetic failure, a self-sustaining spiral of augmented mitochondrial free radical generation and injury and progressive tissue aging. Our model accounts for the observation that mosaicism for specific mitochondrial DNA mutations in the normal aging human brain is most pronounced in areas particularly rich in intracellular iron such as the substantia nigra, caudate, and putamen. Moreover, the effect of CSH and dopamine on astroglial iron homeostasis recapitulates the discordant pattern of iron–transferrin receptor localization observed in the PD nigra (see the previous discussion). Our findings raise the possibility that potentiation of stress-related trapping of nontransferrin derived iron by astroglial mitochondria may be a pivotal mechanism underlying the pathological accumulation of this harmful metal in the basal ganglia of PD subjects. Such iron could conceivably originate from degenerating neurons, glia, or myelin, or from the cerebrospinal fluid (CSF). Others have shown that micromolar quantities of low molecular weight iron are present in normal CSF and that the concentration of this metal in CSF increases under neuropathological conditions. In addition, as described previously, low molecular weight iron may be derived from HO-1–mediated breakdown of cellular heme within oxidatively challenged neural tissues. Consistent with our model are reports that a significant proportion of the excess iron in PD brain may indeed be localized to astroglial mitochondria and that deficiencies of mitochondrial electron transport are prevalent in the brains of PD subjects. As has been previously postulated for MAO-B, the glial mitochondrial iron may promote the conversion of dopamine and environmentally derived xenobiotics to a host of neurotoxic intermediates and thereby perpetuate nigrostriatal degeneration initiated by still-to-be-defined genetic and epigenetic factors in patients with PD. If the latter is true, attempts to pharmacologically block transition metal trapping by "stressed" astroglial mitochondria (e.g., using HO-1 inhibitors and centrally active iron chelators) may constitute a rational and effective strategy in the management of PD and other aging-related neurodegenerative disorders.

ACKNOWLEDGMENTS

The author thanks Mrs. Kay Berckmans and Mrs. Adrienne Liberman for assistance with the preparation of this manuscript.

FURTHER READINGS

Abraham, N. G., Drummond, G. S., Lutton, J. D., & Kappas, A. (1996). The biological significance and physiological role of heme oxygenase. *Cell. Physiol. Biochem., 6,* 129–169.

Aschner, M., & Kimelberg, H. K. (Eds.). (1996). *The role of glia in neurotoxicity.* Boca Raton, FL: CRC Press.

Beal, M. F. (1995). *Neuroscience intelligence unit: Mitochondrial dysfunction and oxidative damage in neurodegenerative diseases.* Austin, TX: Landes.

Calne, D. B. (Ed.). (1994). *Neurodegenerative diseases.* Philadelphia: Saunders.

Kettermann, H., Ransom, B. R. (Eds.). (1995). *Neuroglia.* New York: Oxford University Press.

Langston, J. W., & Young, A. (1992). Neurotoxins and neurodegenerative disease. *Ann. N. Y. Acad. Sci., 648,* 1–385.

Richardson, D. R., & Ponka, P. (1997). The molecular mechanisms of the metabolism and transport of iron in normal and neoplastic cells. *Biochim. Biophys. Acta, 1331,* 1–40.

Schipper, H. M. (1996). Astrocytes, brain aging and neurodegeneration. *Neurobiol. Aging, 17,* 467-480. Author's response to commentaries. (1996). *Neurobiol. Aging, 17,* 488–490.

INDEX

Note: *f,* figure; *t,* table

α-Ca^{2+}-calmodulin kinase II, 69, 73, 77
α-synuclein gene, 147
Aβ-fibrillogenesis, 114, 137
Aβ-peptides
  in Alzheimer's diseased brain, 188
  apolipoprotein isoform interactions, 194
  and cholinergic deficit/hypofunction,
    182–185, 189
  differential uptake astrocytes vs. neurons,
    196
  and glucose uptake, 186
  and muscarinic receptor signaling, 186
A-type K$^+$ currents, 74–75
acetylcholine synthesis, 185
adherens junctions and desmosomes, 174
adrenocorticotropic hormone, 20, 22
advanced glycation end products, 163–165
  receptor on neurons for, 188
age-related changes
  brain structures, 4
  corticosteroid receptors, 22
  dopamine binding, 7
  expression of S1 and EF-1α, 237
  hippocampus, 4, 55
  induction and maintenance of LTP, 69, 72
  interdependence of cells, 54
  memory, 1, 56

neurofilament mRNA, 94–95
normal aging, 83, 87
place cell firing patterns, 55, 60
S1 expression, 224
spatial navigation, 52
aging and glucocorticoids, 21, 30
aging, polymorphisms vs. rare mutations,
  130
Alzheimer's disease
  and Aβ secretion, 136, 172, 188
  animal models, 111–115
  apolipoprotein E risk factor, 190
  astroglial compartment, 249
  cholinergic neurotransmission, 186
  cingulate gyrus involvement, 12
  and cortisol levels, 31–32
  drug response and ApoE, 192
  familial forms, 111, 135
  impaired glucose metabolism, 186
  late-onset, putative gene for, 136
  and memory changes, 7
  MRI studies, 10, 41
  neurofilament pathology, 94–95
  and the presenilins, 112–113, 135–142
  tau preparations, modifications, 155
amino acid substitution, 77, 97, 206–207
amyloid plaques, 84–85, 204
amyloid precursor protein, 111–112
  metabolism, 136–137, 189